Science and Philosophy of Behavior

Science and Philosophy of Behavior

Selected Papers

William M. Baum

This edition first published 2022

Registered Office
John Wiley & Sons, Inc., 111 River Street, Hoboken, NJ 07030, USA

Editorial Office
111 River Street, Hoboken, NJ 07030, USA

For details of our global editorial offices, customer services, and more information about Wiley products visit us at www.wiley.com.

Wiley also publishes its books in a variety of electronic formats and by print-on-demand. Some content that appears in standard print versions of this book may not be available in other formats.

Library of Congress Cataloging-in-Publication Data

Name: Baum, William M., author. | John Wiley & Sons, publisher.
Title: Science and philosophy of behavior : selected papers / William M. Baum.
Description: Hoboken, NJ : Wiley, 2022. | Includes bibliographical references and index.
Identifiers: LCCN 2022017517 (print) | LCCN 2022017518 (ebook) | ISBN 9781119880868 (paperback) | ISBN 9781119880882 (adobe pdf) | ISBN 9781119880875 (epub)
Subjects: LCSH: Behaviorism (Psychology). | Psychology—Methodology. | Social sciences—Methodology.
Classification: LCC BF199 .B32 2022 (print) | LCC BF199 (ebook) | DDC 150.19/43—dc23/eng/20220603
LC record available at https://lccn.loc.gov/2022017517
LC ebook record available at https://lccn.loc.gov/2022017518

Cover Design: Wiley
Cover Image: Courtesy of Tom Smith

Set in 10/12pts and Warnock Pro by Straive, Chennai, India

To my mentor, Dick Herrnstein, and my friend and fellow traveler, Howie Rachlin.

Table of Contents

Preface

This book records almost 50 years' worth of developing and explaining a new way of thinking about behavior. It represents a journey more than a goal. I came to see that the science of behavior needed a new paradigm, and two sorts of changes were required. First, the old molecular view inherited from the nineteenth century, based on discrete responses and contiguity, had to be replaced by a view based on the dynamics of behavior. Behavior is manifestly process and cannot be contained in little momentary packets. Behavior is flow and requires measures and theories that acknowledge its temporal extendedness. Such a new ("molar") view offers plausible and elegant explanations both of laboratory results and of phenomena in the everyday world. The molecular view falls short on both counts.

The molar view derives from two innovations introduced by Skinner: (a) measuring behavior as rate and (b) stimulus control. Thinking in terms of rate of activity allowed theory and explanation to adopt a dynamic approach to behavior. Stimulus control replaced S—R bonds and elicitation as precepts, by substituting "modulation" of activity rate. Although intended to replace simple connectionism, though, stimulus control was described in vague terms. Skinner (1938, 1953) and others said that it "set the occasion" for activity to produce consequences. One might have inferred some hidden connection between context (S^D) and activity, but rarely, if ever, was this made explicit. Some sort of mechanism remained to be specified, and, to my way of thinking, induction filled that lacuna.

Before induction became paramount, Herrnstein (1961, 1970) took another major step: measuring outcomes as rates. I will never forget the first time Herrnstein drew on the blackboard a schematic of the dependence of outcome (reinforcer) rate on activity (response) rate—a curvilinear relation approaching an asymptote. I took this to be a feedback function and the organism and environment to constitute a feedback system.

The second change required for behavior analysis to stand on a sound scientific footing is integration with evolutionary theory. The biological sciences and increasingly the social sciences take evolutionary theory as the conceptual framework for thinking about life and society. Shockingly few behavior analysts think in evolutionary terms or even evince any knowledge of evolutionary theory.

The link to evolution became clear to me when I found I could abandon the notions of reinforcement and strength and instead rely on the concept of *induction* put forward by Eve Segal (1972). She had already laid the groundwork, and all I had to do to build on her work was to make explicit the idea of induction of operant behavior by so-called "reinforcers," "punishers," and "unconditional stimuli." Stimulus control was already implicitly covered by induction; I only needed to make the connection explicit. I was then able to see that these effective events were really *inducers*, inducing activities that enhanced or mitigated their effects. All inducers have one feature in common: they all directly or indirectly affect the likelihood of reproducing—reproductive success or fitness. Thus, I coined the term *Phylogenetically Important Event* (PIE), pointing to the connection to natural selection.

The result I offer here is a basis for a true natural science of behavior that links to biology and anthropology through evolutionary thinking. On one hand, it brings behavior analysis together with ethology, behavioral ecology, and evolutionary biology. On the other hand, it brings behavior analysis together with studies of culture and cultural evolution.

I included in the book only articles that put forward the philosophical and theoretical framework for an effective natural science of behavior. The only quantitative paper is the analysis of avoidance (Baum, 2020; Chapter 10), which I included because it presents the strongest support for induction and the molar view and the greatest challenge to the molecular view. A great deal of empirical work that supported the new framework is omitted. The interested reader will easily find those works in the references and online.

The book is organized into three parts. The first, "Multiscale Behavior Analysis," concerns primarily the framework based on the flow-like nature of behavior and the link to evolution. The second part, which necessarily overlaps with the first part, offers a pure form of behaviorism, building on B. F. Skinner's work, but correcting flaws in Skinner's thinking. For want of a better name, I called this "Molar Behaviorism," but it is really just an update of what Skinner called "radical behaviorism." The third part contains some articles on culture from a behavioral point of view, linking to anthropology.

I am indebted to others for help and support as these ideas unfolded. Most importantly, my mentor, Richard J. Herrnstein guided me, challenged me, and encouraged me. Also of invaluable help was my friend and peer Howard Rachlin. Often Phil Hineline contributed too.

February 10, 2022, Walnut Creek, CA.

Acknowledgments

Chapter 1
Source: Originally published in *Journal of the Experimental Analysis of Behavior,* 20 (1973), pp. 137–153. Reproduced with permission of John Wiley & Sons. Preparation of this manuscript was supported by grants from the National Science Foundation, National Institutes of Health, and National Institute of Mental Health to Harvard University. The author thanks R.J. Herrnstein and H. Rachlin for their helpful advice and criticism.

Chapter 2
Source: A version of this paper was presented at the meetings of the Association for Behavior Analysis, Philadelphia, May 1988. This work was supported by Grant BNS 84-01119 from the National Science Foundation. Originally published in *The Behavior Analyst,* 12 (1989), pp. 167–176. Reproduced with permission of Springer Nature and Association for Behavior Analysis International.

Chapter 3
Source: This paper was originally given in a symposium at University of Nevada, Reno. It was published as a chapter in L.J. Hayes and P.M. Ghezzi (eds.), *Investigations in Behavioral Epistemology*. Context Press, an imprint of New Harbinger Publications, 1997. Reproduced with permission of New Harbinger Publications, Inc.

Chapter 4
Source: Originally published in *Journal of the Experimental Analysis of Behavior,* 78 (2002), pp. 95–116. Reproduced with permission of John Wiley & Sons.

Chapter 5
Source: Originally published in *The Behavior Analysis Today,* 4 (2003), pp. 78–81. © 2003 William M. Baum. Reproduced with permission of American Psychological Association.

Chapter 6
Source: Originally published in *Behavioural Processes,* 66 (2004), pp. 349–359. Reproduced with permission of Elsevier. The author thanks Michael Davison for many helpful comments on earlier versions.

Chapter 7
Source: Originally published in *Journal of the Experimental Analysis of Behavior,* 97 (2012), pp. 101–124. Reproduced with permission of John Wiley & Sons. I thank Howard Rachlin, John Staddon, and Jesse Dallery for helpful comments on earlier versions of this paper.

Chapter 8
Source: Originally published in *Managerial and Decision Economics*, 37 (2016), pp. 239–248. Reproduced with permission of John Wiley & Sons. The author thanks Howard Rachlin for many helpful comments.

Chapter 9
Source: Originally published in H.L. Miller (ed.), *The SAGE Encyclopedia of Theory in Psychology*, pp. 795–798. Thousand Oaks: Sage Publications, Inc., 2016. Reproduced with permission of Sage Publications, Inc.

Chapter 10
Source: Originally published in *Journal of the Experimental Analysis of Behavior*, 114 (2020), pp. 116–141. Reproduced with permission of John Wiley & Sons.

Chapter 11
Source: This paper contains portions of the English version of a book chapter published in Portuguese in D. Zillo and K. Carrara (eds.), *Behaviorismos*. Vol. 2. Sao Paulo, Brazil: Paradigma. The author thanks Howard Rachlin and Tim Shahan for thoughtful comments on earlier drafts. Originally published in *Journal of the Experimental Analysis of Behavior*, 110 (2018), pp. 302–322. Reproduced with permission of John Wiley & Sons.

Chapter 12
Source: Originally published as "Behavior, process, and scale: Comments on Shimp (2020), 'Molecular (moment-to-moment) and molar (aggregate) analyses of behavior'," in *Journal of the Experimental Analysis of Behavior*, 115 (2021), pp. 578–583. Reproduced with permission of John Wiley & Sons.

Chapter 13
Source: Originally presented as the B.F. Skinner Memorial Address at the 1995 Convention of Behaviorology and published in *Behaviorology*, 3 (1995), pp. 93–106. Reproduced with permission of *Behaviorology* journal.

Chapter 14
Source: Originally published in *Behavior and Philosophy*, 35 (2007), pp. 57–60. Reproduced with permission of Cambridge Center for Behavioral Studies.

Chapter 15
Source: Earlier versions of this paper were presented at Association for Behavior Analysis, May 1995, and American Psychological Association, August 1995. The author thanks Howard Rachlin for thoughtful comments on an earlier draft of the paper. Originally published in *The Behavior Analyst*, 34 (2011), pp. 185–200. Reproduced with permission of Springer Nature.

Chapter 16
Source: Originally published in *Behavior and Philosophy*, 45 (2017), pp. 63–78. Reproduced with permission of Cambridge Center for Behavioral Studies.

Chapter 17
Source: Originally published as "Berkeley, realism, and dualism: Reply to Hocutt's 'George Berkeley resurrected: A commentary on "Baum's Ontology for Behavior Analysis"'," *Behavior and Philosophy*, 46 (2018), pp. 58–62. Reproduced with permission of Cambridge Center for Behavioral Studies.

Chapter 18
Source: Originally published as "What is Suicide? Comments on 'Can Nonhuman Animals Commit Suicide?' by David Peña-Guzmán" in *Animal Sentience*, 20 (2018). © William M. Baum.

Chapter 19
Source: Originally published in *Perspectives on Behavior Science*, 42 (2019), pp. 283–289. Reproduced with permission of Springer Nature.

Chapter 20
Source: This paper is gratefully dedicated to my teacher, Richard J. Herrnstein. A version was presented at the Association for Behavior Analysis meeting in Atlanta, May 1994. The author thanks P.N. Hineline, A.S. Kupfer, J.A. Nevin, H. Rachlin, M.E. Vaughan, and G.E. Zuriff for helpful comments on earlier drafts. Originally published in *The Behavior Analyst*, 18 (1995), pp. 1–21. Reproduced with permission of Springer Nature.

Chapter 21
Source: The author thanks F. Tonneau, N. Thompson, and R. Hinde for many helpful comments. Originally published in N.S. Thompson and F. Tonneau (eds.), *Perspectives in Ethology: Evolution, Culture, and Behavior* (Vol. 13), pp. 181–212. New York: Kluwer Academic/Plenum, 2000. Reproduced with permission of Springer Nature.

Chapter 22
Source: I thank M. Ghiselin, S. Glenn, K. Panchnathan, P. Richerson, and H. Rachlin for helpful comments on an earlier draft. Originally published in M. Tibayrenc and F.J. Ayala (eds.), *On human nature: Psychology, ethics, politics, and religion*, pp. 397–415. New York: Academic Press, 2017. Reproduced with permission of Elsevier.

Part I

Multiscale Behavior Analysis

1

The Correlation-Based Law of Effect

Foreword

This article was my first attempt to lay out the molar view of behavior. Reading it in the year 2021, I see both strengths and shortcomings. The paper explains the molar view clearly, but with one error and omission of two essential concepts that came later: the link to evolution and the centrality of induction.

If I were writing this now, my vocabulary would be different. In particular, the word *reinforcer* should be reserved for an event, and the word *reinforcement* should not be used to denote an event but a process of behavior change.

The error I made was in failing to recognize that the molecular view of behavior constitutes a different paradigm from the molar view and, consequently, differs ontologically. This error led me to use "molecular" as synonymous with "small scale" and "molar" as if it meant "large scale." I distinguished discrete responses from activities extended in time, but lacked the concepts of scale and individual. Thanks to meeting with a philosophy of biology group at the California Academy of Science from 1999 onwards, I learned about the ontological category of *individual* and the necessity of part–whole relations. I did not begin applying these concepts until 2001.

Having reviewed the book in which Eve Segal's paper on induction appeared in 1972, I had the concept but did not realize its full significance until later, when I realized that not only non-operant activities but also operant activities were induced.

As I thought critically about the concept of reinforcement that I had inherited via my undergraduate education, I came to realize that it contained a fundamental deficiency. Whether one adopts the concept of reinforcement or utility or maximization or matching, any of these concepts requires a mechanism. Herrnstein (1970) put the matter well when he pointed out that "adaptation" is a question, not an answer.

The implicit requirement of a mechanism to explain how behavior increases or decreases led behavior analysts to propose "strength" as an unseen driver of response rate. Response strength, however, has all the defects of any hypothetical explanation, particularly circularity and untestability. Worse than strength, however, was the unacknowledged appeal to agency. Economic models were particularly prone to leaving a void on mechanism that agency would fill implicitly. If one claims that an organism behaves so as to maximize utility, the burden of mechanism falls on the organism; the organism maximizes. In the absence of any other mechanism, the agency of the organism fills the

Science and Philosophy of Behavior: Selected Papers, First Edition. William M. Baum.

void, and we are left with some sort of inner self choosing and perceiving and deciding. We have failed to escape pre-scientific folk psychology.

I finally realized that induction filled the mechanism gap. The present paper proposed a feedback system including environment and organism—E-Rules and O-Rules. Much could be said about E-Rules—feedback functions and so on—but the paper contains almost nothing about O-Rules. Nowadays I put induction in as the primary O-Rule.

Induction deals with two long-standing problems: how reinforcement and punishment work and what accounts for their efficacy. Adopting the idea of the behavior–environment feedback system, induction, as an O-Rule, completes the loop, because a reinforcer induces operant activity, the operant activity produces the reinforcer, and the reinforcer induces the operant activity—on and on.

Realizing that the events called reinforcers, punishers, and unconditional stimuli induce non-operant activities implies a role for evolution to account for their efficacy. Activities that mitigate fitness-threatening events—encountering a predator, sustaining an injury, and so on—play a role in natural selection, because organisms that fail in that regard leave fewer offspring. Similarly, activities that enhance fitness-promoting events—encountering a mate, gaining food, and so on—allow organisms that deal with them effectively to leave more offspring. Thus, I refer to these fitness-affecting events as *Phylogenetically Important Events.* Instead of value—a term that fails to escape agency—I now call *V competitive weight,* recognizing that activities compete for the time available, which is always limited. Competitive weight depends on induction and is measurable. These ideas appeared in later publications.

Abstract

It is commonly understood that the interactions between an organism and its environment constitute a feedback system. This implies that instrumental behavior should be viewed as a continuous exchange between the organism and the environment. It follows that orderly relations between behavior and environment should emerge at the level of aggregate flow in time, rather than momentary events. These notions require a simple, but fundamental, change in the law of effect: from a law based on contiguity of events to a law based on correlation between events. Much recent research and argument favors such a change. If the correlation-based law of effect is accepted, it favors measures and units of analysis that transcend momentary events, extending through time. One can measure all consequences on a common scale, called *value.* One can define a unit of analysis called the *behavioral situation,* which circumscribes a set of values. These concepts allow redefinition of reinforcement and punishment, and clarification of their relation to discriminative stimuli.

Keywords: molar view, molecular view, correlation, reinforcement, punishment, behavioral situation

Source: Originally published in *Journal of the Experimental Analysis of Behavior,* 20 (1973), pp. 137–153. Reproduced with permission of John Wiley & Sons. Preparation of this manuscript was supported by grants from the National Science Foundation, National Institutes of Health, and National Institute of Mental Health to Harvard University. The author thanks R.J. Herrnstein and H. Rachlin for their helpful advice and criticism.

Introduction

The traditional view of the law of effect makes contiguity between a response and a reinforcer central. It holds that reinforcement strengthens whatever response is contiguous with it, and that a response must be contiguous with reinforcement to be strengthened. Accordingly, a contingency would operate by ensuring contiguity of certain responses with reinforcement.

Recently, a number of authors have criticized this reliance on sheer response-reinforcer contiguity, and have tried to restate the law of effect in more global terms (Herrnstein, 1969, 1970; Seligman, Maier, & Solomon, 1971; Staddon & Simmelhag, 1971; Bloomfield, 1972). The present paper attempts to show that such a reinterpretation of the law of effect follows directly from the understanding that the organism and its environment constitute a feedback system. It attempts also to elucidate a notion of *correlation* that can replace mere contiguity. Finally, it attempts to show some of the implications of the new view.

Instrumental Behavior and Feedback

The opening sentence of *Schedules of Reinforcement* (Ferster and Skinner, 1957) reads: "When an organism acts upon the environment in which it lives, it changes that environment in ways which often affect the organism itself." This statement implies that an organism's relations with its environment can be treated as a closed chain of events: the environment affects the organism's behavior, the organism's behavior changes the environment, the environmental changes again change the organism's behavior, and so on. Although it has long been recognized that behavior, through its consequences, feeds back to the organism—even before Thorndike (Morgan, 1894)—an obvious, but fundamental, implication of the relation has been overlooked until recently.

The Organism-Environment System

Consider how the organism and its environment can be likened to a feedback system. Figure 1.1 diagrams the interactions in the organism-environment system. The experimenter manipulates the E-rules by which the organism's behavior (output) affects the environment, and attempts to discover the O-rules (functional relations) by which the environmental consequences (feedback) affect the organism's behavior. Some of the organism's output is also fed back directly by somesthesis. This loop within the organism is a logical necessity in characterizing the system, because variables such as effort expenditure produce important consequences internal to the organism. These variables are measurable, however, because they can produce effects (e.g., force or work) in the environment as well as within the organism. When a procedure differentially reinforces effort expenditure, the external consequences (reinforcement) will tend to increase effort expenditure, while the internal consequences will tend to keep it from increasing. An internal loop is no different in principle from an external loop. It can be studied from its external effects.

A complete description of the system will include quantitative specification of both 0-rules and E-rules. The 0-rules are familiar as Skinner's (1938) "functional relations". They describe the control of feedback over output. A possible example might be a dependence of response rate on rate of reinforcement. The E rules are feedback loops or

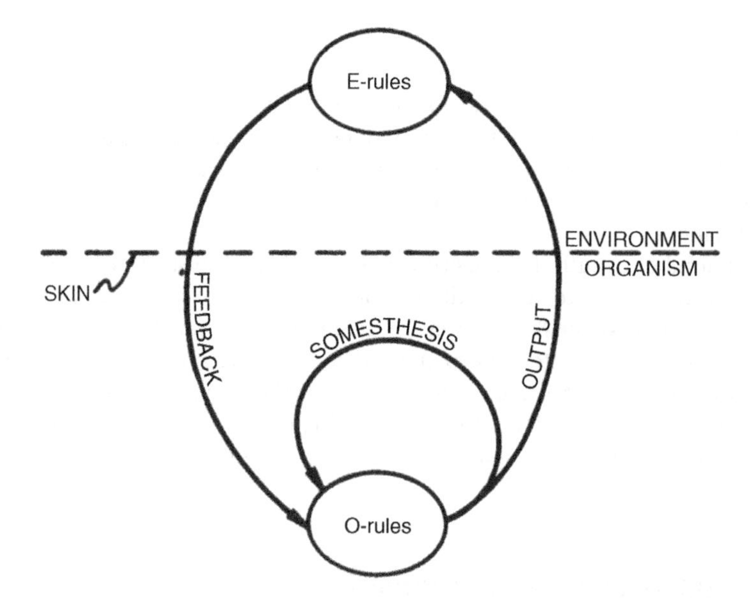

Figure 1.1 Schematic view of the organism-environment system, showing the E-rules that determine feedback to the organism and the 0-rules that determine output from the organism.

feedback functions. By governing feedback, which in turn governs the output (behavior), they cause the output to control itself.

Feedback Functions

A feedback function does not correspond exactly to the usual meaning of a contingency of reinforcement. A contingency is often thought of as a verbal statement of the conditions necessary for reinforcement to occur. A feedback function expresses the quantitative relation a contingency imposes between output and feedback. The contingency "fixed-ratio 5", for example, specifies that every fifth response will be reinforced. A possible feedback function imposed by FR 5 would be: r = 0.2B, where B is response rate and r is rate of reinforcement. Since there will be five responses for every reinforcement, the rate of reinforcement will always be one fifth of the response rate.

Another example would be the function imposed by a variable-interval schedule. A simple version might be:

$$r = \frac{1}{t + .5\left(\frac{1}{B}\right)} \tag{1.1}$$

where r is rate of reinforcement, t is the average scheduled interval, and B is the response rate. The equation states that the rate of reinforcement equals the reciprocal of the average interreinforcement time, which equals the scheduled interval plus half the average inter-response time (1/B). This estimate depends on the assumption that the scheduling of the intervals is unrelated to the distribution of responses in time—that the scheduling of a reinforcement is equally probable at any point within the interresponse time (1/B). This holds as long as the scheduled intervals are not too short or the response rate is not too low. If the scheduled intervals are frequently shorter than the interresponse time, the interreinforcement intervals can exceed the scheduled intervals by more than half the interresponse time. As the scheduled intervals become shorter and shorter, or the interresponse times become longer and longer, the schedule must become functionally more and more similar to continuous reinforcement (CRF or FR 1). When the shortest interresponse time equals or exceeds the longest scheduled interval, Equation 1.1 no longer holds. The feedback function changes to r = B, the relation specified by FR 1.

The broken curves in Figure 1.2 represent some functions from the family produced by varying *t* in Equation 1.1. Each curve rises rapidly to its asymptote, the scheduled rate of reinforcement. Above some low response rate—lower for lower rate of reinforcement—increases in response rate produce no discernible increase in rate of reinforcement. The function for FR 1, the line r= B, appears also. Where each curve intersects this line, the curve ceases to hold, and the line takes over. The circles show the performances of a typical pigeon [Catania & Reynolds (1968), Pigeon 279]. The solid curve was fitted to these data by Herrnstein (1970).

Although any feedback function is derived from properties of the schedule employed, it is more than a description of the apparatus. Whereas the conditions of reinforcement imposed by the apparatus usually can be expressed exactly, specifying feedback functions

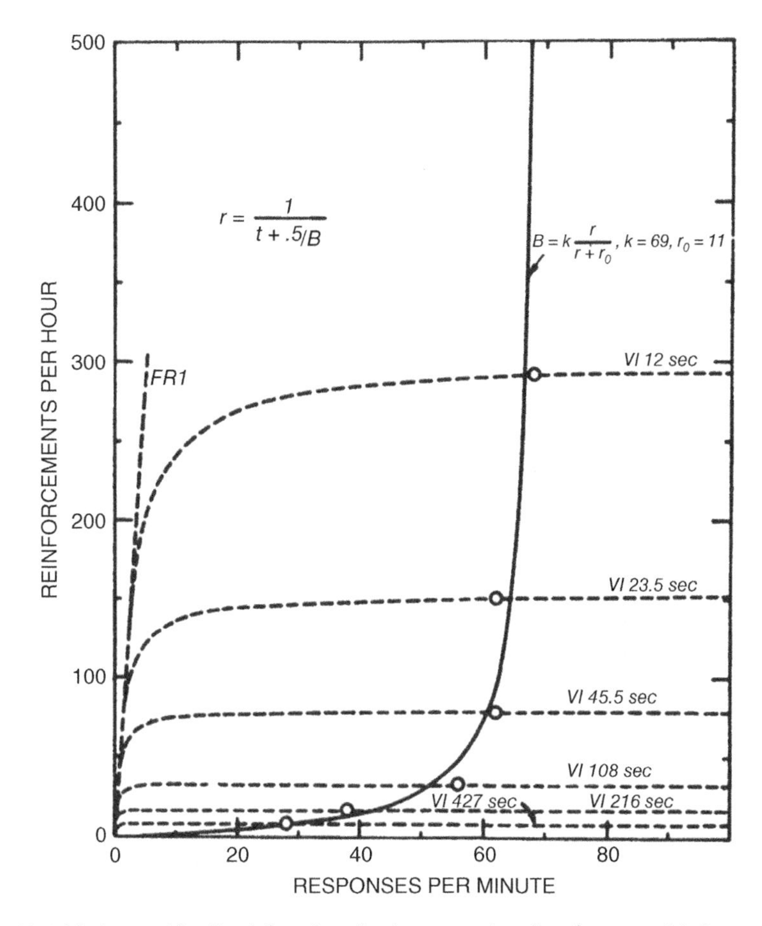

Figure 1.2 Variable-interval feedback functions (broken curves) and performance (circles and solid curve).
Note: The broken line represents the reinforcement function of FR 1. The data are from Catania and Reynolds (1968, Figure 1.1, Bird 279). The solid curve was fitted to the points by Hermstein (1970).

is partly an empirical problem, because the variables that will produce orderly description of performance cannot always he determined in advance. Since it must meet a criterion of conformity to data, a proposed feedback function constitutes part of a theory of performance. In general, it summarizes a number of assumptions about the system. In particular, it states the parameters of feedback and output, and specifies how they are to be scaled. For example, if food intake through time is important to the organism, then both rate of food presentation and amount of food at presentation must be important. How can these two parameters be combined into one scale? When a pigeon's pecking produces grain, one may need only to multiply rate and duration of reinforcement (Neuringer, 1967; Ten Eyck, 1970; Rachlin & Baum, 1969, 1972). A feedback function would summarize the rule for such combination.

The Primacy of Time

An implicit assumption underlies the foregoing discussion: that time is a fundamental dimension of all interactions between behavior and environment. The control depicted in Figure 1.1 occupies time. Performance of the system can be assessed only as it extends through time. This means that no particular momentary event should be seen in isolation, but rather, as part of an aggregate, a flow through time. The relations suggested in Figure 1.1, then, are not relations among momentary events, but a continuous exchange.

Continuous flow is measured as a rate. That emphasis on time inevitably leads to measurement of rates may be understood from the dependence of behavior on physiological needs. Under normal circumstances, the maintenance of the body requires energy expenditure. As time passes, energy resources are depleted. The rate at which the energy reserve is restored must be crucial, because continual failure to offset output will result in death. Rate of energy utilization must, therefore, govern required rate of food and water intake. These basic physiological needs only exemplify, however, the importance of rate in all reinforcement. In normal exchanges, no reinforcer fails to increase in efficacy as deprivation increases or to decrease in efficacy as satiation proceeds. Emphasis on rate of feedback leads inevitably to emphasis on rate of output, as well. If feedback is characterized as a flow through time, then the output governing and sustaining this flow must be similarly continuous. This is not to say that feedback and output undergo no temporary interruptions. It is to say that order in the interactions between behavior and environment appears at the level of aggregate flow in time, rather than momentary events.

A substantial body of research points to more orderly description on such a molar level, rather than on a momentary, molecular level. Herrnstein (1970) summarized evidence that many phenomena of positive reinforcement can be understood as dependencies between rate of responding and rate of reinforcement. It appears that aversive control also can be understood best on this molar level, as a relation between rate of responding and rate of punishment (Schuster & Rachlin, 1968; Herrnstein & Hineline, 1966; de Villiers, 1972). In general, then, this analysis takes rate as a basic dimension of all feedback.

Maintenance and Acquisition

An engineer studying a physical equilibrium system—an amplifier, for example—varies the input or feedback, and measures the output. He asks two kinds of question. First, when the system changes, what new equilibrium becomes established? Second, what is the course of change in the output as it approaches equilibrium? The first asks about stable performance; the second asks about transient performance going from one stable performance to another.

When the feedback to the organism changes, its output, like that of the amplifier, goes through a transient phase, moving, quickly or slowly, directly or with oscillation, toward a new equilibrium. Sometimes a researcher calls the observed change in performance learning; sometimes he accepts it simply as a change from one condition to another. Usually, the nature of the change in situation determines whether or not the transient is called learning. If a new response is made available, it is called learning; if the level of deprivation is changed, it is not. When amount of reinforcement or a schedule parameter is changed, sometimes the transient is called learning, and sometimes not. In all cases, however, the transient results from a change in parameters of either the E-rules or the 0-rules (Figure 1.1)—of either the contingencies (reinforcement or punishment) or the physiological state of the organism. One may question whether the understanding of these performance transients is in any way aided by labelling some of them "learning".

Since more is understood of the organism-environment system at equilibrium than in transition, we are concentrating on stable performance—that is, the maintenance of behavior. It is important to remember, however, that the approach to behavior described in this paper is no way limited to stable performance. The full understanding of the organism-environment system depends on the study of both equilibrium and transition.

Correlation *versus* Contiguity

Correlation and Contingency

Although the flow of feedback may be virtually endless, some finite sample of feedback must control output. Events remote in the past have little influence, but present behavior never depends solely on present circumstances. Since the samples controlling behavior must be finite, they must be subject to error.

A feedback function specifies a regression curve around which the samples of feedback and output vary. A contingency, therefore, establishes a *correlation* between output and feedback. This correlation determines performance.

Staddon and Simmelhag (1971) proposed the following statement of the law of effect:

> If, in a given situation, a positive correlation be imposed between some aspect of an animal's behavior and the delivery of reinforcement, that behavior will generally come to predominate in that situation. (p. 17)

Their use of "correlation" suggests a conception of the effects of contingency similar to the one in this paper. They distinguish two types of principles governing performance: principles of variation, which include all the effects of reinforcers and punishers not contingent on behavior—mainly the result of phylogeny—and principles of reinforcement, which govern the selection of behavior by reinforcement and punishment. Procedures such as autoshaping and classical conditioning operate only through the principles of variation, because they impose no contingency between behavior and consequence: "for all practical purposes, classical conditioning may be defined operationally as a class of reinforcement schedules that involve presentation of reinforcement independently of the subject's behavior (p. 27)." The effects are "due in part to a reinforcement schedule that happens to prescribe no correlation between the delivery of reinforcement and the subject's behavior (p. 27)." In the present context, such non-correlation procedures omit feedback. They are open-loop systems. Since this paper concerns the law of effect, or feedback, it is limited to consideration of the effects of procedures that impose a correlation between behavior and reinforcement or punishment.

Figure 1.3 illustrates, with hypothetical data, the relationships in a correlation. Part A shows the kind of temporal relation that characterizes a typical schedule of reinforcement. Suppose that it shows performance on a VR schedule. Time proceeds from left to right. The upper line shows the distribution of responses, the lower the distribution of reinforcements. Note that although only some responses produce reinforcement, each reinforcement follows immediately upon a response. The broken vertical lines delineate two equal time samples. The response rate in the first sample is higher than in the second. Appropriately, the rate of reinforcement also is higher in the first sample than in the second, because ratio schedules make rate of reinforcement always directly proportional to the response rate. A VR feedback function appears in Part C. Since the time samples are limited in size, they conform only approximately to the function imposed by the schedule. When one plots the rates of reinforcement and responding from various time samples like those in Part A, they produce points like those in Part C that cluster around the ideal feedback relation, which is really the regression line of the correlation between rate of reinforcement and rate of responding.

According to the traditional view of reinforcement, responding on such a schedule depends on the close contiguity between some responses and reinforcement. The

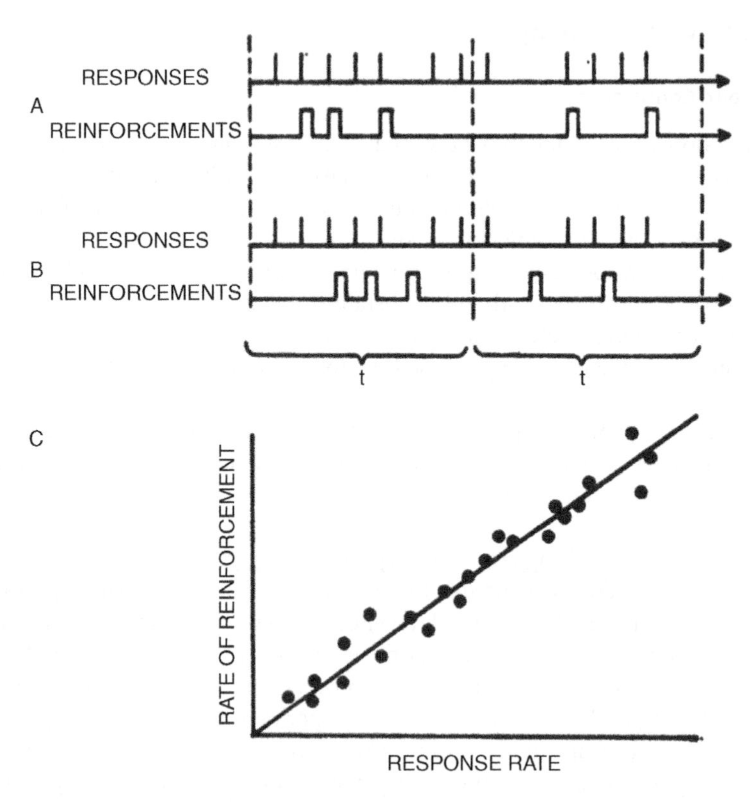

Figure 1.3 Contingency as correlation.
Note: **A**: Close contiguity between responses and reinforcement, as in a usual schedule of reinforcement. Time progresses from left to right. The broken vertical lines delimit two equal time samples. **B**: Correlation without close contiguity. The rates of responding and reinforcement in the two time samples are the same as in A. **C**: An example of a correlation—the relation between rate of reinforcement and response rate imposed by a variable-ratio schedule. See text for explanation.

correlation-based law of effect suggests that simple response-reinforcer contiguity cannot account for instrumental behavior—that the molar relation between responding and reinforcement is crucial. It does imply a definite role for contiguity, however. We will discuss the relationship between contiguity and correlation a little later. First, since most writing on the law of effect has emphasized response-reinforcer contiguity, whereas the correlation-based view calls for its de-emphasis, we must assess the adequacy of the contiguity-based law of effect.

The Contiguity-Based Law of Effect

Thorndike's (1911) original statement of the law of effect reads:

> Of several responses made to the same situation, those which are accompanied or closely followed by satisfaction to the animal will, other things being equal, be more firmly connected with the situation, so that, when it recurs, they will be more likely to recur. (p. 244)

Two underlying notions about behavior are apparent: first, that strengthening depends on close contiguity between response and reinforcement ("satisfaction"), and second, that response and reinforcement should be conceived as discrete events, occurring at certain moments in time. The two ideas are logically connected. Temporal contiguity between events presupposes a point in time at which they coincide. These notions have long historical momentum, because contiguity and momentary events are basic to description of behavior as composed of reflexes. It is not surprising, therefore, that psychologists after Thorndike (e.g., Hull, Skinner, Spence, and Mowrer) conceived of instrumental behavior as the outcome of pairing of momentary events.

This approach has been criticized recently along three lines. One line would focus on the phrase "of several responses" in Thorndike's statement. It suggests that all responses are equally susceptible to strengthening by a given reinforcer. Recent work has shown, however, that a reinforcer usually is more effective in strengthening some responses than others (Breland & Breland, 1961; Seligman, 1970; Staddon & Simmelhag, 1971). Although phylogenetic biases may constrain the law of effect in important ways, they bear little on the theme of this paper, because our concern is not to delimit, but better to describe.

A second line of criticism attacks the concepts of discrete response and reinforcement. We will consider this in the next section, when we discuss the molar view of behavior and consequences.

The third line of criticism, perhaps the most fundamental, questions the adequacy of contiguity itself. The arguments have arisen from the study of two separate phenomena: avoidance and conditioned reinforcement.

Avoidance

The observation that organisms avoid unfavorable events poses a serious problem for the requirement of reinforcement contiguity in the law of effect. A man will not only flee a fire in his house; he will take precautions against fire. A rat will not only jump out of a chamber in which it is being shocked; it will jump out of a chamber in which it has been shocked in the past, if by doing so it avoids the shock. In both examples, no obvious reinforcement follows the behavior to maintain it. How then is the law of effect to account for avoidance?

Two-Factor Theory

Although the contiguity-based law of effect cannot explain avoidance, it does provide a simple explanation for escape. The way to account for avoidance was to recast it as escape: to suppose some unobservable reinforcer contiguous with the response. If an animal were exposed to electric shock, the stimuli associated with the shocks might, through Pavlovian conditioning, come to evoke much the same autonomic reaction *(e.g.,* "fear") as electric shock. The responses that terminated the conditioned aversive stimuli and the reaction they produced would be maintained by negative reinforcement. Thus, avoidance could be thought of as escape maintained by conditioned negative reinforcement ("fear" reduction).

This theory of avoidance, known as two-factor theory, because it appealed to both classical and instrumental conditioning, has been thoroughly reviewed and criticized by Herrnstein (1969). It quickly encountered difficulty in the observation that avoidance could be maintained in the absence of termination of any obvious exteroceptive conditioned aversive stimuli (Sidman, 1963). To account for such performances, two-factor theorists had to postulate unobserved conditioned aversive stimuli in addition to unobserved reinforcement *(e.g.,* Anger, 1963). This modification, Herrnstein said, made the theory irrefutable, because it could no longer be subjected to empirical test. As an alternative explanation of avoidance, Herrnstein offers the simple proposition that the behavior can be acquired and maintained by reduction in frequency of aversive stimulation. An experiment by Herrnstein and Hineline (1966), in which rats were trained to press a lever solely in order to reduce the frequency of unavoidable shocks, supports his thesis. Herrnstein views avoidance, therefore, as acquired and maintained through negative reinforcement, which he defines as reduction in frequency of aversive stimulation. It follows that the conditioned stimulus in discriminated avoidance and the passage of time in free-operant avoidance (Anger, 1963) need not be thought of as conditioned aversive stimuli. Instead, Herrnstein suggests, these stimuli serve simply as cues or signals to the organism. They are discriminative stimuli correlated with the contingency between responding and reduction in frequency of aversive stimulation.

Conditioned Reinforcers

Not every response in a sequence need be reinforced for the behavior to be acquired or maintained. This commonplace observation poses difficulty for the requirement of response-reinforcement contiguity. The carpenter building a house does not demand food after every nail he drives. He does not require reinforcement at each new stage of construction. And when he is done, he will accept inedible money in place of food.

Although the problems posed by behavioral chains and avoidance may appear to differ, at least superficially, the solutions accepted have been highly similar. If stimuli paired with aversive stimulation can become conditioned aversive stimuli, then stimuli paired with reinforcement can become conditioned reinforcers. An associative process like Pavlovian conditioning played a crucial role in both explanations.

A neutral stimulus, according to this view, after numerous pairings with a primary reinforcer, acquires reinforcing capability of its own, through a purely associative process similar to Pavlovian conditioning. Indeed, the basic notion was originally Pavlov's (1927), since secondary reinforcement was implicit in secondary conditioned reflexes.

The explanation for the maintenance of behavior chains proposes that the stimuli correlated with each link in the chain serve as a reinforcer for the behavior in the previous link that produces them. Since the stimuli produced are discriminative stimuli for the next link in the chain, conditioned reinforcers are generally, if not always, discriminative stimuli.

This conception of chaining has been so widely accepted that some authors have used it to explain complex behavioral sequences in which no exteroceptive discriminative stimuli appear. Ferster and Skinner (1957), for example, explained performances on both fixed interval (FI) and fixed-ratio (FR) schedules as chains. They suggested that FI performance, in which responding is absent for a period after reinforcement and then accelerates to a moderate response rate, was mediated by a chain of behavior, only some of which was recorded by the apparatus. After reinforcement, a pigeon might, for example, walk around in a circle several times before pecking the response key, then circle a few more times, peck again, circle again, peck again, and gradually circle less and peck more, until it is only pecking.

The unrecorded behavior (circling here), Ferster and Skinner suggested, could function as a crude "clock" to mediate the apparent temporal discrimination in FI performance. In a similar vein, they proposed that FR performance, in which the ratio requirement is met by a rapid run of responses up to reinforcement, depends on the animal's "counting" its responses, each response producing a change in an interoceptive discriminative stimulus analogous to the reading of a counter. The progress of the stimulus along a continuum from its state at zero pecks to its state at reinforcement, they asserted, maintains the run of responses through conditioned reinforcement: "... at any point during a fixed ratio, a response may be reinforced because it increases the number and advances this stimulus toward the reinforced end of the continuum." (1957, p. 40).

In contrast, they explained performance on variable-interval (VI) or variable-ratio (VR) schedules, in which reinforcement occurs irregularly in relation to time and behavior, without appeal to conditioned reinforcement. They suggested instead that the rate of reinforcement produced by a VI schedule serves as a discriminative stimulus for responding (1957, p. 362), and that VR performance results simply from differential reinforcement of high response rates (1957, p. 391). The notions that a rate of reinforcement is a dimension of the environment that can acquire stimulus control and that rate of responding is a dimension of performance that can be differentiated accord well with the viewpoint of this paper. We will discuss such non-momentary (molar) dimensions a little later.

More recent work on FI and FR performance has sought explanations beyond the level of momentary response and reinforcements. The explanation of FI performance as a chain has been discredited. Dews (1962, 1965) has shown that the characteristic pattern of responding persists when periods of timeout break up the interval and interrupt responding. B. A. Schneider (1969) showed that, after extensive training, a fixed-interval becomes functionally equivalent to a period of extinction followed by variable-interval reinforcement. Both Dews and Schneider suggest that FI performance depends on more molar aspects of the experimental situation than momentary response-generated stimuli.

The move from momentary control of behavioral sequences parallels the move from momentary control of avoidance. Just as one may question the necessity of attributing hedonic value to the stimuli for avoidance, one may question the necessity of attributing hedonic value to the stimuli in a chain (positive two-factor theory). If the stimuli in avoidance serve only a discriminative function, and the behavior is maintained by the correlation between responding and shock-rate reduction, then the stimuli in a chain also may serve only a discriminative function, and the behavior be maintained by the correlation between responding and rate of reinforcement.

An experiment by Schuster (1969) directly attacked the notion of conditioned reinforcers. He gave pigeons a choice between two equal VI schedules of food reinforcement, one of which provided extra presentations of the stimuli paired with food delivery, on a superimposed FR schedule. The birds developed an aversion for the schedule with the extra presentations of the stimuli paired with food. If these stimuli acted as a conditioned reinforcer—if they had hedonic value—the birds should have preferred the schedule with the extra presentations. Instead, they chose the schedule in which the stimuli were more reliably paired with food. Schuster concluded that stimuli correlated with reinforcement exert control over behavior that produces them, not because they acquire reinforcing properties of their own, but only because they signal the availability of reinforcement[1].

J. W. Schneider (1972) cast still more doubt on the notion of conditioned reinforcers. He gave pigeons a choice between two alternatives, each consisting of two chained VI schedules. He maintained one chain constant, and varied the lengths of the components in the other. The birds' responding on the varied chain changed appropriately as the components changed: longer and earlier components sustained lower rates of responding. Although the behavior was clearly under control of the chain stimuli, the birds were indifferent in their choice between the two chains as long as the chains provided the same overall rate of primary reinforcement. Regardless of the lengths of the components, the chains were equivalent when the sums of their components were equal. If the response-produced stimuli in the chains possessed reinforcing value of their own, one would expect their value to vary with the lengths of the components, even if the overall time to primary reinforcement were constant. Since the pigeons remained indifferent, one cannot suppose the stimuli had reinforcing value that simply added to the value of the food. A theory of conditioned reinforcers could explain the simple summing of times that Schneider found only with great complexity.

It seems, therefore, that the stimuli in a chain serve a discriminative function, but not a reinforcing function. As with conditioned aversive stimuli, conditioned reinforcers can be treated simply as discriminative stimuli without hedonic value.

If we are to do without the concept of conditioned reinforcer, how can we account for the observation that originally inspired the concept? Long sequences of behavior remain intact, even though only the terminal response is actually reinforced. Schuster (1969) pointed out that the nature of the stimuli controlling behavior in the sequential links of a chain can be understood in terms of the functions that these stimuli serve in the maintenance of the organism. That is, the stimuli in a chain exert control by virtue of their relation to reinforcement. They signal that reinforcement is either closer in time or imminently available. The entire chain is organized around its ultimate outcome; the stimuli act as the cement of this organization. We will return to this question when we consider redefinition of reinforcement and punishment.

The attacks on positive and negative two-factor theory have a similar thrust. They suggest that the strict requirement of contiguity between responses and reinforcement

---

[1]A recent review by Gollub (1970) overlooked the significance of this experiment. Since the extra stimulus presentation produced a higher response rate, Gollub concluded: "This experiment does not . . . invalidate any particular theory of reinforcement, but rather corroborates Fantino (1968) that higher response rates in a terminal link produce lower preferences in the initial links" (p. 367). Herrnstein (1964) and Autor (1969) showed, however, that when high response rates are not *required,* there is no relationship between response rate and choice. Schuster's experiment does, therefore, invalidate theories of conditioned reinforcers, because the added stimulus presentations failed to enhance preference.

espoused by Thorndike, Skinner, and others unnecessarily complicates the accounts of such basic phenomena as avoidance and chaining. Greater flexibility and simplicity prevails when the law of effect is stated in terms of correlation: behavior increases in frequency if the increase is correlated with an increase in rate of reinforcement or a decrease in rate of aversive stimulation.

Contiguity and Correlation

The notion that close contiguity is necessary to the law of effect has been supported by the recognition that delay substantially reduces the effectiveness of a reinforcer. It must be understood, however, that greater delays of reinforcement usually ensure lower rates of reinforcement, as well. Chung and Herrnstein (1967), for example, studied pigeons' choices between two delayed reinforcers. Responses at each alternative produced a blackout followed by food. The experimenters varied the durations of the blackout. They found that the relative responding at the alternatives matched their relative immediacies of reinforcement, when immediacy was defined as the reciprocal of the delay. It can be seen, however, that the reciprocal of the delay of reinforcement is a rate of reinforcement: the rate of reinforcement during the delay stimulus conditions. The alternatives could be viewed as two chains, each having a terminal link consisting of response-independent reinforcement at a rate specified by the duration of the blackout. Autor (1969) showed that response-independent reinforcement is functionally equivalent to the usual response-contingent reinforcement in such a choice situation. One can say, therefore, that the pigeons' choices were governed not by the delays of reinforcement but by the rates of reinforcement in the terminal links of the alternative chains. Many studies of delay of reinforcement lend themselves to a similar analysis (e.g., Logan, 1960).

Some few studies of delay, however, do require interpretation in terms of response-reinforcement contiguity. Such procedures omit exteroceptive stimuli signaling the delay. Dews (1960), for example, trained pigeons to peck a key for reinforcements that occurred at a delay, unsignaled and independent of intervening responses. With such a procedure, the interval between a response and a reinforcement can vary from zero, if a response occurs just at the same moment as a delayed reinforcement, up to the scheduled delay, if no responses intervene between the reinforced response and reinforcement. Although reinforcement depends on responding, the contiguity between responses and reinforcements is poor—the longer the delay, the poorer the contiguity. Since Dews found, in general, that the longer was the delay the lower was the response rate, it seems clear that response-reinforcement contiguity can affect performance.

At first glance, this observation might appear incompatible with the correlation-based view. It only illustrates, however, that contiguity plays an important role within a correlation. Figure 1.3 makes it clear that some temporal grouping of responding and reinforcement is necessary for a correlation to exist. The smaller the duration of a sample, the closer the temporal grouping must be to maintain a correlation. The poorer the grouping, the poorer the correlation. The usual method of scheduling reinforcement, which makes each instance of reinforcement contiguous with a response (Figure 1.3A), makes for close temporal grouping of responding and reinforcement. Response-reinforcement contiguity, therefore, ensures a good correlation between output and feedback. It minimizes the variability around the regression function (Figure 1.3C).

Figure 1.3 suggests that contiguity may act through its effect on correlation, because poorer contiguity means more variability around the feedback (regression) function. It is possible that all effects of varying contiguity are due solely to the resulting variation in goodness of correlation (as measured by the variance around the regression function). In this way, although correlation determines performance, contiguity still retains a role as a parameter.

How important is close response-reinforcement contiguity? Temporal grouping is still possible without it. Part B of Figure 1.3 illustrates performance produced by a schedule in all respects like a variable-ratio, except that reinforcements do not necessarily follow immediately upon a response. The situation can be likened to pressing a slightly faulty elevator button. The button has to be pressed several times to summon the elevator reliably. The elevator comes only when one presses the button, but only after a variable delay. It is also true, within limits, that the more often one summons the elevator the more often it comes. Note that, just as in Part A, the higher response rate of the first time sample in Part B is associated with a higher rate of reinforcement. The points on the graph of Part C would be exactly the same from the schedule in Part B as from the schedule in Part A.

Is performance on the schedule of Figure 1.3B comparable to usual VR performance? Such schedules do indeed maintain responding, but poorer response-reinforcement contiguity produces lower and more erratic rates of responding. A computer samples response rates in equal successive intervals, and adjusts the rate of reinforcement in each interval to be directly proportional to the response rate in the one before. When the duration of the sampling interval is changed, the proportionality, and therefore the regression line, is held fixed. As a result, increasing the sampling interval loosens the correlation, whereas decreasing the sampling interval tightens it.

Figure 1.4 shows some sample performances of two pigeons, drawn from the last days of exposure to the conditions, during which day-to-day variation in response rate appeared stable. Three basic features of the procedure can be seen in the records. First, comparison of the slopes with the densities of reinforcement marks in various segments reveals the positive relation between responding and reinforcement. Second, as the sampling interval increased, reinforcement during a period of pausing became more frequent, and the pauses preceding such reinforcements grew longer. Loosening the correlation, therefore, had the side-effect of decreasing response-reinforcer contiguity. Third, as the duration of the sampling interval increased, and goodness of correlation decreased, response rate decreased. The less precise the correlation, the less it controls responding.

Herrnstein (personal communication) has done an experiment that further illustrates the lack of need for close response-reinforcer contiguity. In the absence of responding, reinforcements occurred at a low rate. Responding switched the animals to a higher rate of reinforcement, but still reinforcements occurred with irregular temporal relation to responses. The schedule produced acquisition and stable rates of responding.

The performances shown in Figure 1.4, Herrnstein's results, and Dew's (1960) results all support the conclusion that poor response-reinforcer contiguity reduces responding, but cannot eliminate it. As long as it produces food, responding persists. The correlation between responding and reinforcement stands out as the essential ingredient in instrumental behavior.

Can one do an experiment to separate the effects of contiguity and correlation? Since contiguity cannot vary without affecting goodness of correlation, it would be necessary to hold contiguity constant while varying correlation. One might, for example, study different correlations within a family (e.g., Figures 1.2 and 1.4), or correlations from different families, trying to find or create situations in which delay of reinforcement, though greater than zero, remained invariant.

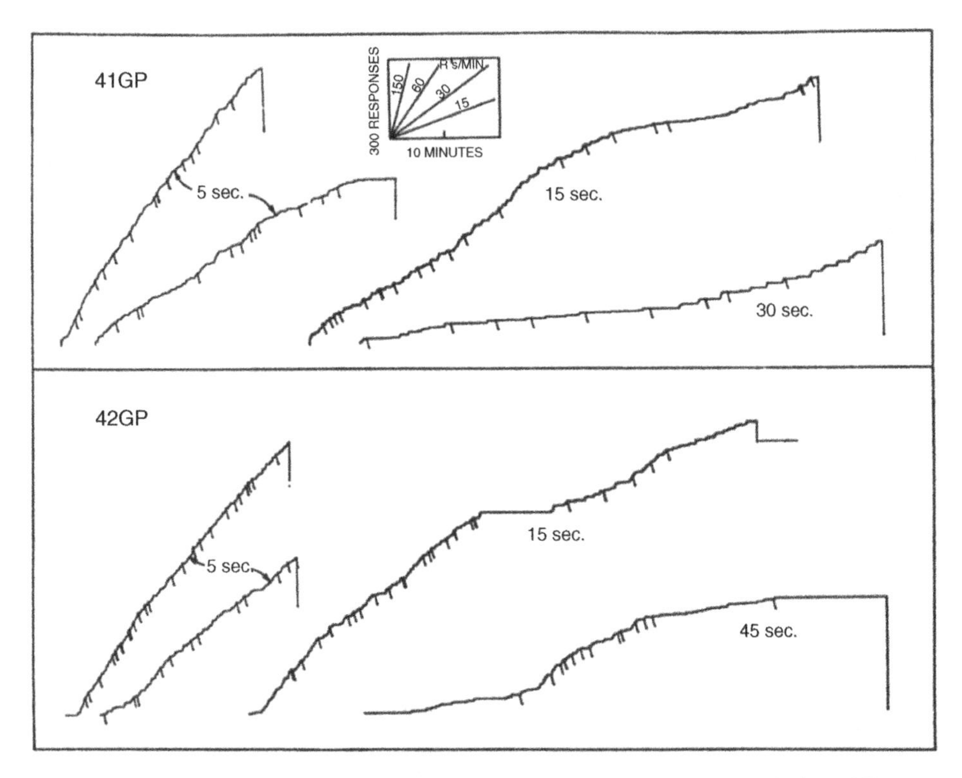

Figure 1.4 Cumulative records of performances of two pigeons, 41GP and 42GP, with three different sampling intervals (durations given in seconds) for adjusting rate of reinforcement to be proportional to response rate.
Note: The proportionality maintained is that of a VR 40. Each record shows an entire day's session. Sessions ended either after 40 reinforcements or 48 minutes. Downward deflections of the pen indicate reinforcements (4 sec access to grain). For further explanation, see text.

Although it may prove difficult to distinguish correlation and contiguity empirically, there are non-empirical reasons for favoring a law of effect based on correlation. The concept of correlation as embodied in feedback functions can be useful for describing contingencies, particularly in complex and natural situations, and as a means of developing a comprehensive theory of instrumental behavior. It applies readily to procedures such as avoidance, in which the absence of discrete consequences contiguous with responses requires awkward theorizing about unobservable events for the assumption of contiguity to hold. The concept of correlation has the additional advantage that it draws together apparently diverse procedures into a single conceptual framework. Positive reinforcement, punishment, avoidance, negative reinforcement, and DRO (punishment of responding by imposing a negative correlation between response rate and rate of reinforcement) can all be described in terms of feedback functions. Discrete-trial procedures and free-operant procedures, classical conditioning and instrumental conditioning, continuous reinforcement and intermittent reinforcement, all can be described and related within this framework. Even if it should prove impossible to distinguish experimentally between contiguity and correlation, the weight of conceptual power and simplicity seems to lie with correlation.

The Molar View

Molar Behavior

According to the contiguity-based law of effect, an organism's behavior consists of a sequence of the various responses that the organism can make. Since the responses are discrete and distinguishable from one another, the most direct method for assessing the composition of this sequence over any particular period of time (e.g., an experimental session) is to count the number of instances of each response under study.

For the requirement of response-reinforcement contiguity, it is sensible, even necessary, to assume discrete momentary responses. When we recognize that responding enters into a more molar relation with reinforcement, that contiguity is not essential, the need for assuming discrete responses disappears. The notion of correlation and the description of instrumental behavior as part of a feedback system require instead that we characterize both behavior (output) and consequences (reinforcement, punishment, and response cost: feedback) on a more molar level, transcending the momentary. As noted earlier, the concept of continuous exchange between organism and environment implies measurement that extends over time.

Procedures have begun to appear in which the conditions of reinforcement preclude the assumption of discrete responses. Brownstein and Pliskoff (1968) presented reinforcers to pigeons at variable intervals simply for being in the presence of either of two different colored lights. Each light was correlated with a particular rate of reinforcement. As long as a bird remained in the presence of a light, it continued to receive reinforcers at the light's rate of reinforcement. It could change from one color to the other by pecking a response key. Brownstein and Pliskoff found that the times spent in the presence of the lights depended in a simple and orderly way on the rate of reinforcement: the ratio of the times equaled the ratio of the rates of reinforcement. Baum and Rachlin (1969) found a similar relation between the times spent in two locations and the rates of reinforcement for being in those two locations. In both experiments, the pigeons apportioned their time between two alternatives according to the same matching law that applies to concurrent reinforcement for pecking; the proportion of time allocated to each alternative matched the relative rate of reinforcement for that alternative.

Baum and Rachlin (1969) argued that such laws of time allocation are more generally applicable than laws of response allocation, because the collective response times of discrete responses, such as key pecks or lever presses, which are highly constant in duration, are directly proportional to the number of responses. Laws of response allocation, therefore, are directly convertible to laws of time allocation. The reverse does not hold, however, because in experiments like those of Brownstein and Pliskoff (1968) and Baum and Rachlin (1969), no empirical basis exists for defining a discrete response to make the conversion from time to responses. Extending the argument, one would state the law of effect as a relation between the time spent in an activity (e.g., key pecking, being in a certain location, lever holding) and the rate of reinforcement produced by that activity.

Regardless of how we measure response frequency, whether as response rate or as proportion of time spent responding, it is a variable that must be sampled and averaged over time. Since it transcends particular instances of discrete responses, it can be called a molar variable that enters into a molar relation (correlation) with another molar variable, the consequence (e.g., rate of reinforcement, rate of aversive stimulation, *etc.*).

Molar Consequences

Much of the recent interest in rate of reinforcement as an independent variable stems from Herrnstein's (1961) finding that, in a choice between two concurrent variable-interval schedules, pigeons matched the proportion of their responses to each alternative to the relative rate of reinforcement produced by the alternative. This matching law, now well substantiated, has been obtained in a wide variety of situations (e.g., Herrnstein, 1964; Catania, 1963a, 1963b; Shull & Pliskoff, 1967; Brownstein & Pliskoff, 1968; Baum & Rachlin, 1969; see Herrnstein, 1970, for overview).

Rate of reinforcement, like response frequency, is a variable that, by definition, must be sampled over a substantial period of time. Any variable that fails to reduce to a discrete event in time implies an averaging or integrating capability on the part of the organism. Such integrating must be commonplace in an organism's reactions to its environment, because responding rarely, if ever, depends solely on a present situation. Past experience almost always plays a large role. Averaged variables like rate of reinforcement or rate of punishment (Schuster & Rachlin, 1968), therefore, suggest an analogy to a complex system, in which input data are collected over intervals of time into aggregates, and then processed as aggregates, rather than individually. Computer systems, for example, typically treat data in this manner. Such integration occurs commonly in mechanical systems as well. The continuous movement of an automobile, for instance, depends on a succession of discrete explosions in its engine. In a like manner, an organism can be viewed as collecting time samples of the significant events in its environment (e.g., reinforcers and punishers), which it integrates and utilizes to control its behavioral output. The exact nature of this integrating or averaging process has been the subject of some recent research (Killeen, 1968; Davison, 1969; Duncan & Fantino, 1970; Schneider, J. W., 1970).

Feedback to a behaving organism is more than just reinforcers and punishers. Other stimuli, perhaps "neutral" in themselves, but correlated with reinforcement or punishment—discriminative stimuli—also control behavior.

We usually characterize discriminative stimuli solely in terms of their presence or absence. They can be viewed, however, as integrated feedback in just the same manner as reinforcers or punishers. Schuster (1969), for example, showed that rate of presentation of a discriminative stimulus can have a strong effect on pigeons' preferences in a choice situation.

The suggestion of Ferster and Skinner (1957, p. 326) mentioned earlier, that responding on a variable-interval schedule depends on the rate of reinforcement acting as a discriminative stimulus, implies that the rate of occurrence, not only of a discriminative stimulus, but of a reinforcer (or punisher), can control a discrimination. One type of evidence that supports this conception is the observation that discrimination of extinction from reinforcement, measured by the number of responses made in extinction, improves with repeated extinction and reinstatement of reinforcement (e.g., Bullock & Smith, 1953). A similar improvement occurs with repeated removal and reinstatement of aversive stimulation in avoidance (Boren & Sidman, 1957).

Whatever other dimensions might characterize an event, it always will possess a rate of occurrence. Rate is the universal dimension. The rate of occurrence of a reinforcer, punisher, or discriminative stimulus can control behavior, just as the other attributes of the event can control behavior. The organism, in other words, integrates all feedback over time.

To understand the implications of this idea for our conceptions of reinforcement, punishment, and behavioral chains, we must develop two preliminary notions: value and the behavioral situation.

Consequences as Value

The notion of molar consequences suggests that all the various parameters of reinforcement and punishment can be drawn together into a single scale called *value*. At least two approaches to construction of such a scale have been suggested (Baum & Rachlin, 1969; Premack, 1965, 1971). Baum and Rachlin (1969) suggested that the proportion of time spent in an activity equals the relative value of the activity, that is, its value relative to the sum of the values of all the sources (e.g., all possible activities) in the situation:

$$\frac{t_1}{T} = \frac{v_1}{\sum_{i=1}^{n} v_i} \tag{1.2}$$

where v_i is the value of Activity i (there are n such), t_i is the time spent in Activity i, and T is the total time (implicitly assumed to be exhausted by the n activities). The absolute value (v_1) of an activity is a function of the feedback it produces. It is directly proportional to rate of reinforcement (Herrnstein, 1961) and duration of reinforcement (Catania, 1963b; Neuringer, 1967). It may be inversely proportional to delay of reinforcement (Chung and Herrnstein, 1967). The value of an activity that decreases the frequency of electric shock appears to be directly proportional to the resultant reduction in rate of shock (de Villiers, 1972). Punishing responses with electric shock, on the other hand, reduces absolute value (Holz, 1968; Schuster & Rachlin, 1969).

For a situation in which a single activity is studied alone, Equation 1.2 may be simplified as follows (*cf.* Herrnstein, 1970):

$$\frac{t_1}{T} = \frac{v_1}{v_1 + v_0} \tag{1.3}$$

where v_0 is the sum of the values of all the activities other than v_1. In a choice situation in which two responses are studied, Equation 1.2 can be written:

$$\frac{t_1}{T} = \frac{v_1}{v_1 + v_2 + v_0} \tag{1.4}$$

$$\frac{t_2}{T} = \frac{v_2}{v_1 + v_2 + v_0} \tag{1.5}$$

Note that whereas v_0 must be the same in Equations 1.4 and 1.5, v_0 in Equation 1.3 can vary from one activity to another and is not necessarily the same as in Equations 1.4 and 1.5, because different situations will produce different values in the alternative activities (grooming, walking about, *etc.*) that are not directly controlled by the contingencies of reinforcement and punishment of the experiment. The ratio of Equations 1.4 and 1.5 produces the matching equation (Herrnstein, 1970):

$$\frac{t_1}{t_2} = \frac{v_1}{v_2} \tag{1.6}$$

This states that the relative time spent in two activities equals the relative value of the two activities.

Value as Molar

Since the value of an activity depends on variables like rate of reinforcement and rate of punishment, which cannot be assessed at any particular point in time, but must be averaged over a period of time, the value of an activity must likewise be conceived to extend through time. In other words, since value depends on integrated feedback, an activity has value and changes value only over extended periods of time, and behavior varies with changes in value only over extended periods of time. A concept like *momentary* value could be meaningful only as the temporal derivative of value expressed as a function of time (just as momentary velocity is the derivative of distance with respect to time).

From moment to moment, however, the organism engages in one activity or another and switches from one activity to another. These moment-to-moment relationships among activities have little to do with value, because at any moment the organism may be engaging in an activity of any value; it simply engages more often in high-valued activities. The momentary fluctuations in an organism's activities result from momentary fluctuations in variables that have a constant average effect over extended periods of time (e.g., deprivation).

Premack (1971) pointed out that averaged behavioral measures fail to capture momentary fluctuations that are often of interest to psychologists. During the course of a session, for example, satiation may reduce an originally high level of drinking below a low, but constant, level of wheel-running. Two activities can also differ in their temporal distribution, one producing frequent small satisfactions (e.g., eating), one producing less frequent large satisfactions (e.g., copulation), and yet still be of equal value and take up equal average times. In general, the study of momentary relations is compatible with the study of average relations, however; the two complement each other.

Behavioral Situations

A set of activities will have a corresponding set of values only under certain specified conditions. Let us call these conditions the *behavioral situation*.

A behavioral situation consists of a set of possible activities, a set of possible events or stimuli, and a set of feedback functions determining the effects of the activities on the events. Figure 1.1 attempted to diagram the salient features of a behavioral situation. A variable-interval schedule, for example, provides a feedback function governing reinforcement for key pecking (Figure 1.2). We can imagine another feedback function indicating response cost, in terms of energy expenditure and loss of opportunity for alternate reinforcement (v_0 in Equation 1.3). As response rate increases, response cost increases. The interaction of these two feedback functions would determine performance. This description implicitly specifies the important activities and events: key pecking, the activities that reduce response cost, and reinforcement. To complete the description, other events and stimuli, such as response feedback, keylight, and chamber size, although perhaps of lesser important, must be specified also.

Up to now we have assigned values only to activities. We can assign values also to situations. The value of a situation usually equals the sum of the values of all its possible activities. In a chain schedule, for example, an activity in one situation (link) leads to the next situation, which contains a higher-valued activity in addition to or in place of the activity in the first situation, and therefore has higher total value. The maintenance of a behavioral chain depends on this succession of situations from lower to higher value.

When a situation contains response-independent reinforcement or response-independent punishment, then the value of the situation exceeds or falls short of the sum of the activity values. Studies of chained schedules illustrate these effects of response-independent events. Autor (1969), for example, showed that the same performance holds in the initial link of a chain if the value (rate of reinforcement) of the terminal link is the same, regardless of whether the reinforcement in the terminal link is response-contingent or response-independent. Schuster and Rachlin (1968) found a similar equivalence for punishment. They studied a concurrent-chain schedule in which the two terminal links were identical, except that in one, every response produced an electric shock, whereas in the other, electric shocks occurred at a regular rate, independent of the animal's responding. They found that preference in the initial choice link for the terminal link with response-independent shock was an inverse function of the rate of these shocks. Furthermore, the animals were indifferent between the two terminal links when the rates of shock were equal, regardless of whether the shock was response-dependent or response-independent. Assuming that the responding in the initial link reflects the values of the terminal links, these results suggest that the value of a situation depends simply on rate of punishment and rate of reinforcement, regardless of whether the punishment and reinforcement arise from behavior. The value of an activity, on the other hand, depends entirely on its producing reinforcement and punishment. In the terminal link with response-produced shocks, for example, the rate of responding was low, whereas in the terminal link with response-independent shocks, the rate of responding was high and largely independent of the rate of shock. The denominator of Equation 1.2 should represent the value of the situation, rather than the sum of the values of the activities (Rachlin & Baum, 1972).

Definition of Reinforcement and Punishment

Now that we have developed the notions of value and situation, we can redefine reinforcement and punishment in terms consistent with the correlation-based law of effect. Reinforcement can be viewed as a transition from a lower-valued situation to a higher-valued situation. A simple schedule of reinforcement, for example, periodically produces a situation in which eating is possible—a maximal-valued situation for a hungry organism. Rate of reinforcement, therefore, could be thought of as rate of situation transition—that is, rate at which, in a lower-valued situation, transitions into a higher-valued situation occur.

Punishment can be viewed as the converse of reinforcement—that is, transition from a higher-valued situation to a lower-valued situation. Punishment by timeout from reinforcement arranges such a situation transition. Electric shock or another noxious event can· be thought of as a low-valued situation, even if briefly presented.

A behavioral situation can be likened to a room with several exit doors. The value of the situation depends on what other situations lie behind the doors. The feedback functions control parameters such as frequency and duration of exits. The organism might, for example, find itself thrust briefly through the door to a situation including electric shock, and then returned. Or it might be moved into another situation, remain there for a time, and then exit from that one into a third situation. This conception of reinforcement and punishment depends on a generalized notion of chaining. It suggests that instrumental behavior can be viewed as moving the organism through *chains of situations.* Where we conceive a chain to stop may be a matter of convenience. We generally stop at the last measurable event outside the organism's skin. Food presentation to a hungry animal, for example, we usually consider a terminal situation, even though it makes eating possible. Electric shock is seen as a brief presentation of a highly unfavorable situation, even though it stimulates pain receptors and produces autonomic responses. These stopping places are probably arbitrary, at least in principle.

Describing instrumental behavior in terms of chains of situations accords well with the view that conditioned reinforcers have no hedonic quality of their own, but serve rather as signals providing information about availability of reinforcement and punishment. In fact, we can define a *discriminative stimulus* as a signal of a situation transition—that is, a stimulus correlated with a situation transition. Examples would be the CS in discriminated avoidance, as well as the change of stimulus in chained schedules.

On this view, behavioral chains are maintained by reinforcement, because transition from a situation further from the terminal situation to one nearer the terminus, and thus higher-valued, constitutes reinforcement. This would be *conditioned* reinforcement insofar as the development of the chain depends on experience. Note, however, that to call a situation transition conditioned reinforcement is not to call the stimulus signaling it a conditioned reinforcer. The new situation might be a conditioned reinforcer, but the discriminative stimulus would not (Schuster, 1969; Schneider, J. W., 1972). The situation transitions, not the discriminative stimuli, maintain a behavioral chain.

Schuster's (1969) experiment, in which pigeons preferred the situation with the more reliable signal, suggest that discriminative stimuli play a purely informative role. Other experiments also have shown that, in the absence of any difference in primary reinforcement, animals prefer a situation with informative stimuli (e.g., Bower, McLean, & Meacham, 1967; Wilton & Clements, 1971). Such stimuli may be of use to the organism in permitting it to perform most efficiently; that is, with no surplus of energy expenditure or sacrifice of reinforcement.

Conclusion

If, as recent research (e.g., Herrnstein, 1969, 1970; Staddon & Simmelhag, 1971) suggests, we drop the contiguity-based law of effect in favor of a law based on molar correlation, many benefits ensue. We can arrive at an integrated understanding of various procedures, such as avoidance, chained schedules, superstition, and classical conditioning. And we can define the concepts of reinforcement, punishment, and discriminative stimulus in a manner that clarifies the relationships among them. This view has been, and promises still to be, highly productive.

2

Quantitative Prediction and Molar Description of the Environment

Foreword

This paper, like some others, argues that behavior and environment constitute a feedback system. This insight was implicitly recognized early, in Skinner's (1938) *Behavior of Organisms* and Ferster and Skinner's (1957) *Schedules of Reinforcement*. It compels a molar view of behavior and environment. Here and in 1973 ("The correlation-based law of effect"), I portrayed the feedback loop as consisting of E-rules that characterize the environment and O-rules characterizing the organism's response to the feedback from the environment. In this paper, I focus on E-rules and try to explain how they may be specified. Later publications focused on E-rules in the form of matching theory and power-function induction. I would use a slightly different vocabulary today: "reinforcer rate" or "food rate" for "rate of reinforcement," for example. The final version of the paper included a longer section on hypothetical constructs, but the editor thought it took up too much space and removed it.

Abstract

Molecular explanations of behavior, based on momentary events and variables that can be measured each time an event occurs, can be contrasted with molar explanations, based on aggregates of events and variables that can be measured only over substantial periods of time. Molecular analyses cannot suffice for quantitative accounts of behavior, because the historical variables that determine behavior are inevitably molar. When molecular explanations are attempted, they always depend on hypothetical constructs that stand as surrogates for molar environmental variables. These constructs allow no quantitative predictions when they are vague, and when they are made precise, they become superfluous, because they can be replaced with molar measures. In contrast to molecular accounts of phenomena like higher responding on ratio schedules than interval schedules and free-operant avoidance, molar accounts tend to be simple and straightforward. Molar theory incorporates the notion that behavior produces consequences that in turn affect the behavior, the notion that behavior and environment together constitute a feedback system. A feedback function specifies the dependence of consequences on behavior, thereby describing properties of the environment. Feedback functions can be derived

Science and Philosophy of Behavior: Selected Papers, First Edition. William M. Baum.

for simple schedules, complex schedules, and natural resources. A complete theory of behavior requires describing the environment's feedback functions and the organism's functional relations. Molar thinking, both in the laboratory and in the field, can allow quantitative prediction, the mark of a mature science.

Keywords: molar description, feedback function, behavior–environment system, operant behavior, hypothetical constructs

Source: A version of this paper was presented at the meetings of the Association for Behavior Analysis, Philadelphia, May 1988. This work was supported by Grant BNS 84-01119 from the National Science Foundation. Originally published in *The Behavior Analyst*, 12 (1989), pp. 167–176. Reproduced with permission of Springer Nature and Association for Behavior Analysis International.

Like any experimental science, the science of behavior is judged by its ability to allow prediction and control. Sometimes qualitative prediction and control suffice: in clinical settings, classrooms, and situations where one requires only that some behavior decrease and other behavior increase. Some applications, however, require quantitative prediction. In the field of organizational behavior management (OBM), for example, particularly in deciding whether some technique of behavioral change is cost effective, one wishes to know not only whether behavioral output will increase but *how much* it will increase. Similar questions arise in behavioral ecology, which has applications in wildlife management. There, one wishes to predict how much of a resource an organism is likely to consume, and how much might be left at the end of its predation (see, e.g., Taylor, 1984; Stephens & Krebs, 1986).

Apart from these practical concerns, quantitative prediction is generally more satisfying than qualitative, because quantitative prediction is widely considered the mark of a mature science. If we can tell only what sorts of changes should occur, we are at a more primitive scientific stage than if we can also tell how much change should occur.

The basic requirement for quantitative prediction is that one be able to write mathematical formulas having the general form $B = f(x)$, where B stands for response rate and x stands for an environmental independent variable. Since reinforcers and schedules of reinforcement are qualitative descriptions, they cannot stand for x. Rather, x must be a measure of some quantitative dimension of reinforcement, such as magnitude in grams or delay of reinforcement in seconds.

Molar *Versus* Molecular Explanation

Magnitude and delay exemplify variables that can be called *molecular*, which here will mean that they can be measured on any one occurrence of an event, such as the presentation of a reinforcer. Each time a grain magazine operates, one can measure the number of grams eaten or the delay since the last response. Molecular variables contrast with *molar* variables, which can only be measured over an aggregate of many events (e.g., presentations of a reinforcer). Rate of reinforcement, for example, cannot be measured on any one presentation of grain, but must be calculated by counting the number of presentations over some substantial period of time (Baum, 1973; Rachlin, 1976).

In what follows, I will argue that molecular variables cannot suffice for quantitative prediction. Moreover, attempts to rely on them exclusively not only fall short of quantitative prediction, but necessitate hypothetical constructs of questionable validity.

Molecular Theories

To see the inadequacy of molecular theories, let us consider explanations of two phenomena: (1) the higher response rates on ratio schedules than on interval schedules, and (2) free-operant avoidance.

The standard molecular account of the ratio-interval rate difference appeals to two factors, the strengthening effect of reinforcement on the response immediately preceding it and the differential reinforcement of inter-response times (IRTs) (e.g., Mazur, 1986). In interval schedules, the longer the time since the last response—the longer the IRT—the higher the probability of reinforcement. Thus, longer IRTs are differentially reinforced and become more frequent, lowering response rate. Acting in the opposite direction, reinforcement tends to increase response rate in both interval and ratio schedules. Since its effect is opposed by differential reinforcement of IRTs only in an interval schedule, response rate is higher on a ratio schedule. One shortcoming of this theory is that without further specification of how response rate depends on reinforcement, one can make no quantitative prediction. As it stands, the theory tells us that the ratio rate will be higher, but not how much higher. To make a quantitative prediction while doing without molar variables like rate of reinforcement, one would have to specify how each instance of reinforcement increments *response strength,* a hypothetical construct that would in turn determine response rate.

Explanation of the second phenomenon, free-operant avoidance, by molecular theory also requires hypothetical constructs. The molecular theory is based on the two-factor theory of signaled avoidance. Stimuli preceding an aversive event like electric shock are said to elicit "fear," and the avoidance response is reinforced by fear reduction when those stimuli are removed (Solomon & Wynne, 1954). In free-operant avoidance, where there are no exteroceptive stimuli, appeal is made to temporal regularity in the presentation of shock, which is considered to produce conditioned aversive temporal stimuli (CATS) that are supposed to elicit "fear" and allow its reduction by the avoidance response (Anger, 1963).

The molecular explanation of free-operant avoidance has the same two short comings as the molecular explanation of the higher response rates on ratio than interval schedules. First, the theory predicts only that avoidance responses will occur; it makes no quantitative prediction about their rate of occurrence. Second, neither "fear" nor CATS are observable, and their properties are unknown. They are hypothetical constructs, necessary only if one must avoid referring to frequency of shock, a molar variable.

Molecular theories require hypothetical constructs because the explanation of any response lies not only in events at the moment of its occurrence but in an aggregate of events over a span of time, often loosely referred to as a "history of reinforcement." A rat presses a lever when a light is turned on because in the past when the light was on presses were reinforced with some frequency and were extinguished when the light was off. Historical variables are often molar variables; to the extent that one must explain behavior historically, one will need often to refer to molar variables.

What's Wrong with Hypothetical Constructs?

Hypothetical constructs have earned a bad name, not because they are bad in principle—atoms are, after all, hypothetical constructs—but because in practice they are usually vague and ill-defined. If they were well-defined, we would know the mathematical relations they embody, and we could make quantitative predictions. Vague hypothetical constructs, however, allow no quantitative prediction and are probably worse than none at all.

The hypothetical constructs characteristic of molecular theories, like "strength" and "fear," serve the same sort of explanatory purpose as mental constructs like "anxiety" and "memory," to invent present causes for historically caused behavior. They arise from a common prejudice that insists on placing causes in the present. Staddon (1973) suggested that the prejudice may have originated in the course of evolution; it may be adaptive to perceive the event that immediately precedes some occurrence as the cause of the occurrence. To say that the causes of behavior are historical is to say that there is a gap, that they cannot be found immediately before the behavior, but that they can be found in the history, and that there is neither need nor use to fill the gap with fictional mediators.

To say that the causes of behavior are historical is to suggest also that they constitute an aggregate—that historical causes translate into molar environmental variables. Like mental constructs, the hypothetical constructs of molecular theories function only as *surrogates*, surrogates of molar environmental variables, such as rate of shock or rate of reinforcement. Like mental constructs, too, they are superfluous, because adequate molar theories are possible without them.

Molar Theories

In contrast to molecular theories, molar theories refer to environmental variables that are physical and readily specified. Questions may arise about how to calculate rate of reinforcement—over how long a time period or by using an arithmetic or a harmonic mean of intervals between reinforcers (Killeen, 1968)—but the properties of any computation are well known, and no hypothetical constructs are needed. Moreover, whereas molecular theories exclude molar variables, molar theories in no way exclude molecular variables. Magnitude and delay may be important, as well as rate of reinforcement. And molar theories make quantitative prediction of response rate possible.

A molar theory can explain free-operant avoidance, for example, relatively simply. Once we are allowed to refer to the frequency of the aversive event, we can see that avoidance responding is maintained by the reduction it produces in that frequency (Hermstein, 1969). Hermstein and Hineline (1966), for example, found a direct relation between rate of avoidance responding and the amount of reduction in shock rate that responding produced, whereas they found an inverse relation between number of responses to extinction and the amount of reduction that had been possible in prior training.

The Behavior-Environment Feedback System

Molecular variables by themselves fail to provide adequate accounts of behavior because behavior produces results in the environment that in tum affect behavior. That is, behavior and environment together constitute a closed feedback system. To explain behavior one must think of both behavioral output (e.g., response rate) and environmental input (e.g., rate of reinforcement) as continuous variables—that is, as flows through time.

Figure 2.1 depicts the behavior-environment feedback system in extremely simple form. The environment provides functions of the form $r = g(B)$, in which we see the dependence of some environmental variable r (e.g., rate of reinforcement) on a behavioral variable B (e.g., response rate). These functions, called "E-rules" in an earlier paper (Baum, 1973), nowadays are called feedback functions. By themselves, they tell nothing about how an organism will behave; they tell only what sort of an environment the organism is behaving in.

To form a theory of behavior, we must know what the organism brings to the situation. The organism provides functions of the form $B = f(r)$, in which we see the dependence of the behavioral variable B on the environmental variable r. These functions, called "O-rules" earlier, are the same as Skinner's functional relations (Baum, 1973).

The figure emphasizes that in operant relations feedback should be understood as axiomatic: B affects r just as much as r affects B. It illustrates also that in a feedback system the distinction between independent and dependent variables be comes arbitrary; strictly speaking, all variables depend on one another.

A reasonable research program might begin by trying to specify feedback functions and then trying to discover functional relations. Once these two goals are met, the result is a quantitative theory that allows quantitative predictions of behavior. In practice, these two attempts go on in parallel, but there is a sense in which the description of the environment—the feedback functions—might need to come first. To learn about the organism, the experimenter systematically manipulates the environment. To specify these manipulations exactly, the feedback functions need to be worked out. Once we can be precise about the ways in which the environment varies, then we can isolate the invariances that the organism contributes to the system. For that reason, I am focusing here on feedback functions.

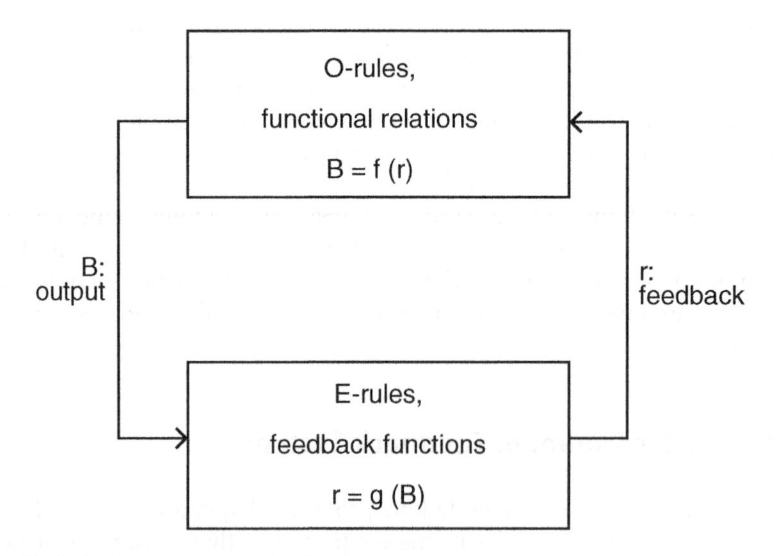

Figure 2.1 The behavior-environment feedback system.

Simple Feedback Functions

Figure 2.2 shows two methods of depicting the dependence of reinforcement on behavior in interval and ratio schedules. On the left, the requirements for reinforcement are represented in cumulative-recorder coordinates. In the ratio schedule, when the cumulative record hits the horizontal line corresponding to the number of responses required by the schedule, reinforcement occurs. In the interval schedule, when the cumulative record crosses the vertical line corresponding to the time at which reinforcement becomes available, the next response produces reinforcement.

The right-hand side of Figure 2.2 shows feedback functions for ratio and interval schedules. Those for *ratio* schedules appear as lines through the origin, because rate of reinforcement is directly proportional to response rate in a ratio schedule; if ten responses are required for each reinforcer, then the rate of reinforcement must be one-tenth of the response rate. The proportionality varies inversely with the ratio; the larger the ratio, the flatter the line. The feedback function for an *interval* schedule approaches the programmed rate of reinforcement as an upper limit; if reinforcers are scheduled only once a minute, then they can be obtained no more frequently than once a minute. The concave curvature reveals the "diminishing returns" characteristic of interval schedules; that is, as response rate grows, increases in response rate produce smaller and smaller increases

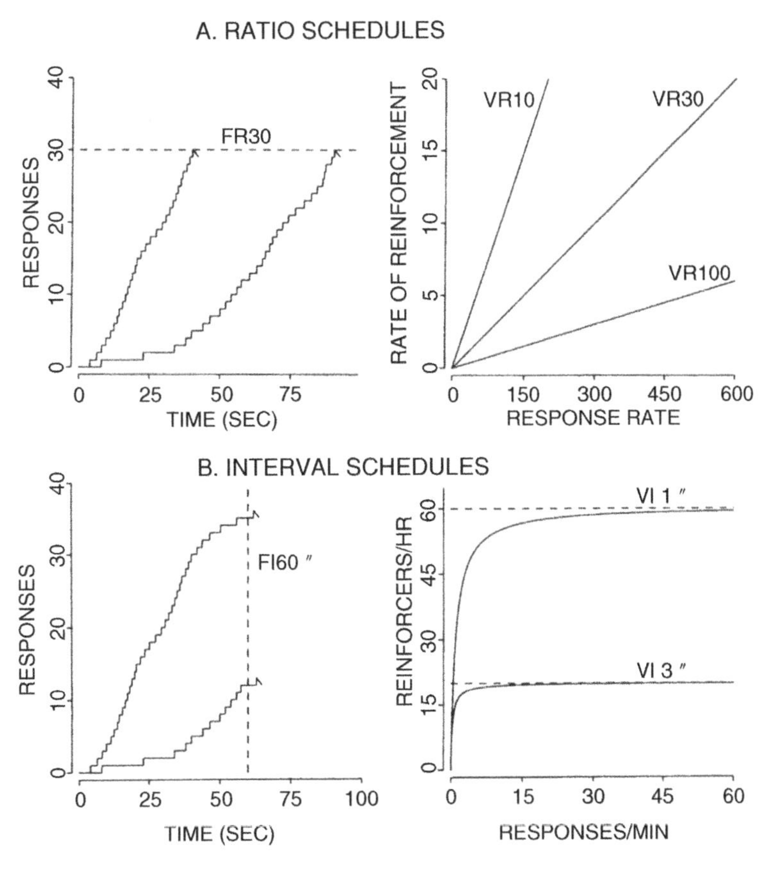

Figure 2.2 Cumulative-record representations (left) and feedback functions (right) for ratio and interval schedules.

in rate of reinforcement, and beyond a certain response rate further increases produce virtually no increase in rate of reinforcement. You may check your mailbox ten times a day, but if the mail is delivered once, your efforts will only produce mail once.

The difference between the feedback functions of ratio and interval schedules provides a straightforward account of the differences in performance, including the difference in response rate. The linear feedback function of the ratio schedule may be thought of as differential reinforcement of high response rate, because the higher the response rate, the higher the rate of reinforcement; hence the high rates characteristic of ratio schedules. A larger ratio gives a lower slope to the line and less differential reinforcement in the sense that each increment in rate of reinforcement is more costly with the larger ratio. If the ratio is large enough, ratio strain occurs, and responding will drop from a high level to zero; intermediate rates should not (and do not) occur (Baum, 1981b). This explanation accounts for the high work rates maintained by piecework wages (an example of a ratio schedule) and also for one of the chief objections to them: the employer is tempted to maximize profit by requiring output that just falls short of ratio strain. Molecular analysis offers no comparable account, because ratio schedules provide no differential reinforcement of IRTs; the probability that the next response will be reinforced is unaffected by the passage of time.

In the interval-schedule feedback functions shown at the lower right in Figure 2.2, differential reinforcement of response rate occurs only at low response rates; if you only check for mail once a month, then checking twice a month would probably produce mail twice as often. As the curve flattens out, differential reinforcement ceases at a relatively low rate; hence the lower rates characteristic of interval schedules. The higher the upper limit of the feedback function, the more slowly the curve approaches the limit and the higher the response rate at which the curve becomes virtually flat. This shift is usually small and predicts only moderate increases in response rate with increasing rate of reinforcement when the interval schedule is shortened, in keeping with those usually observed (Baum, 1981b). If, however, the scheduled upper limit were very high indeed, as in an extremely short interval schedule (e.g., variable-interval 2 s), then the rising portion of the curve would continue even into high response rates, and the result should be a transition to extreme response rates like those normally maintained by ratio schedules. In at least one experiment, I have observed such an effect (Baum, 1986).

Feedback Functions and Compound Schedules

Pure examples of ratio and interval schedules are difficult to find in the every day world. One of the ways in which the complexity of everyday life can be approached is to combine ratio and interval requirements into compound schedules. Three types of compound schedules are conjunctive, alternative, and interlocking schedules.

Conjunctive interval-ratio schedules. Figure 2.3A shows, the same two ways as before, the characteristics of conjunctive interval-ratio schedules. The cumulative-record depiction of the schedule (left) shows how both the interval and the ratio requirement must be met before reinforcement can occur. If the organism responds at a high rate, it will meet the ratio requirement (horizontal line) early, but must persist until the interval requirement (vertical line) is met. If it responds at a low rate, it will fail to have met the ratio requirement when it satisfies the interval requirement, and must continue until it has met the ratio requirement. There is one response rate that satisfies both requirements simultaneously.

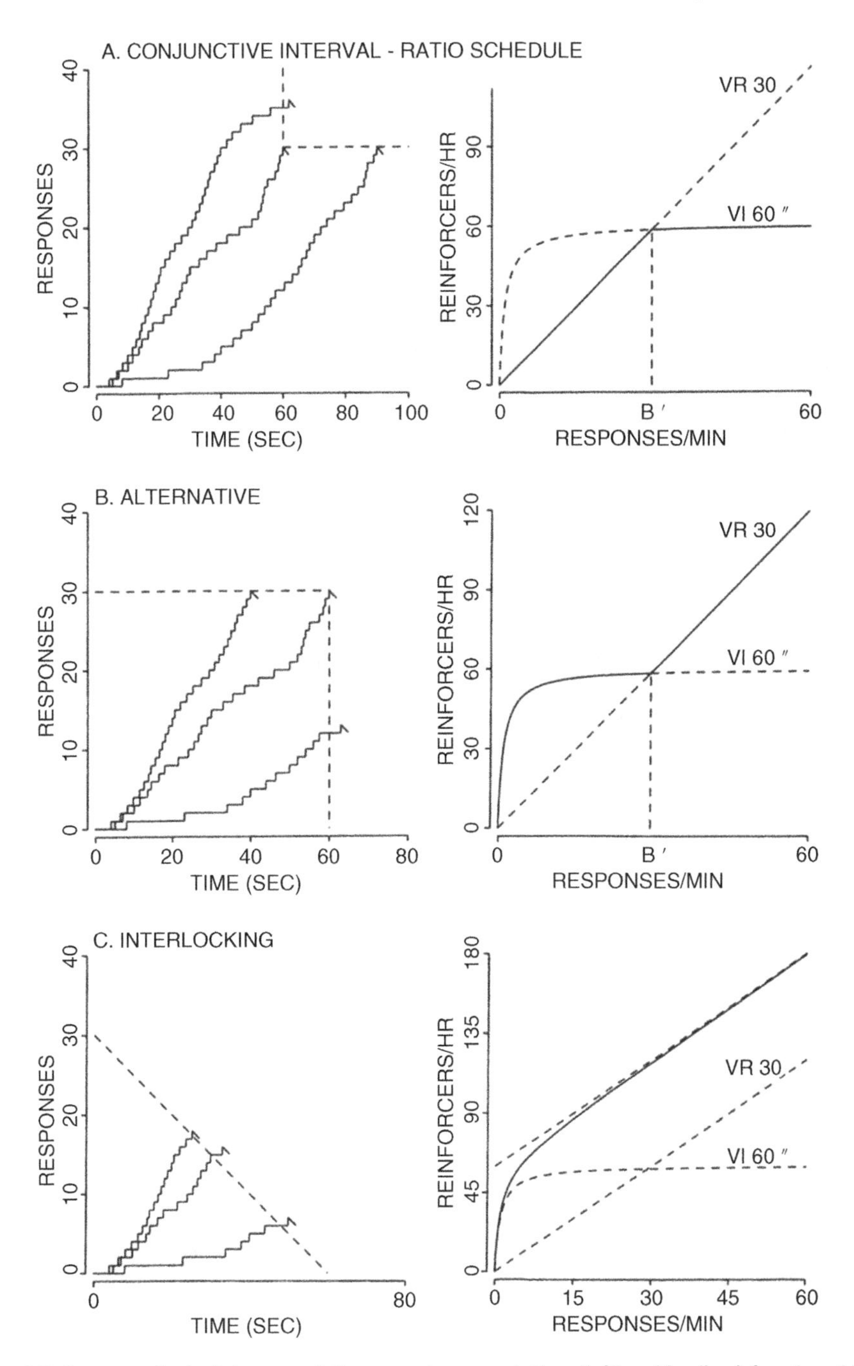

Figure 2.3 Compound schedules: cumulative-record representations (left) and feedback functions (right).

The feedback function (right) shows how the requirements affect the dependence of rate of reinforcement on response rate. Low response rates pay off according to the ratio schedule. High response rates pay off according to the interval schedule. The response rate that satisfies both requirements simultaneously, *B*,′ where the two functions intersect, can, with a few additional assumptions, be shown to be the optimal performance.

This schedule might be a realistic representation of hourly wages. The interval requirement alone falls short, because employers typically expect some minimal performance even when they pay by the hour. The result is that the employee must meet a work requirement within a time schedule. Without other incentives (i.e., modification to the feedback function), the employee's optimal response rate is *B'*.

Alternative interval-ratio schedules. Figure 2.3B shows the characteristics of alternative interval-ratio schedules. The cumulative-record depiction shows how either the ratio requirement may be met by a high rate or the interval requirement may be met by a low rate. The feedback function reveals that low rates pay off according to the interval schedule, whereas high rates pay off according to the ratio schedule. There is a response rate *B'* that meets both requirements simultaneously, but here that rate has no special advantage. The situation offers an implicit choice: respond at high rates, above *B',* and take control of the rate of payoff, or respond at low rates, below *B',* and go easy.

Interlocking interval-ratio schedules. Figure 2.3C shows the nature of interlocking interval-ratio schedules. The cumulative-record depiction shows that various possibilities exist, high rates paying off sooner but with more effort and low rates paying off with less effort but later. The schedule allows a whole range of compromises between working and waiting. The feedback function shown on the right (solid curve) is the sum of the ratio and interval feedback functions (dashed curve and lower line). It illustrates that the situation is a smooth blend of ratio and interval-like payoffs. At relatively lower rates, the curvature indicates that the schedule tends toward what may be called the *corrective* tendency of interval schedules—the lower the rate, the better the payoff. At relatively higher rates, the curve approaches the upper dashed line as an asymptote, which means that the situation shifts gradually toward ratio payoff. The feedback functions reveal that interlocking and alternative schedules resemble one another in offering choice between taking control and going easy.

Natural Resources

A different kind of complexity enters when we consider the situation organisms face in exploiting a resource in the natural environment. As one eats the berries in a berry patch, there are fewer and fewer to be found: the patch *depletes.* Depletion means that as the resource is consumed, it becomes progressively more difficult to obtain. For schedules, this means that as reinforcers are delivered, the schedule requirement changes so as to make them less available.

Adjusting schedules as depleting resources. Figure 2.4 shows how adjusting schedules can model depleting resources. Since hunting, searching, and otherwise exploiting resources share with ratio schedules the property that rate of reinforcement depends directly on behavior al output, the schedule that models a natural resource is the adjusting ratio. Figure 2.4A shows adjustment of ratio schedules. Either the ratio requirement increases with number of reinforcers delivered or the probability of reinforcement decreases. Figure 2.4B shows the effect of the adjustments on the rate of reinforcement. With increasing time spent exploiting the resource, the rate of reinforcement falls, perhaps quickly, and then slowly or perhaps slowly and then quickly, depending on the pattern of use. For example, systematic search for berries, working from a starting point and never retracing, will result in little or no decline in rate of finding berries until the patch has been completely covered, at which point rate of finding berries will decline precipitously. If a bird were to hunt for seeds in a patch of grass completely at random, then the rate of capturing seeds would decline rapidly at first and then ever more slowly as the number of seeds declined all over the patch (Baum, 1987).

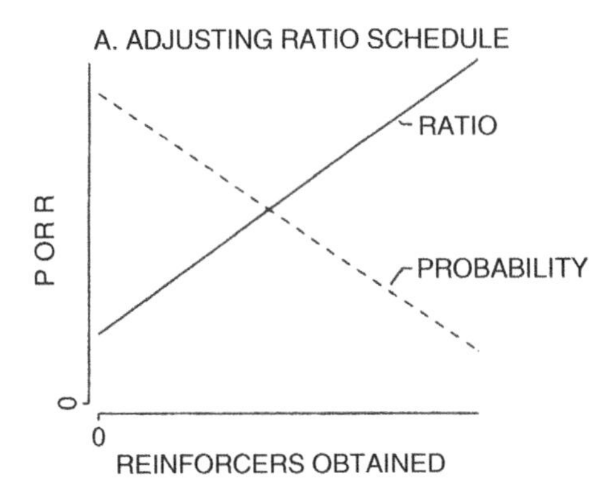

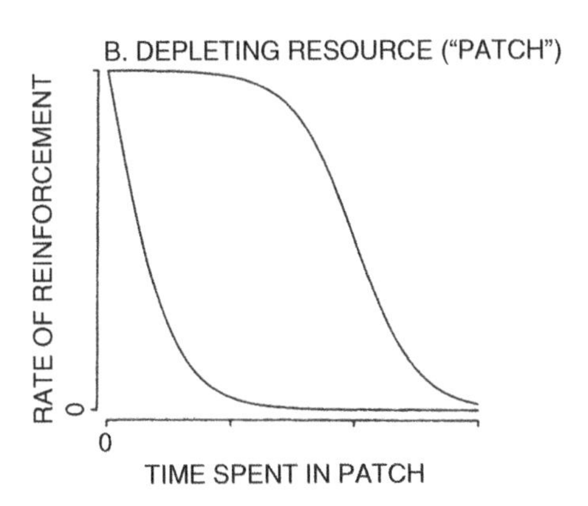

Figure 2.4 Adjusting ratio schedules as models of depleting resources or "patches."
Note: A: As more reinforcers are obtained, the ratio requirement increases or the probability that a response will be reinforced decreases. B: Depending on how the ratio or probability is adjusted, different patterns of depletion can result.

The complexity of this type of situation arises because, unlike the schedules we considered before, adjusting schedules cannot be considered *stationary;* their parameters shift in time. This means that not only response rate but also time must be taken into account. For deriving feedback functions, at least two additional performance variables have to be considered: time spent responding ("time in") and time away from the patch. When one is searching for berries, interacting with the patch, that is time in; when one goes home or takes a nap under a nearby tree, no longer interacting with the patch, that is time away.

Interval schedules as depleting. Looked at this way, even an ordinary interval schedule shares some of the character of a depleting resource (Staddon, 1980). Figure 2.5 illustrates this. Figure 2.5*A* shows differential reinforcement of IRTs in the usual molecular view of a variable-interval schedule. In such a view there is no time in, because responses are treated as if they have no duration, and the only time considered is the time between responses; so the IRT is considered time away. Figure 2.5B shows the feedback function of the molar view. Here, there is no time away, because rate of reinforcement and response rate are computed using all the available time. Hence, Figure 2.5B shows the effect of response rate during time in. Figure 2.5C shows how the two views combine when we consider both time away and time in. After some time away, there is a higher probability of reinforcement—hence a higher rate of reinforcement—at the beginning of time in. The height of this peak depends on the duration of time away (Figure 2.5A). As

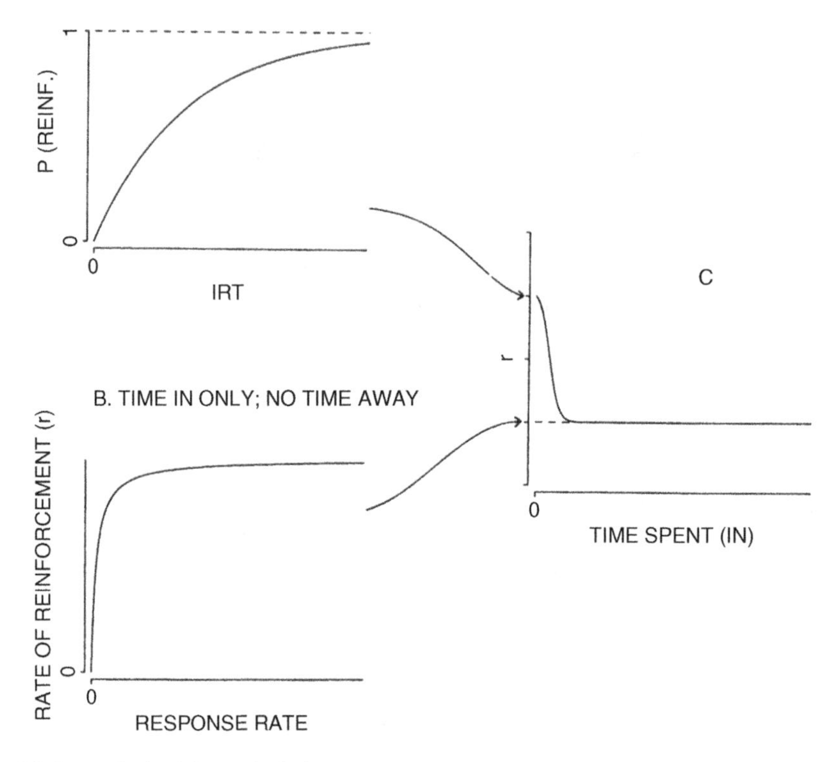

Figure 2.5 Interval schedules as depleting.
Note: The most molecular view (A) includes only time away. The most molar view (B) includes only time in. Combining the two (C) reveals that interval schedules may share the property of depletion with natural resources.

time in proceeds, the rate of reinforcement drops to a horizontal asymptote, the height of which depends on the response rate during time in, as given in Figure 2.5B.

Feedback function for a patch. The feedback function for a depleting resource (patch) must take into account three performance variables: response rate while in the patch, time in the patch, and time away from the patch. This means the feedback function will be four dimensional. Figure 2.6 shows only how overall rate of reinforcement would vary with time in, for a given time away and response rate. The curve indicates that there is an optimal duration of time in. The exact height and position of the maximum depend not only on time away, but on how the patch replenishes—whether quickly or slowly.

Conclusion

Molecular theories of behavior—those that rely solely on momentary events for explanation—have two great drawbacks. First, because present behavior arises from a history of events, extended through time, molecular theories require invention of hypothetical constructs like response strength and conditioned fear to represent the extended history in momentary events of the present. Since these hypothetical constructs are

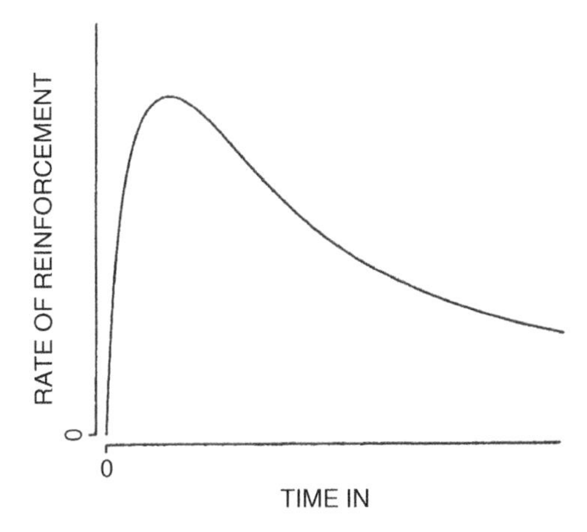

Figure 2.6 Feedback function for a depleting patch.
Note: Effects of varying time spent responding ("time in").

unmeasurable and vaguely defined, their explanatory power is more illusory than real. Second, molecular theories, even with the constructs, make only qualitative predictions. Were the properties of the constructs precisely defined, quantitative prediction might become possible, but the constructs would stand merely as surrogates of well-understood molar historical variables that can be specified in physical terms. Once defined, the hypothetical constructs become superfluous, because one can proceed to a direct molar analysis, relating molar behavioral variables to molar environmental variables.

With maturity, a science of behavior should be able to make quantitative predictions. Since quantitative predictions are possible only with molar laws, behavioral analysis can progress toward this goal only by looking beyond momentary events to molar variables and molar relations.

In the laboratory and in the field, a better understanding of contingencies can be had by thinking of their molar properties, by thinking of the impact of contingencies on history over a period of time. Instead of asking, "Does this contingency change behavior?" one can ask, "If we make rate (or amount) of reinforcement depend on response rate according to this relation, how much does behavior change?" I have emphasized here the tactically prior problem of describing the environment in molar terms, relying on the concept of feedback functions (E-rules; Figure 2.1). The specification of functional relations (0-rules; Figure 2.1) will follow or accompany development of such molar descriptions. The result will be a science of behavior that is both powerful and practical.

3

The Trouble With Time

Foreword

This paper aimed to show not only how a molar view would be useful in understanding behavior but also how an adequate understanding of behavior requires a molar view. A view based on contiguity and discrete events has fundamental weaknesses, such as inability to deal with less-than-perfect contiguity (gaps of time) and impossibility of identifying behavior at a moment. Comparisons with Schrödinger's writings about physics and Whorf's writings about linguistics reveal that these inadequacies arise from a culturally derived concept of time as moving from the future through the present into the past. This concept of time is by no means intuitive, and escaping from it is essential to an adequate science of behavior. Instead of efficient causes, final causes would be more appropriate for defining and explaining behavior, because behavior consists of patterns or activities that exist in extension through time.

Abstract

Events in the remote past affect behavior in the present. Focus on moments in time offers no explanation of how the past can affect the present. Physics solved the problem of action at a distance by giving up corpuscular mechanics. Behavioral science needs to do the same by giving up discrete responses and momentary relations in favor of extended patterns of behavior and extended controlling relations. The smaller the timeframe in which we study behavior, the greater the uncertainty about the identity of the behavior and its control. Both the physicist Erwin Schrödinger and the linguist Benjamin Whorf pointed out the pitfalls of limited causal thinking. Rachlin approached the problem by identifying extended behavioral patterns with final causes instead of efficient causes. Behavior may be thought of as hierarchical patterns of activities that transcend moments of time.

Keywords: extended patterns, extended control, final causes, Erwin Schrödinger, Benjamin Whor

Science and Philosophy of Behavior: Selected Papers, First Edition. William M. Baum.

Source: This paper was originally given in a symposium at University of Nevada, Reno. It was published as a chapter in L.J. Hayes and P.M. Ghezzi (eds.), *Investigations in Behavioral Epistemology*. Context Press, an imprint of New Harbinger Publications, 1997. Reproduced with permission of New Harbinger Publications, Inc.

Time presents a problem for a science of behavior because events in the remote past often affect behavior in the present. The problem parallels the challenge of action at a distance in physics. To someone committed to viewing the universe as composed of moving particles that affect one another only by collision (i.e., corpuscular mechanics), the idea that one body might affect another at a distance would seem mysterious and even impossible. For one object to affect the motion of another, the two would have to touch. Similarly, in a science of behavior, to someone committed to viewing actions as driven by immediate events (i.e., behavioral mechanics), the idea that events might affect behavior across a gap of time would seem mysterious and impossible. For an event to affect action, the two would have to be in temporal contiguity.

Twentieth-century physics solved the problem of action at a distance by replacing corpuscular mechanics with notions like field theory, relativity, and quantum mechanics, notions that see wholes and relations where corpuscular mechanics saw disconnected particles. Behavior analysis is still in the process of solving its problem of action at a (temporal) distance. In this paper, I shall argue that, as the problems are parallel, so the solutions are parallel. Where behavioral mechanics would see disconnected events across gaps of time, a molar view sees whole patterns and relations. As developments in twentieth-century physics seemed counterintuitive, so developments in molar behavior analysis may seem counterintuitive, until they become familiar.

The trouble with mechanical explanations of behavior is not so much that they are wrong as that they are incomplete. If you say that reinforcement of good manners in childhood produces polite behavior in an adult, listeners may respond that the reinforcement in childhood must have had effects in the brain that persist in the adult and that these effects in the brain produce the polite behavior in the present. This presumes a notion of brain function—that information or memories are stored in the brain and later retrieved for use—that is probably false, but even if it were true that some state in the brain produced polite behavior by its immediate presence, the explanation begs the question of where the brain state came from. The mechanical explanation is altogether too silent about the events in childhood.

Evolutionary biologists affirm the incompleteness of mechanical explanations and resolve any apparent contradiction between them and more global or historical explanations by making a distinction between *proximate* and *ultimate* explanations (e.g., Alcock, 1993). Proximate explanations concern the physiology of individual organisms within their lifetimes, whereas ultimate explanations concern evolution of populations across generations by natural selection. Ultimate explanations span great gaps of time, because they refer to countless ancient events to account for the populations of today. As someone would assume that events earlier in an individual's lifespan would be preserved in the brain, so evolutionists assume that ancient events are preserved in the gene pool. As a brain state might result in a particular immediate action, so a genotype is expressed in a particular phenotype. Proximate explanations tell about particulars and short-term events, whereas ultimate explanations tell about tendencies and history. Neither gives a complete picture by itself; both are required for a full understanding.

The ultimate explanations of behavior analysis span, not generations, but significant portions of an individual's lifespan, which, though short by evolutionary standards, are long in comparison with momentary events. The common property of temporal extendedness makes many parallels between explanations. The ultimate explanations of evolutionary biology refer to three types of influences in a species's past: (1) uncontrollable environmental factors, such as droughts, ice ages, and asteroid impacts; (2) accidental mutations; and (3) selective pressure (e.g., cryptic coloration surviving better or large males reproducing more). Similarly, the ultimate explanations of behavior analysis refer to three types of influence in an individual's past: (1) uncontrollable environmental factors, such as death of a parent, presence of a skillful teacher, and a move to California; (2) behavioral accidents (e.g., making a wrong turn because of poor visibility or a distraction); and (3) selective pressure (e.g., reinforcement of good manners or punishment for aggression). Just as biological ultimate explanations apply to populations of organisms, so behavior-analytic explanations apply to populations of actions. Evolutionary theory cannot say why a particular individual moth is colored black, but might say why that type is common or rare in the population and why it is increasing or decreasing in frequency. Similarly, behavior analysis cannot explain why a rat presses a lever at just the moment when it does or why a driver runs a stop-light on a particular occasion, but might explain why that type of action is rare or common and why it is increasing or decreasing in frequency.

Speaking more generally, understanding populations is an example of understanding *patterns*. To understand a population is to understand a pattern of types and their frequencies. Since patterns extend either in time or in space, they always raise problems about sampling. If a population of moths contains thousands of individuals, one might try to record a sample of a hundred. To measure choice between two response keys, one collects a sample of perhaps a thousand responses. The larger one's sample, the more certainty one has about having measured the pattern, and the smaller one's sample the more uncertainty one has. The situation resembles Heisenberg's *uncertainty principle* in physics, which states that the more accurately one measures one aspect of a particle's motion, the less certain one can be about other aspects. For example, the more one tried to be accurate about a particle's position, the more one would be uncertain about its velocity. The point applies just as well to a ball flying from a bat. It is easy enough to imagine the ball having a particular velocity at a particular point in its trajectory, but in practice the more accurately we measure its position, the less we know about its velocity. For example, a high-speed photograph would reveal exact position, but a picture of a ball hovering in the air reveals nothing about its velocity, because measuring velocity requires some portion of the trajectory to be sampled. The larger the portion of the trajectory sampled, the more certainty one may have about the measure of velocity, but the less certainty one has about which point in the trajectory to assign to that velocity.

A principle analogous to Heisenberg's uncertainty principle invariably applies to the study of patterns. A stop-action snapshot of a rat or a person tells with certainty the position of the organism and the disposition of its limbs, but reveals almost nothing of what behavior is occurring, because that requires a pattern. A picture of a rat in front of a lever cannot tell with certainty that the rat is pressing the lever, let alone at what rate the

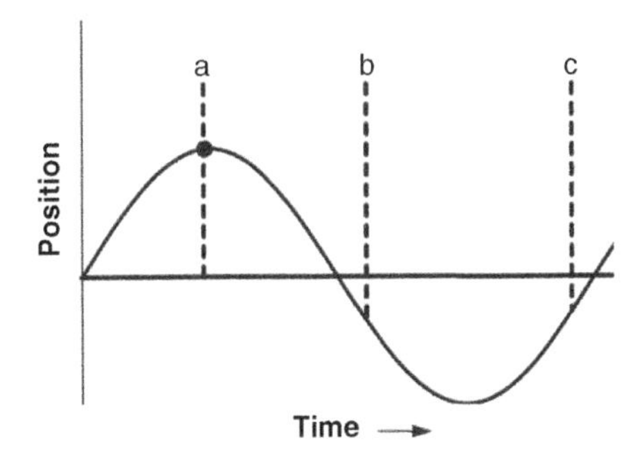

Figure 3.1 A curve representating a pattern of variation transcending moments of time.
Note: The larger the sample, the less uncertainty about the pattern. A single point a gives almost no certainty, an interval ab gives more, and a longer interval ac gives still more.

lever is being pressed. Although one may conceive of behavior at a moment, as one may conceive of instantaneous velocity, in practice there is no such thing as behavior of the moment. We cannot decide what a creature is doing without an adequate sample of its movements. There is no way to tell if a person bent over a book is reading or daydreaming except by observing subsequent (and possibly prior) behavior: Does the person speak and otherwise behave afterwards in ways that could only imply having read the book? (And have such sequences occurred before?)

Figure 3.1 illustrates the situation. A single momentary measure of amplitude at point *a* would leave almost complete uncertainty about the wave form shown. A sample spanning an interval like *ab* would allow the form to be better known, and an interval like *ac* would reduce uncertainty still further. Even if one had a full cycle, however, some uncertainty would remain, because there still might be variability from cycle to cycle. The whole pattern can never be known with absolute certainty, but the larger the sample, the more certainty about the pattern. When we have enough of a sample, even if some uncertainty remains, we say we have observed a pattern. Thus, behavior, insofar as it consists of populations of actions, is observed only in patterns extended in time. Only from such extended patterns are we able to draw conclusions with any certainty about: whether response rate is characteristic of a fixed-ratio schedule, whether there is a preference for the left response key, whether John knows about current events, whether Sally is religious, whether Ira's intentions toward Martha are pure, or whether Martin thinks about life at all.

Schrödinger on Causal Thinking

By no means do such considerations occur only in the study of behavior. Erwin Schrödinger (1956), whose work on quantum mechanics won a Nobel prize, pointed out limitations to what he called *causal thinking*—the idea that the universe can be understood as sequences of events linked into causal chains. This idea, which he considered to be peculiar to the scientific worldview, sees the universe in terms of repeatable conditions (causes) and subsequent events (effects). Such relations are selected for scrutiny and relied on for explanation, particularly when thought of as a chain in which causes alternate with effects, each link being a cause-effect relation. All other coincidences of events are dismissed as chance. Such a view, Schrödinger remarked, seems to capture little of what is interesting in life:

> ... in reflecting on the course of events one always brings together in one's mind such perceptions or observations as stand in the relation of necessity to each other. Causal chains are selected and designated as the only thing of consequence. In real life, however, hundreds of causal chains constantly cross each other, and so it happens that events constantly coincide which are not related in an intelligible way and whose coincidence passes for chance in the eyes of those who are scientifically minded. They are matters like a solar eclipse and a lost battle; a black cat which crosses my path from the left and a business misadventure on the same day. But also matters like a train connexion missed in Basle because a dog was run over, which makes me meet an acquaintance from Istanbul the same evening and thereby (logical subject is still the dead dog) turns all my future life into a new direction. (Schrödinger, 1956, p. 201)

The story of the dead dog and the missed train illustrates how, in real life, "hundreds of causal chains constantly cross each other." To Schrödinger ,the search for strict causality seemed to omit most of what counts in life:

> ... anybody who looks closely at a path of life familiar to him, such as his own, will gain the impression that random coincidences of events or circumstances which are not directly causally connected play a very large, actually the most interesting, part in it; compared to them the role of the transparent causal chains appears rather trivial... And this may lead to the conclusion that [causal thinking], reasonable as it may appear, still explains only a small and at that the most trivial part of the interconnexions that really interest us, while the essentials remain uncomprehended. (Schrödinger, 1956, p. 202)

These essentials, these uncomprehended interconnexions, include matters like the runover dog, the missed train, the friend from Istanbul being in Basle, and one's life taking a whole new direction. Their nature is captured more by the notion of an extended pattern into which all the pieces fit together to make a comprehensible whole, like a tapestry in which causal chains are like the threads woven together to make the pattern.

The idea of hundreds of causal chains crossing each other in a way that presents a potentially intelligible pattern suggests both the limitations of causal thinking and the possibility of an alternative sort of analysis. Talking about extended patterns raises difficulties, however, because our language, so rich in apparatus for causal thinking, lends itself poorly to discussion of temporal patterns. Schrödinger complained how our language has become completely adapted to causal thinking— what he called the Greek outlook on nature:

> Dozens of vital particles, like *because, as, although, in order to, so that, why, nevertheless, suppose, let alone, all the same,* etc., have assumed a definite logical significance, they have their exact counterparts in all languages that spiritually (not necessarily etymologically) are derived from the Graeco-Roman Mediterranean civilization and make it simple to translate from one language into another. Non-Greek thinking appears from our standpoint and expressed in our language not only as strange, not only as distorted and wrong, but easily as drivel without rhyme or reason. (Schrödinger, 1956, p. 201)

Whorf on the Awkwardness of English

Schrödinger's complaint that languages like English raise obstacles to the analysis of patterns was echoed by the chemist-turned-linguist Benjamin Whorf. Whorf (1956) went further than Schrödinger in his analysis of the limitations of English. He contrasted English with the Hopi language, arguing that Hopi was oriented toward talking about events, whereas English was oriented toward talking about objects. The unwieldiness of English as a language for science Whorf blamed on the way it treats objects— particularly the distinction between *form* and *substance*. We may speak easily of water, for example, as substance without form; English has many mass nouns like this. When we need to talk about some specific bit of water, we resort to locutions like "a cup of water" or "the water in that cup." The cup gives the form to the substance. Similarly, for any object, we name it with substance and a form. Driving home the point that language can be otherwise, Whorf comments on the Hopi approach to objects:

> It has a formally distinguished class of nouns. But this class contains no formal subclass of mass nouns. . . . Nouns translating most nearly our mass nouns still refer to vague bodies or vaguely bounded extents. They imply indefiniteness, but not lack, of outline and size. In specific statements, 'water' means one certain mass or quantity of water, not what we call "the substance water." Generality of statement is conveyed through the verb or predicator, not the noun. . . . The language has neither need for nor analogies on which to build the concept of existence as a duality of formless item and form. (Whorf, 1956, pp. 141–142).

On this "duality" of formless substance and substanceless form Whorf blamed the dichotomy between an inner world and an outer world— what behaviorists call *mentalism*. It is altogether too easy in English to talk about substanceless forms as if they had a separate reality, with the result that there seems to be an inner world containing forms without substance and an outer world containing forms with substance. Whorf argued that Hopi illustrates a possible alternative:

> It is no more unnatural to think that thought contacts everything and pervades the universe than to think, as we all do, that light kindled outdoors does this. And it is not unnatural to suppose that thought, like any other force, leaves everywhere traces of effect. Now, when **WE** think of a certain actual rosebush, we do not suppose that our thought goes to that actual bush, and engages with it, like a searchlight turned upon it. What then do we suppose our consciousness is dealing with when we are thinking of that rosebush? Probably we think it is dealing with a "mental image" which is not the rosebush but a mental surrogate of it. But why should it be **NATURAL** to think that our thought deals with a surrogate and not with the real rosebush? Quite possibly because we are dimly aware that we carry about with us a whole imaginary space, full of mental surrogates. To us, mental surrogates are old familiar fare. Along with the images of imaginary space, which we perhaps secretly know to be only imaginary, we tuck the thought-of actually existing rosebush, which may be quite another story, perhaps just because we have that very convenient "place" for it. (Whorf, 1956, pp. 149–150)

This resembles Skinner's (1969c, 1974) objections to introspection and copy theory. Skinner also used the word *surrogate*, although nowadays the word *representation* is more common. Whorf complained that the very ease of such talk in English, couched as it is in mentalistic terms, makes it almost impossible to speak nondualistically:

> From the form-plus-substance dichotomy the philosophical views most traditionally characteristic of the "Western world" have derived huge support. Here belong materialism, psychophysical parallelism, physics— at least in its traditional Newtonian form— and dualistic views of the universe in general. Indeed here belongs almost everything that is "hard, practical common sense." Monistic, holistic, and relativistic views of reality appeal to philosophers and some scientists, but they are badly handicapped in appealing to the "common sense" of the Western average man— not because nature herself refutes them (if she did, philosophers could have discovered this much), but because they must be talked about in what amounts to a new language. (Whorf, 1956, p. 152)

Whorf on Time

This parallels Shrödinger's complaint given earlier, and Whorf's discussion, like Schrödinger's, extended to the way we talk about time. He pointed out that speakers of English— just as they take for granted the view of objects as formless substance combined with form— take for granted a certain view of time: "as a smooth flowing continuum in which everything in the universe proceeds at an equal rate, out of a future, through a present, into a past; or, in which, to reverse the picture, the observer is being carried in the stream of duration continuously away from a past and into a future." (Whorf, 1956, p. 57) Although this view is often considered "intuitive," Whorf argued that it is simply a part of our culture:

> It is sometimes stated that Newtonian space, time, and matter are sensed by everyone intuitively, whereupon relativity is cited as showing how mathematical analysis can prove intuition wrong. . . .[L]aying the blame upon intuition for our slowness in discovering mysteries of the Cosmos, such as relativity, is . . . wrong. . . Newtonian space, time, and matter are no intuitions. They are recepts from culture and language. That is where Newton got them. (Whorf, 1956, p. 152–153)

Whorf argued that many other views of time and space are possible, "descriptions of the universe, all equally valid, that do not contain our familiar contrasts of time and space." (Whorf, 1956, p. 58) He suggested as examples both relativity theory in modern physics and the Hopi worldview. To give a sense of other possibilities, he again compared English with Hopi:

> . . . the Hopi language. . . contain(s) no words, grammatical forms, constructions or expressions that refer directly to what we call "time," or to past, present, or future, or to enduring or lasting, or to motion as kinematic rather than dynamic (i.e., as a continuous translation in space and time rather than as an exhibition of dynamic effort in a certain process), or that even refer to space in such a way as to exclude that element of extension or existence that we call "time," and so by implication leave a residue that could be referred to as "time." Hence, the Hopi language contains no reference to "time," either explicit or implicit. (Whorf, 1956, pp. 57–58)

How does Hopi describe the universe without space and time? Whorf argued that, like our naive view of space and time and like relativity theory, "the Hopi language and culture conceals a metaphysics." It relies on new concepts and abstractions, "abstractions for which our language lacks adequate terms." Instead of dividing the world into past, present, and future, Hopi distinguishes what Whorf called *manifested* and *manifesting*, or *objective* and *subjective*. He explained thus:

> The objective or manifested comprises all that is or has been accessible to the senses, the historical physical universe, in fact, with no attempt to distinguish between present and past, but excluding everything that we call future. The subjective or manifesting comprises all that we call future, **BUT NOT MERELY THIS**; it includes equally and indistinguishably all that we call mental. . . The subjective realm . . . embraces not only our **FUTURE**, much of which the Hopi regards as more or less predestined in essence if not in exact form, but also all mentality, intellection, and emotion, the essence and typical form of which is the striving of purposeful desire, intelligent in character, toward manifestation— a manifestation which . . . in some form or other is inevitable. . . It is in a dynamic state, yet not a state of motion— it is not advancing toward us out of a future, but **ALREADY WITH US** in vital and mental form, and its dynamism is at work in the field of eventuating or manifesting, i.e. evolving without motion from the subjective by degrees to a result which is the objective. (Whorf, 1956, pp. 59–60)

For behavior analysts struggling with the limitations of the concept of time in Indo-European language, this description of the Hopi view may be suggestive, particularly if they are talking, as Whorf put it, "in what amounts to a new language." One may take the Hopi ideas that events *manifest* and that events manifesting are *already with us* to suggest a coherent pattern of events that becomes clearer as we gain experience with it, instead of moving at us or our moving through it. Instead of the standard metaphors of motion, likening the flow of time to a road or a river, with the result that one event seems to follow another in a sort of chain, one might use metaphors of clarification and revelation, as when we sometimes speak of events *unfolding*— a mode of talking that suggests, instead of discrete events, extended patterns that gain in definition with study or experience.

Although such a view of events in no way rules out consideration of mechanism—i.e., proximate explanations—it lends itself well to considering ultimate explanations, because definition of patterns is the explanatory mode in which a phenomenon is understood by seeing how it fits into a larger pattern. The characteristics of a population of mice, for example, are understood by the way they fit with the characteristics of the environment—resources, predators, weather, and so on. If the mice in one habitat have lighter coloration than the mice in another habitat we look for a difference in the color of the ground that might affect their visibility to avian predators. Defining patterns is a key mode of explanation in evolutionary biology, modern physics (e.g., Capra, 1983), and behavior analysis.

Rachlin's Discussion of Final Causes

Rachlin (1992; 1994) treats the difference between proximate explanations of momentary events versus ultimate explanations of patterns within populations in the light of Aristotle's distinction between *efficient* causes and *final* causes. Efficient causes are prior events conceived as connected directly (i.e., mechanically) to the subsequent event (the effect to be explained). Behavior-analytic and evolutionary proximate explanations are of the efficient-cause variety, because they propose prior events to be necessary and sufficient to explain present behavior. Even when a gap of time separates cause from effect, since behavior analysts acknowledge that something must fill the gap— either in the physiology of the organism or in a theoretical variable [e.g., Staddon's (1993) state variables]— the explanations conform to the efficient-cause mode. Even history of reinforcement is sometimes treated as if it were some kind of aggregate efficient cause, although, as we shall see, it would be more proper to treat it as a final cause.

Final causes may be thought of as extended patterns. Actions are explained by final causes, Rachlin (1994) explains, by *fitting into* them, as the notes of a tune fit into the tune. It is a final-cause explanation to say that Jane is eating an appetizer because she is eating dinner. On a larger scale, it is a final-cause explanation to say that Jane is eating dinner because she eats dinner every day at 7:00 PM (i.e., it is her pattern). On a still larger scale, it is a final-cause explanation to say that Jane eats dinner every day at 7:00 PM because she maintains her health by eating regularly.

When thought of as extended through time, final-cause explanations become historical explanations (Baum & Heath, 1992). Particularly when thought of as repeating patterns, final causes make sense of explanations of behavior that refer to "training." When we say that a rat presses a lever now because it was trained to do so in the past, we are implicitly appealing to a pattern of events (chamber, stimuli, lever-pressing, food deliveries) into which the lever-pressing fits. It is a pattern that is manifest (now and in the past) and, with no superordinate change (e.g., alterations of contingencies), continues to manifest (in the future).

Generalizing further, we see that any ultimate explanation is an explanation by final causes. When it is proposed that behavior is to be explained by environmental contingencies, the explanation fits into this category, because the contingencies impose a temporally extended pattern of contexts, actions, and consequences. To be effective, a contingency must operate over and over through time and must affect behavioral variants differentially through time. The same line of reasoning holds for evolutionary theory. A species' environment, which includes exploitable resources and challenges to survival such as predators and fires, imposes differential reproductive success among variants within a population— that is, it imposes a pattern of differential consequences that holds stable within and to some extent across generations. As some variants out-reproduce others, the more successful ones increase in frequency in the population. If darker colored mice are less susceptible to predation, then they produce more descendants. Eventually, the population stabilizes, because the most successful variants come to dominate and any variants that deviate are selected against. At this stage it can be said (and at every earlier stage it could have been said) that the population *fits into* the larger pattern (of varying reproductive success) defined by the environment, because the pattern defined by the environment embraces all possible variants, whereas within the possible variants the actual population constitutes only a subset. Thus, the pattern defined by the environment is the final cause of the population pattern. Returning to the contingencies that shape operant behavior, we draw a parallel conclusion. Out of all the variants that might occur, only the most successful— most reinforced and least punished— dominate; the rest drop away. The population of actions that actually occurs fits into the larger pattern of differential reinforcement defined by the contingencies.

By way of illustration, Figure 3.2 shows two reinforcement feedback functions (Baum, 1973; 1981b). The top one, imposed by a variable-interval schedule, shows the dependence of reinforcement rate on response rate. All possible pairs of rate of reinforcement and response rate lie along this curve. Actual performance is represented by a range along the curve; only some of the possible performances actually occur. The frequency distribution in the lower left of the graph represents the pattern of variation in response rate,

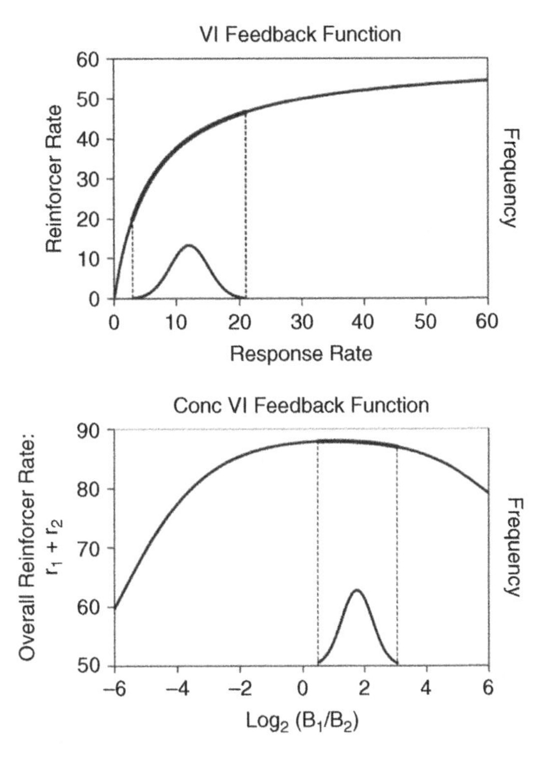

Figure 3.2 Feedback functions illustrating how the pattern actual performance fits into the larger pattern imposed by a contingency.
Note: The frequency distributions show patterns of variation in response rate (upper graph) and preference (lower graph). The broken lines and thicker sections of curve show the range of actual performances as a subset of those included in the larger pattern, the entire curve. Upper graph: a feedback function for a VI 1 min. Lower graph: a feedback function for two concurrent variable-interval schedules, VI 1 and VI 2.

aggregated across minutes, hours, or sessions. The range of response- and reinforcement-rate pairs that actually occurs is indicated by the broken vertical lines and the thickening of the feedback curve. The lower feedback function, imposed by two concurrent variable-interval schedules, relates overall rate of reinforcement to the distribution of responding between the two schedules, measured as the logarithm (base 2) of the ratio of response rates (B_1/B_2). If all responses occur at schedule 1 (VI 1 min), then the rate of reinforcement is that obtained from that schedule alone. The same holds if all responses occur at schedule 2, which is indicated as the leaner schedule (VI 2 min). If responding is distributed between the two schedules, a higher rate of reinforcement occurs, because reinforcers are delivered by both schedules. As with the variable-interval feedback function, actual performance appears as a frequency distribution of preferences and a range along the curve, because out of all the possible performances shown by the curve, the actual performance consists of a small subset.

Rachlin(1992; 1994) emphasizes the hierarchical nature of final causes. Except for the all-encompassing category, which might be called "living one's life,"[1] every activity fits

[1]Someone who believed in reincarnation would hold that even living one's present lifetime fits into a still larger (Karmic) pattern across many lifetimes.

into a larger pattern of activities, and every activity constitutes a pattern into which other activities fit. Tooth-brushing fits into the larger activity of maintaining oral hygiene, which also includes flossing. Maintaining oral hygiene fits into maintaining bodily hygiene, which also includes bathing. Maintaining bodily hygiene fits into maintaining health. And so on.

If living one's life were broken down into constituent activities, these might be family life, professional life, having friends, and personal satisfaction. Let us say that we keep track of Tom, who is a married accountant in his mid-thirties. Out of his waking hours, we might find that the pattern of time spent in an average week was as shown in the leftmost chart in Figure 3.3. Tom spends most of his time working (say, 50 hours out of 112), a fair amount of time connected with his wife and two children (householding), some time on personal goals, and a small amount of time relating to friends. That is the pattern of his life in his mid-thirties, which might fit into a larger pattern spanning his whole lifetime. In his mid-thirties, any one of the categories shown in Figure 3.3 might be examined in more detail to reveal a subordinate pattern, as illustrated by the second chart, which shows the pattern of Tom's personal activities. This chart shows a pattern of time spent in maintaining health (exercise, tooth-brushing, bathing, and so on), reading, meditating, studying cooking, and working on his car. Since the activities included here constitute personal satisfaction, they exclude reading done for professional reasons, cooking done for the household, or working on the car as an activity with friends. Every one of these personal activities exhibits a pattern itself. As an example, Figure 3.3 shows the pattern of Tom's meditating. The time is divided among three different techniques: Most is spent in breath work, some is devoted to chanting a mantra, and some is spent in focus on a specific object. Each of these activities could be examined in further detail for its pattern.

Accepting Figure 3.3 for the moment as a representation of activity patterns, we may see the logic of final-cause explanations and their relation to what are loosely called "goals" and what behavior analysts call reinforcement. Tom chants because he meditates, the goal of chanting is meditation, and the reinforcement of chanting is successful meditation. Tom meditates because it is personally satisfying, the goal of meditating is personal satisfaction, and the reinforcement for meditating is personal satisfaction. When we come to the leftmost chart, we face the question, "What is the goal of life?" This has a variety of answers; I suggested elsewhere that evolutionary theory leads to the

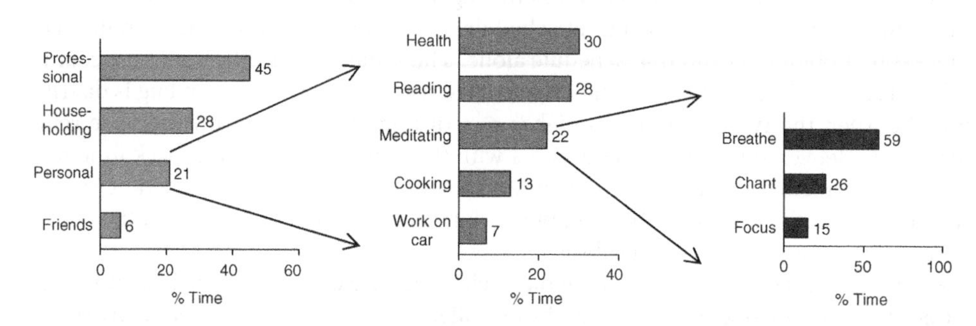

Figure 3.3 The hierarchical nature of patterns and final causes.
Note: Leftmost: pattern of frequencies at a very general level of analysis. Middle: pattern of frequencies of activities grouped under "personal activities." Rightmost: pattern of frequencies of activities grouped under "meditating."

conclusion: health, resources, relationships, and reproduction (Baum, 1995b). The four categories in the leftmost chart of Figure 3.3 map roughly into these four goals.

The same hierarchical view applies to the more quantitative conceptions embodied by the feedback functions in Figure 3.2. The concurrent-VI function (Figure 3.2, bottom) incorporates two functions like the VI function (Figure 3.2, top); it could be said that the two alternatives' VI functions *fit into* the concurrent-VI function. More molecular relations, such as differential reinforcement of interresponse times (Morse, 1966), fit into the VI feedback function, and more molar relations characterizing broader arrangements, such as multiple concurrent schedules or concurrent chain schedule, would encompass the concurrent-VI function. I suggested elsewhere the way feedback functions fit into a herarchy of behavioral situations (Baum, 1973).

How should we talk about patterns of behavior? What terms— what metaphors— should apply? Figure 3.3 is less an answer to this question than a statement of the problem. Many questions remain. It remains to be seen how to categorize behavior into mutually exclusive activities or whether this is even necessary. It remains to be seen how sequential dependencies should be treated. And so on. Such questions will be addressed as this pattern of thinking unfolds.

Conclusion

The science of behavior has reached a maturity that permits a reassessment of its early reliance on mechanical, efficient-cause models of the universe. Although such models lent clarity to the science early on, in the second half of the twentieth century their limitations have become apparent. Their crucial flaw lies in their reliance on the conception of time as a succession of moments. They fall short both on empirical grounds and logical grounds, because behavior cannot be measured or defined at moments of time. A viable alternative may be to view behavior as organized into hierarchical patterns of activities that transcend moments of time and manifest or unfold rather than move out of a future, through a present, and into a past.

4

From Molecular to Molar: A Paradigm Shift in Behavior Analysis

Foreword

A true natural science of behavior is possible, a science that treats behavior as composed of natural processes, understood and explained by other natural processes. A molar view releases the science from the thralldom of discrete events, moments in time, and contiguous (efficient) causes, because it sees behavior as composed of temporally extended processes, which may be called *activities.* In a molar view, all processes are concrete, observable, and measurable. Its explanations require invoking no hypothetical processes like response strength.

This paper begins to present the molar view and its advantages compared with the molecular view. As I write this 20 years later, two points that are implied now seem clear and worthy of emphasis: induction and ontology. Induction begins to feature with the 2012 paper, "Rethinking Reinforcement." The ontological implications of conceptualizing of behavior as process are profound, because they not only change how we observe and measure behavior, they change how we think about two old thorny problems: variation and disposition.

Skinner attempted to deal with variation in behavior by defining the units in terms of their function—e.g. getting a lever pressed. He made the mistake, however, of calling the population a class, instead of an individual, thereby creating confusion that persists to this day. A disposition is the tendency to behave one way rather than another, and the problem it raises is the question of its ontological status. Skinner's attempted solution was to refer to an organism's "repertoire" as composed of discrete units with different probabilities, and disposition was probability. Probability, however, is not what we measure; we measure rate. Neither a probability nor a rate can be assessed at a moment. Thus, probability, like strength, recedes into some hypothetical realm. In the molar view, dispositions, if spoken of at all, are real and measurable, because a disposition is a rate of occurrence. A response rate or a behavioral allocation is a disposition, and either we measure dispositions or we eliminate the term, depending on whether the word seems useful. In a molar view, all behavior is actual. No need exists for works like "latent" or "potential." When I am watching television, other activities are neither latent nor potential—they take up time before and after I watch television. The various activities of my day constitute an allocation and may be aggregated across many days if the allocation needs to be more general. The same holds for activities measured in the laboratory. Like a disposition, an

Science and Philosophy of Behavior: Selected Papers, First Edition. William M. Baum.

organism's repertoire is actual and measurable; it consists of the activities that actually occur through time—an allocation. In the molar view, the science of behavior requires no hypothetical entities.

Abstract

A paradigm clash is occurring within behavior analysis. In the older paradigm, the molecular view, behavior consists of momentary or discrete responses that constitute instances of classes. Variation in response rate reflects variation in the strength or probability of the response class. The newer paradigm, the molar view, sees behavior as composed of activities that take up varying amounts of time. Whereas the molecular view takes response rate and choice to be "derived" measures and hence abstractions, the molar view takes response rate and choice to be concrete temporally extended behavioral allocations and regards momentary "responses" as abstractions. Research findings that point to variation in tempo, asymmetry in concurrent performance, and paradoxical resistance to change are readily interpretable when seen in the light of reinforcement and stimulus control of extended behavioral allocations or activities. Seen in the light of the ontological distinction between classes and individuals, extended behavioral allocations, like species in evolutionary taxonomy, constitute individuals, entities that change without changing their identity. Seeing allocations as individuals implies that less extended activities constitute parts of larger wholes rather than instances of classes. Both laboratory research and everyday behavior are explained plausibly in the light of concrete extended activities and their nesting. The molecular view, because it requires discrete responses and contiguous events, relies on hypothetical stimuli and consequences to account for the same phenomena. One may prefer the molar view on grounds of elegance, integrative power, and plausibility.

Keywords: resistance to change, individual, class, concrete/abstract, molar, molecular, atomism

Source: Originally published in *Journal of the Experimental Analysis of Behavior*, 78 (2002), pp. 95–116. Reproduced with permission of John Wiley & Sons.

> *No man is an* Iland, *intire of it selfe; every man is a peece of the* Continent, *a part of the* maine; *if a Clod bee washed away by the* Sea, Europe *is the lesse . . . any mans* death *diminishes* me, *because I am involved in* Mankinde; *And therefore never send to know for whom the* bell *tolls; It tolls for* thee.
>
> John Donne (1572–1631).

[1]Every scientific paradigm includes both epistemological claims—claims about knowledge, such as what it is and how it is obtained—and ontological claims—claims as to what we are to know about (Kuhn, 1970). In paradigm clashes, ontological claims often

[1]I thank Michael Ghiselin and Howard Rachlin for helpful comments on an earlier draft of this paper.

matter most. In the Ptolemaic view of the universe, the planets, like other heavenly bodies, revolve around the earth. Their irregular movements in the sky were explained with the use of epicycles, circles within the circular orbits around the earth. In the modern view, the sun, moon, and planets constitute a solar system. The concept of epicycle makes sense in one paradigm but is absent from the other. The concept of solar system exists in the other paradigm but is absent from the first. Neither concept is wrong. Each makes sense within one paradigm but is nonsense in the other. That is why, in contrast to theoretical disputes, paradigm clashes cannot be settled by data. Any particular set of data may be meaningful to one paradigm and meaningless to another or may have different interpretations according to different paradigms. The interpretations will be "incommensurate"—that is, each will make sense only within its paradigm (Kuhn, 1970).

The purpose of this paper is to describe and support a paradigm that has developed within behavior analysis over about the last 30 years. I will call it the molar view, because *molar* carries the connotation of aggregation or extendedness, and the molar view is based on the concept of aggregated and extended patterns of behavior. Its roots may be traced back to the 1960s, but it became clearly visible in the 1970s (e.g., Baum, 1973; Rachlin, 1976), and it was articulated explicitly in the 1980s and 1990s (e.g., Baum, 1997; Chiesa, 1994; Lee, 1983; Rachlin, 1994, 2000). Like the heliocentric view of the solar system, the molar behavior-analytic view clashes with an older paradigm (Baum, 2001; Dinsmoor, 2001; Hineline, 2001), which I will call the molecular view, because it is based on an atomism of discrete events at moments in time. I will focus on the contrasting ontological claims made by the molar and molecular views, even though they also clash on epistemological grounds (e.g., the uses of cumulative records vs. digital counters), because the ontological clash, though more fundamental, is less obvious.

The Molecular View

In the 19th century, many psychologists sought to put psychology on a sound scientific basis by focusing on the association of ideas, sensations, and movements. These units of consciousness were conceived of as discrete events that could be "hooked" together or associated according to certain principles. Chief among these principles was the law of contiguity, which stated that two events that occurred close together in time (i.e., in temporal contiguity) would tend to recur together. In particular, if the idea of food happened to follow closely upon the idea of a musical tone, then when the idea of the tone recurred, the idea of the food would recur. This seemed a way to account for both the stream of consciousness and for the build-up of complex ideas from simpler ideas, as molecules are built up from atoms.

Although the association of ideas became less popular in the 20th century, the original atomism persisted in the concepts of stimulus and response. A stimulus was a discrete event in the environment, and a response was a discrete event in behavior. The principle of association by contiguity persisted in the concept of the conditional reflex.

In a classic paper, "The Generic Nature of the Concepts of Stimulus and Response," Skinner (1935/1961) attempted to create definitions of stimulus and response that would serve as the basis for a science of behavior. One can hardly overstate the importance of this paper to the development of behavior analysis. Skinner proposed a solution to the problem of particularity that plagues behavior analysis as it does any science: If each event (stimulus or response) is unique, how does one achieve the reproducibility required for scientific study? His answer was that a stimulus or a response was not a unitary event but was a class of unitary events. Although any particular event might be described with great precision, the goal for defining a stimulus or response, as a class, was to specify the class's defining properties. The lever press, for example, would consist of the class of acts, all of which achieved the necessary movement of the lever. The nondefining properties could be ignored or could serve as the means for further differentiation. One would know if one's defining properties were correct by the consistency of one's results when the class is so defined—that is, in "smooth curves for secondary processes" (p. 366), what he was later to call functional relations. In this way, Skinner made possible a science of behavior—that is, behavior, as opposed to physiology or consciousness.

Skinner's stimulus and response, however, were classes of discrete events, the same sort of events as the previous century's ideas, situated at moments in time and explained by contiguity between events in time. A reflex for Skinner (1935/1961) was a correlation between two classes, meaning that when a member of the stimulus (as class) occurred, it would be followed by a member of the response (as class). Conditional reflexes were created by the repeated contiguity of members of the two classes. Later, he treated the law of effect in similar fashion: The response (as class) was strengthened by repeated contiguity between its members and the members of the reinforcer (as stimulus class).

Skinner (1938) equated response strength to probability. He proposed to measure probability as response rate. He saw response rate as an expression of response probability or strength, often writing as if this were the true dependent variable (Skinner, 1938, 1950, 1953/1961, 1957/1961). Response rate would be the outcome of probability acting moment to moment, as if at every moment a probability gate determined whether a response would occur just then or not. Changes in response rate were an outcome of changes in response strength, possibly acting locally, as in the fixed-interval scallop (Ferster & Skinner, 1957). Stimulus control occurred as a result of modulating response strength, as if in the presence of a discriminative stimulus the probability gate became more or less liberal in its moment-to-moment decisions.

A Molar View

In 1969, Baum and Rachlin proposed a different view of response rate. We drew on T. F. Gilbert's (1958) suggestion that responses like lever presses and key pecks might occur in bouts at a constant rate, which he called the tempo of responding. Variation in response rate, in this molar view, would result from variation in the duration of bouts and the pauses between bouts (e.g., Shull, Gaynor, & Grimes, 2001). The time spent responding (T) at the tempo (k) would deter mine the number of responses (N): $N = kT$. If S is the duration of the sample (usually the session duration), then response rate (B) is given by

$$B = \frac{kT}{S} \quad (4.1).$$

We suggested that behavior might be thought of as divided among activities that lasted for periods of time (i.e., bouts). Taking time as the universal scale of behavior, we proposed that the dependent variable be thought of as time spent responding ($T = N/k$) or proportion of time spent responding:

$$\frac{T}{S} = \frac{N}{kS} \quad (4.2)$$

Accordingly, we wrote of choice as time allocation, the allocation of time among continuous activities, and we characterized the matching law as a matching of relative time spent in an activity to relative reinforcement obtained from that activity. The idea was extended to other changes in responding, such as behavioral contrast (White, 1978).

The elaboration of this idea is the paradigm that I am calling the molar view of behavior. Whereas the central ontological claim of the molecular view is that behavior consists of discrete responses, the central ontological claim of the molar view is that behavior consists of temporally extended patterns of action. I shall call these activities. Besides the concept of an activity, the molar view is based also on the concept of nesting, the idea that every activity (e.g., playing baseball) is composed of parts (batting) that are themselves activities. I shall focus first on activities and discuss nesting later.

Historically, the notion of reproducible discrete events, whether ideas or responses, allowed a scientific approach to the subject matter. The concept of discrete response allowed quantification. Skinner's (1938) preparation, based as it was on the reflex, encouraged researchers to think of the response as momentary, as an event without duration. Skinner even set out one of his requirements for a response to study that it should be brief and easily repeated, because these properties would allow rate to vary over a wide range. A concept like the delay-of-reinforcement gradient depends on the idea that the response occurs at a certain point in time from which delay is measured. The concept of contiguity itself depends on the idea that two discrete events (e.g., response and reinforcer) mark the beginning and end of such a delay, the duration of which, for perfect contiguity, should be zero.

The notion of activity takes for granted the possibility of quantification, extending it beyond discrete responses and contiguity. The key difference lies in the recognition that activities take up time. In an earlier paper, I argued that the reinforcement relation might be thought of as a correlation (Baum, 1973). In the molar view, an activity like lever pressing, extending in time, is seen as *accompanied* by the reinforcers it produces. Many reinforcers may be involved, and consequences, being extended like activities, often consist of changes in reinforcer rate or changes in reinforcer magnitude. The idea that reinforcers accompany an activity might be misinterpreted to mean that delay is irrelevant, a claim that apparently would be easy to refute by experimentation. As in other paradigm clashes, however, in the molecular–molar clash the phenomena observed are simply seen in different terms. A procedure that arranges food delivery to immediately follow depressions of a lever arranges a strong correlation between lever pressing and rate of food delivery. A procedure that arranges food deliveries to follow responses at some delay arranges a lesser correlation, because the food deliveries may fail to accompany the activity. To effect changes in behavior, strong correlations work best; to maintain activities already occurring frequently, weak correlations may suffice (Baum, 1973).

The point might seem trivial, until we turn to other types of activity. Had Skinner chosen to study wheel running instead of lever pressing, his situation would have been different. Experimenters often measure wheel running in revolutions or quarter-revolutions, but these are artifices aimed at creating discrete responses where none are apparent. Premack (1965, 1971) instead proposed time as the measure. If we turn this reasoning onto activities like lever pressing, we see that discrete responses like lever presses are an outcome of instrumentation. "Responses" are momentary events (usually switch closures) contrived by the apparatus (e.g., lever or key). Their momentary character is only an artifact of the use of electrical switches. As a thermometer, whether stuck into a bowl of ice cream or a pile of cow manure, indicates only the temperature, regardless of the stuff it is stuck into, so a lever, when stuck into the activity of a rat, indicates only rate of switch closure, regardless of what sort of activity it is stuck into.

When I was a graduate student, laboratory practice dictated that one attach a lever or key to a pulse former, a device that would generate a uniform pulse on each operation of the lever or key. Laboratory lore claimed this was essential to keep the subjects from holding the switch. I tested this claim by omitting the pulse formers in an experiment on concurrent schedules (Baum, 1976). I measured both number of lever presses (switch operations) and time that the lever was depressed. The measures turned out to be equivalent, because the activity continued to result in jiggling of the lever, producing many switch operations. Even though the duration of the operations varied, the average was consistent enough that time and number of presses could be interchanged, as in Equation 4.2. Just as a rat's jiggling of a lever operates it at certain points in the jiggling, a pecking key operates only at a certain point in the movement of a pigeon's head. When Pear (1985) arranged a system for keeping continuous track of the position of a pigeon's head, key pecking appeared as cyclic motion, a wave form.

Two simple arguments might persuade one to think about behavior in terms of activities rather than discrete responses: (a) Momentary events are abstractions and (b) reinforcement operates by selection. The first, that momentary events cannot be observed, but only inferred, was argued in an earlier paper (Baum, 1997). I shall recapitulate briefly. Suppose I show you a snapshot of a rat with its paw on a lever, and I ask you, "Is the rat pressing the lever?" You will have to reply, "I don't know. I have to see what comes next." You won't know until you see the whole (extended) lever press. In other words, you would never be able to tell from a momentary picture of the behavior what behavior was occurring at that moment. Only after the whole pattern unfolded would you be able to look back and infer that at that moment the rat was pressing the lever. The momentary response is never observable at the moment. It is always inferred afterwards. Like instantaneous velocity in physics, instantaneous behavior cannot be measured at the moment it is supposed to have occurred but must be inferred from a more extended pattern. That is why it might fairly be called an abstraction. That is why one may argue that it is impossible to reinforce a momentary response; one can only reinforce some activity with some duration.

The second argument is that reinforcement consists of selection. Possibly Ashby (1954) was the first to recognize the parallel between reinforcement and natural selection. Campbell (1956) spelled out the idea that reinforcement is a type of selection, and R. M. Gilbert (1970) and Staddon and Simmelhag (1971) elaborated it further. Skinner (1981) himself proposed it eventually. The essential point is that behavior varies as do genotypes within a population of organisms. As differential reproductive success increases some genotypes while decreasing others, so differential reinforcement increases some behavioral variants while decreasing others. This parallel, however, requires competition among variants, in the sense that the increase of one variant necessitates the decrease of others. With genotypes, this occurs because the size of the population tends to remain constant at the carrying capacity of the environment (i.e., the number of organisms that the resources of the environment can maintain). For behavior, the competition requires that the total of behavior in any period of time should, like the carrying capacity, remain constant. In other words, it requires that behavior take up time, and presumably that all the behavior that occurs in a time period take up all the time. Just as the limit to a biological population necessitates that if longer-necked giraffes increase in frequency then shorter-necked giraffes must decrease in frequency, so the temporal limit to behavior necessitates that if pecks at the left key increase in frequency then pecks at the right key must decrease in frequency. At the least, the idea that reinforcement is a kind of selection requires that discrete responses must have duration. Add to that the possibility that the so-called responses may vary in duration, and you have granted that behavior consists of activities. The so-called response becomes indistinguishable from a bout of an activity.

Someone wedded to the molecular view might reply in a number of different ways. One might reject the analogy to natural selection altogether and stick with momentary responses. One might insist that in any time period greater and lesser amounts of behavior may occur. The molarist might reply that sacrificing the elegance of selection is a high cost to pay for the sake of keeping to momentary responses.

Strength Versus Allocation

The difference between the molar and molecular views may be seen in their different guiding metaphors. Whereas the molecular view relies on the idea of strength, the molar view relies on the idea of allocation. These different ideas depend on the different ontological claims of the two paradigms.

In the molecular view, reinforcers strengthen behavior. They do this by following immediately upon or soon after a bit of behavior, a response. The underlying assumption of a central role for contiguity in time entails the ontological claims of the molecular view, because contiguity exists at or around a certain moment in time. For two events to be contiguous, to occur close to the same moment, they must either be momentary or have distinct beginnings and ends. For a response to be followed immediately by a reinforcer, the response and the reinforcer must be discrete events. Thus, the contiguity- based notion of reinforcement espoused by the molecular view entails the ontological claim that behavior consists of discrete events. Strength cannot attach to a single instance of a discrete unit, because each of those is unique. Therefore, in the molecular view, strength attaches to the class and is inferred from the number of members of the class that occur in any given time period (i.e., from the calculated rate). A rat may press a lever 30 times per minute or twice per minute. I may drive to work seven times per week or once per week. A reinforcer may follow each occurrence of a discrete unit, but only some occurrences need be followed by reinforcers for the rate of occurrence of members of the class to be maintained—that is, reinforcement may be intermittent. The more often members of the class are reinforced, the greater the strength of the class.

In contrast, the guiding metaphor of the molar view is allocation. If the molecular view likens behavior to picking numbers out of a hat, the molar view likens behavior to cutting up a pie. Choice is time allocation (Baum & Rachlin, 1969). All behavior entails choice. All behavior entails time allocation. To behave is to allocate time among a set of activities. Such an allocation is a behavioral pat tern. If a pigeon spends 60% of its time pecking at one response key, 30% of its time pecking at another, and 10% of its time in other, unmeasured activities, that is the pattern of its behavior while in the experiment. If a person spends 47% of his or her recreational time watching television, 40% reading, 10% walking, and 3% going to movies, that is a pattern of recreational behavior. Such patterns or allocations are necessarily extended in time.

Because every activity itself is composed of other activities—that is, because every activity is a whole constituted of parts—every activity itself contains an allocation of behavior. In this sense we may say that every activity *is* a behavioral allocation. The pigeon's allocation in the experiment may be called its experimental activity, and the allocation of recreational time may be called the person's recreational activity. We shall discuss this further when we take up the concept of nesting.

In the molar view, the appearance that behavior might be composed of discrete units arises because activities often occur in episodes or bouts. Task completion provides many examples: completion of a fixed-ratio run, of a house, of writing a paper, of reading a book. In the laboratory, the training of response chains originated to try to model such extended units. In the molecular view, the sequence is thought to terminate at least some of the time with a reinforcer and to be held together with conditional reinforcers along the way to ultimate reinforcement. In the molar view, an activity like building a house entails a pattern of activities such as pouring the foundation, framing the structure, insulating, putting in windows and doors, and finishing the interior. House construction seems like a unit only because it is labeled as such, as one may call an episode of napping a nap or a bout of walking a walk. To a building contractor, construction of one house would seem more like an episode of building than a discrete unit. Against the molecular view, one might argue that behavioral chains sometimes bear little resemblance to extended behavior in the real world. Is it plausible to treat obtaining a bachelor's degree as a behavioral chain? Except for repetitive sequences like those of the assembly line, real-life sequences like building a house or baking a cake rarely follow a rigid order. Even the fixed-order chain of the laboratory (e.g., chain fixed interval fixed ratio) may be seen as a sequence of activities (fixed-interval activity, then fixed-ratio activity), and if some activities are maintained better than others, that is a matter to be studied. Completion of any task may be seen as an episode of an activity. Even activities that usually end in a certain way, such as search, last for varied durations; sooner or later the forager encounters prey, sooner or later one finds a parking space.

In particular, the molar view holds that the so-called response is an episode of an activity. Grant that pecks and presses take up time and that the time taken up by each cycle of activity—of body motion back and forth or up and down—takes up about the same amount of time (Baum, 1976; Pear, 1985), and then Equation 4.2 illustrates how switch operations (N) are convertible to time and how that time may be considered relative to the total (T). In other words, rate of pecking or pressing is equivalent to an allocation—relative time spent pecking or pressing. Even if counting responses is convenient, from the molar viewpoint a response rate is a relative time in the activity.

In the molar view, discrete behavioral units are not only illusory but often are simply impossible. As Baum and Rachlin (1969) argued before, many activities lack any natural unit. What is the discrete unit of watching television, reading, sleeping, or driving a car? Applied behavior analysts recognize this when they set goals such as increasing time on task. Even in the laboratory, activities like wheel running and lever holding lack any nonarbitrary unit. Such activities, for which the molecular view must invent "responses," readily lend themselves to the idea of allocation.

Three Examples of Molar Explanation

Although the conflict between two paradigms cannot be resolved by data, the power of a paradigm may be seen in its ability to interpret various phenomena of the laboratory. Three examples of the power of the molar view appear in its ability to treat (a) variation in tempo, (b) asymmetrical concurrent performances, and (c) resistance to change.

Variation in Tempo

Herrnstein's (1970, 1974) formulation of the matching law relied on response rate. He expressed it in the form:

$$\frac{B_1}{\sum_{i=1}^{n} B_i} = \frac{r_1}{\sum_{i=1}^{n} r_i} \quad (4.3),$$

which states that the relative response rate of any of n alternative responses matches the relative reinforcement obtained from those n alternative responses. Herrnstein (1970) further supposed that the sum total of behavior (the denominator on the left side of Equation 4.3) was a constant. On this assumption, Equation 4.3 is rewritten as:

$$B_1 = \frac{kr_1}{r_1 + r_0} \quad (4.4)$$

where r_0 represents all reinforcement obtained from alternatives other than Alternative 1, and k represents the sum total of behavior expressed in the response units of B_1. In keeping with the notion of time allocation that Baum and Rachlin (1969) put forward, however, the constant k in Equation 4.4 may be reinterpreted as the tempo of the activity defining B_1. It equals the response rate that would occur if all behavior were allocated to Alternative 1—that is, the response rate if all the time were spent in Activity 1, sometimes called the asymptotic response rate. Substituting Equation 4.1 into Equation 4.4 allows one to rewrite Equation 4.4 to have proportion of time spent in Activity 1 matching proportion of reinforcement obtained from Activity 1.

One seeming challenge to Equation 4.4 arose from research by McDowell and associates (e.g., Dallery, McDowell, & Lancaster, 2000) that cast doubt on Herrnstein's assumption of constancy of k. They found that several operations, such as varying deprivation or reinforcer magnitude, result in different values of k when the response (of B_1) remains ostensibly the same. McDowell offered equations that predicted variation in asymptotic response rate, but at the cost of assuming discrete responses of invariant duration (e.g., McDowell, 1987). How might one explain variation in k while retaining the molar view?

One may interpret McDowell's findings as showing that the operations that vary k affect the tempo of responding, perhaps by affecting response topography. If k increases with increasing magnitude of reinforcement, that might be because the increased magnitude results in more vigorous responding, which results in less time per response. The higher tempo of the activity would result in more switch closures counted in the same amount of time. Thus the observation of varying k is readily accommodated by the molar view.

Asymmetrical Concurrent Performances

Another challenge to the matching law is the observation that behavior at two choice alternatives may differ qualitatively. The two-alternative version, expressed as

$$\frac{B_1}{B_2} = \frac{r_1}{r_2} \quad (4.5)$$

may be thought of as derived by taking the ratio of Equation 4.3 to the similar equation written for Alternative 2. Such a derivation would be justified only if k were equal for the two alternatives. Suppose that the topography of the two responses differed, resulting in a difference in tempo (k). For example, suppose that one alternative was reinforced according to a variable-interval (VI) schedule and the other was reinforced according to a variable-ratio (VR) schedule (e.g., Baum & Aparicio, 1999; Herrnstein & Heyman, 1979). If the tempo on the VR alternative were higher, the same amount of time spent at that alternative would result in more responses counted (i.e., more switch operations) there than if that time were spent at the VI alternative. Relative "responses" would deviate from matching.

The generalized matching law has been used to estimate such deviations from matching:

$$\frac{B_1}{B_2} = b\left(\frac{r_1}{r_2}\right)^s \quad (4.6),$$

where b is a proportionate bias that is independent of the rates of reinforcement, r_1 and r_2, and s is the sensitivity to variation in the ratio of reinforcement. If b and s both equal 1.0, the strict matching of Equation 4.5 occurs. When deviations from strict matching occur, they are usually estimated as values of b and s different from 1.0. If the tempos of B_1 and B_2 differed, one would expect a value of b different from 1.0, favoring the VR (e.g., Baum & Aparicio, 1999; Herrnstein & Heyman, 1979).

A stronger challenge to the matching law arises from the observation that a difference in topography or tempo may affect s (Baum, Schwendiman, & Bell, 1999). Equation 4.6 is usually fitted to behavior and reinforcer ratios in its logarithmic form,

$$log\frac{B_1}{B_2} = s \cdot log\frac{r_1}{r_2} + logb \quad (4.7),$$

because this form is symmetrical around the indifference point (behavior and reinforcer ratios both equal to 1.0) and because, being linear, it is easier to fit. The equation is fitted to behavior ratios determined for several reinforcer ratios to both sides of equality (i.e., sometimes making Alternative 1 richer, some times making Alternative 2 richer) on the assumption that parameters s and b remain independent of variation in the reinforcer ratio. Baum et al. found that when pigeons were exposed to pairs of concurrent VI schedules long enough for performance to remain stable over a substantial sample, the behavior ratios deviated systematically from Equation 4.7. On closer examination, a simple pattern of behavior appeared: Responding occurred almost exclusively on the rich alternative, interrupted only by brief visits to the lean alternative. This pattern, which we called "fix and sample," predicted two lines, each with slope s equal to 1.0 but with different bias b, depending on whether Alternative 1 or Alternative 2 was the lean alternative (Houston & McNamara, 1981). We found that the two-line model, with the same number of parameters as the one-line model, fitted the data better and with no systematic deviations.

Figure 4.1 illustrates the finding. The two left graphs show the behavior ratios from Pigeon 973, fitted with one line (top) and with two lines (bottom). A casual look at the top graph would lead one to conclude that the results were typical of such experiments: a good fit to Equation 4.7 with a moderate amount of undermatching (s = 0.8). Closer inspection reveals that, going from left to right, the data points first lie above the line, then below the line, then above again, and then below again. Not only is the two-line fit better, but also the variation in choice appears to conform closely to the assumed slopes of 1.0. Indeed, the undermatching shown in the top graph is explained by the inappropriate fitting of one line to data better described by two.

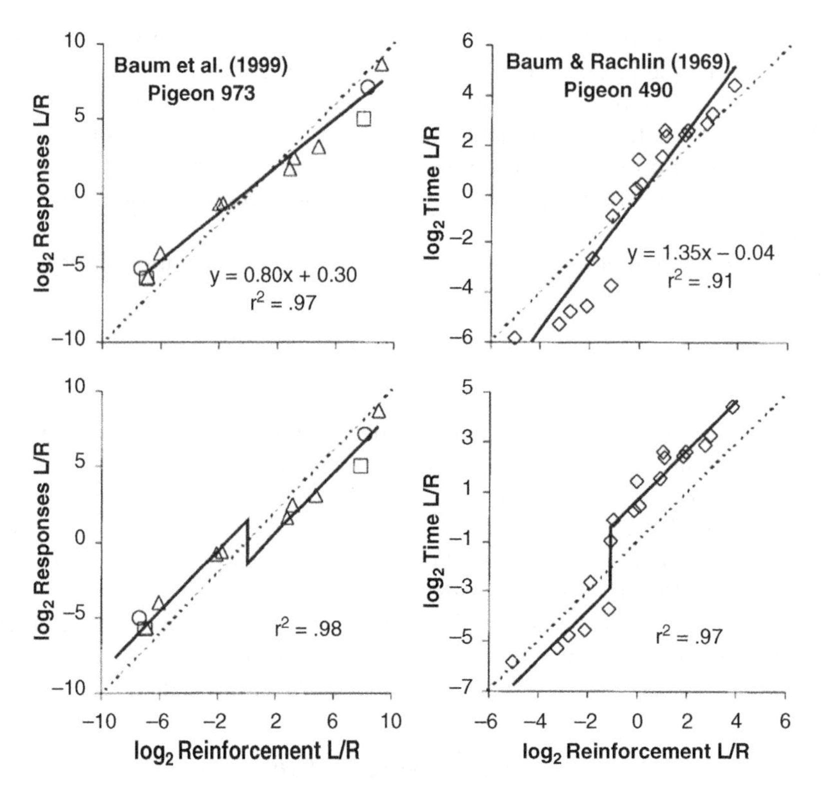

Figure 4.1 Apparent deviations from matching explained by the fix-and-sample pattern
Note: Left: The top graph shows apparent undermatching. Close examination reveals systematic deviation from the fitted line (the generalized matching law). The bottom graph shows the same data fitted with the two lines predicted by fix and sample. Right: an instance of apparent overmatching, with systematic deviation from the generalized matching law. The bottom graph shows the data fitted with the two lines predicted by fix and sample, eliminating the systematic deviation.

The two right graphs in Figure 4.1 demonstrate a similar explanation for an example of apparent overmatching. They show behavior ratios from Pigeon 490 in the Baum–Rachlin (1969) experiment, fitted with one line and with two. Again the two-line fit (bottom) shows less systematic deviation, again the slopes of 1.0 seem appropriate, and again the apparent deviation of slope s from 1.0 may be explained as the result of inappropriate fitting of one line to data better described by two.

Although the two-line fits in Figure 4.1 both show biases that explain the deviations of s from 1.0, these biases differ from the bias b in Equations 4.6 and 4.7. Whereas the usual bias is assumed to be independent of reinforcer ratio, the biases in the two-line model depend on the reinforcer ratio, because they depend on which is the leaner alternative. In the undermatching example, the bias favored whichever was the lean alternative. In the overmatching example, the bias favored whichever was the rich alternative. The overmatching example shows standard position bias too; that is why the vertical crossover occurs at a reinforcer ratio less than 1.0.

The results shown in Figure 4.1 support the idea that behavior at the rich alternative and behavior at the lean alternative constitute different activities that, in turn, comprise parts of a more extended pattern (i.e., the fix-and-sample activity) involving both alternatives. Behavior at the rich alternative consists of staying there—fixing—whereas behavior at the lean alternative consists of brief visiting—sampling. In the undermatching example, the pigeons pecked at two keys, but visits to the lean key nearly always consisted of single isolated pecks. To explain the apparent bias in favor of the lean alternative, one need only assume that less time was spent per peck at that key. For example, a pigeon stationed in front of the rich key would "travel" to the lean key by stretching its neck toward the key, and in that position, make a brief exploratory peck. Measured as time spent, a number of pecks at the lean key would represent less time than the same number of pecks at the rich key, and counting them equally would overestimate the time spent at the lean alternative. In other words, if the subscripts in Equation 4.7 were reinterpreted to mean rich and lean, rather than left and right, then the bias would differ from 1.0 only because the time per peck differed, just as the earlier discussion suggested for concurrent VI VR schedules.

In the overmatching example, Baum and Rachlin (1969) measured only time spent on two sides of a chamber; there were no response keys. We observed informally that the pigeons would station themselves near the middle of the chamber and move rapidly back and forth. Whereas behavior on the rich side consisted of standing or dancing, a visit to the lean side consisted of stepping over there, waiting out the signaled changeover delay (COD), and then either running to the feeder on the lean side or immediately hopping back again to the rich side. Under such circumstances, the visits to the lean side constituted brief episodes of an activity (i.e., sampling) different from that at the rich side (fixing). One need only assume that our procedure underestimated the time spent per visit to the lean side to explain the apparent biases in the lower right graph in Figure 4.1. In particular, we excluded time spent during the COD from our calculations, on the ground that it was signaled timeout from the experiment. Had we included that time, we would have had to decide how to allocate it between the two sides. Owing to premature changeovers to the lean side, most of the changeover time was probably spent on the lean side. Limitations of our electromechanical equipment prevented us from solving this problem.

The important point made by Figure 4.1, however, is that such instances of undermatching and overmatching may be readily explained by adopting the molar view, because it allows choice between the rich and lean alternatives to be seen as time allocation between two different activities. The molecular view presumably would explain Figure 4.1 as the aggregation of many discrete responses made at the two alternatives, combined with assumptions about the reinforcement of switching and delay gradients. A molarist might judge the account logically correct, but would regard it as implausible and inelegant, illustrating once again that a paradigm clash is resolved by such considerations rather than by data (Kuhn, 1970).

Activities, Stimulus Control, and Resistance to Change

A seeming challenge for the molar view comes from research and theory about behavioral momentum, a recent re-expression of the notion of response strength (Nevin, 1974, 1992; Nevin & Grace, 2000). The experiments show that persistence of responding (resistance to change) depends on prior rate of reinforcement. The molar interpretation of the results, which eschews response strength, stems from a molar view of reinforcement and stimulus control.

In an earlier paper (Baum, 1973), I argued that, just as behavior is extended, so too are consequences extended. As response rate is real (i.e., to be known about), so rate of reinforcement and rate of punishment are real. The molar analogue to contiguity is correlation—that is, correlation between extended activities and extended consequences. If I forget to add baking powder when I'm making a cake, the result is disappointing, but my cake baking generally pays off well and is maintained by its high rate of mainly good consequences. In the molar view, reinforcement is like starting and stoking a fire. Special materials and care get the fire going, and throwing on fuel every now and then keeps the fire going.

McDowell's experiments on variation in k (e.g., Dallery et al., 2000) and experiments on asymmetrical concurrent performances support the idea that behavior consists of extended allocations or activities. Reinforcement and punishment may change the time spent in an activity (i.e., allocation), but also may change the allocation among the parts and even the parts themselves. This latter kind might be called change in topography, referring roughly to the way an activity is done. Whatever the change, the molar view attributes it to differential extended consequences.

Once we move away from the atomism of discrete responses, we should expect that the way we talk about reinforcement and stimulus control will change too, even if only in subtle ways. In the molecular view, for example, "continuous" reinforcement is often contrasted with "intermittent" reinforcement.

One may question whether the notions of intermittent and continuous reinforcement have any meaning in relation to extended activities. In the molar view, reinforcers coincide with various parts of the activity in various forms. What matters is the aggregate of consequences that the activity (allocation) produces relative to other activities (allocations) over time. In the simplified environment of the laboratory, concurrent schedules arrange that a pattern of choice (an allocation) produces a rate of reinforcement (Baum, 1981b). As with choice patterns of self-control, in which impulsive behavior is immediately reinforced and self-control pays off better only in the long run, even though the more extended choice pattern would produce more reinforcers in the long run, local reinforcement contingencies may prevent the choice pattern (allocation) from evolving toward the maximum possible (Rachlin, 1995b, 2000; Vaughan & Miller, 1984). An experiment by Heyman and Tanz (1995), however, showed that providing signals allowed changes in reinforcer rate to reinforce changes in choice pattern (allocation). They arranged that when pigeons' choice over a sample of responses deviated toward a more extreme allocation than would be expected from the matching law (Equation 4.5), relative reinforcement would remain unchanged, but that the overall reinforcer rate would increase and a light would come on. They found that deviations from matching were reinforced by the changes in overall reinforcer rate.

The molar concept of reinforcement also implies a molar concept of stimulus control. In the experiment by Heyman and Tanz (1995), the light signaled a relation between an extended pattern (an allocation) and an increase in reinforcer rate. In the molecular view, a discriminative stimulus signals that some responses may be intermittently reinforced, and its presence increases the probability of the response. In the molar view, a discriminative stimulus signals more frequent reinforcement of one activity or allocation than another, and its presence increases the time spent in that activity. Reinforcers that occur in the presence of the stimulus plus the presence of the activity or allocation increase the control of the stimulus over that activity or allocation.

In the molar view, a response rate, whether measured in experiments in which the activities consist of repetitive motions like key pecks or lever presses (e.g., White, 1985) or used to measure general activity (Buzzard & Hake, 1984), is equivalent to an allocation, a pattern of behavior, an activity. We have seen that experiments on variation in k and asymmetrical concurrent performances may be interpreted as changes in response rate resulting from changes in extended patterns of responding. Heyman and Tanz's (1995) experiment embodies the molar idea of stimulus control, in which control over extended patterns of responding entails control over response rate and choice. If extended patterns of behavior (activities or allocations) may be reinforced and controlled by discriminative stimuli, then we should expect that response rates are both reinforceable and subject to stimulus control.

We may apply these expanded notions of reinforcement and stimulus control to experiments on behavioral momentum and resistance to change (Nevin, 1974, 1992; Nevin & Grace, 2000). In a typical experiment, pigeons are exposed to a multiple schedule composed of two components, each consisting of a VI schedule of food reinforcement for pecking at a response key. One VI is richer than the other and thus maintains a higher rate of key pecking. Once the two rates of key pecking have stabilized, a variety of different operations may be used to disrupt the responding; the usual ones are prefeeding, food presentations during timeout periods between components, and extinction. The typical result is that response rate decreases in both components, but by a larger proportion in the component with fewer reinforcers. For prefeeding and food presentations between components, the decreases in response rate might be interpreted as the result of a decrease in magnitude of programmed (VI) reinforcement relative to background reinforcement. In terms of Equation 4.4, these operations would decrease rate of pecking by increasing r_O. How Equation 4.4 might account for the difference in rate of extinction is less clear, but one might suppose that higher rates of key pecking, once established, tend to persist longer in the absence of reinforcement. Nevin, Tota, Torquato, and Shull (1990) reported an experiment, however, that undermined such seemingly straightforward explanations. Pigeons were exposed, as usual, to a multiple schedule, one in which the same VI schedule occurred in both components. In one component, however, it was the only schedule present, whereas in the other component, a second schedule reinforcing pecks on a second key (Experiment 2) or delivering food independently of behavior (Experiment 1) was paired with the constant schedule. Although the overall rate of reinforcement was lower in the single-VI component, the response rate there was higher than the response rate on the same key in the component with the concurrent VI (as would be expected; Rachlin & Baum, 1972). The crucial result was that, comparing the response rates on the constant-schedule key during extinction, the lower response rate decreased more slowly than the higher response rate.

Nevin (1992) interpreted this result and the earlier experiments to mean that reinforcement builds behavioral momentum: The more reinforcement, the more momentum; the more momentum, the less the response is susceptible to disruption. To maintain this theory, however, he had to distinguish between what he called operant and respondent aspects of the components. The difficulty was that pecks at the constant-VI key were reinforced at the same rate in both components; the extra reinforcers that made the difference in resistance to change were associated with the second key or some other behavior, and how reinforcement of other behavior could increase a response's momentum was unclear. Nevin concluded that, because the overall reinforcer rate was higher in the two-VI component, the momentum of all behavior in a component must depend on all the reinforcers in that component. The association of reinforcers for other behavior with the component's stimulus constituted a respondent or Pavlovian aspect to determining momentum.

The ideas of reinforcement and stimulus control of extended patterns of behavior (allocations) open the way to a different interpretation of experiments on resistance to change. In a multiple schedule, we may suppose that if the contingencies of reinforcement differ from component to component, they will generate different allocations of behavior in the presence of the different discriminative stimuli. All the reinforcers in a component serve to reinforce the allocation occurring there, and the stimulus enjoins that allocation. The more reinforcement, the more the stimulus enjoins the allocation. In a multiple schedule with two different VI schedules in two components, the higher reinforcer rate will be associated with the allocation that generates (i.e., is equivalent to) the higher response rate, stimulus control will be stronger over that allocation, and that stimulus will sustain that allocation longer as it disintegrates (i.e., transforms into some other allocation including little or no pecking) during extinction. As the experiments show, the higher rate allocation will take longer to disintegrate than the lower rate allocation. In Nevin et al.'s (1990) crucial experiment, different allocations occur in the one-VI component and in the two-VI component. The allocation in the one-VI component entails a higher response rate on the constant-VI key, but that allocation is less reinforced. The allocation in the two-VI component entails responding on both keys and is more reinforced. Because stimulus control is stronger over the two-VI allocation, that one disintegrates more slowly during extinction. Hence the response rate on the constant-VI key falls more slowly in the two-VI component.

Although this explanation of variation in resistance to change bears some similarity to Nevin's explanation, it has advantages. First, it is arguably simpler. It requires no appeal to separate operant and respondent aspects, because it invokes only the idea that stimulus control depends on rate of reinforcement. Second, it requires only an expansion of the concepts of stimulus control and reinforcement to apply to extended patterns of behavior, instead of the introduction of new concepts, such as behavioral momentum and mass (Nevin & Grace, 2000). These concepts, borrowed by analogy from Newtonian mechanics, seem particularly unlikely to explain the dynamics of behavior, because mechanics offers only nonhistorical immediate causes (Aristotle's efficient causes; Rachlin, 1995b). In the molar view, reinforcement is a process of selection, resembling natural selection—an entirely different sort of causation and fundamentally historical (Baum & Heath, 1992; Baum & Mitchell, 2000; Skinner, 1981).

Ideal Response Classes Versus Concrete Behavioral Patterns

The concept of behavioral momentum, like the concept of response strength, flows from the molecular, atomistic view of behavior. Momentum, like strength, is considered the possession of a class, the members of which are momentary responses. Skinner's (1935/1961, 1938) operant, for example, was a class with discrete responses as members, and when its strength was high its members occurred at a high rate. His ill-fated idea of the reflex reserve depended on just such a notion of strength, and Nevin's notion of momentum—the contemporary equivalent of the reflex reserve—similarly depends for its definition on the idea of response class. In the molecular view, one supposedly specifies the ideal properties required for membership in the class (e.g., a certain force, a certain extent, etc.). Any lever press or key peck that possesses the ideal properties may be recorded. To estimate response rate, one counts a number of instances and divides by the time interval during which they were counted. An increase or decrease in response rate reflects an increase or decrease in strength or momentum of the class. The more the members of the class are reinforced, the more is the class's strength. That response rates on interval schedules fall short of those on ratio schedules, for example, is explained by the differential reinforcement of long inter-response times (IRTs) in interval schedules, the IRT being considered another property of the response or instance (Ferster & Skinner,

1957; Morse, 1966). Reinforcement then selectively strengthens different response classes. The high rates on ratio schedules are attributed to the absence of differential reinforcement of IRTs. In this view, response rate always remains an abstraction, because the concrete particulars are the responses, the class instances.

In the molar view, an activity occupies more time or less time, depending on the conditions of reinforcement. No notion of strength or momentum enters the picture. When behavior is seen as composed of continuous activities or extended patterns (i.e., allocations), response rate is no longer an expression of strength or momentum. A response rate, as an allocation, is seen as concrete.

Increases or decreases in rate of key pecking may or may not indicate increases or decreases in the time spent pecking. The examples of varying k (tempo), asymmetrical concurrent performances, and varying resistance to change show that at least two possibilities exist. First, response rate may increase or decrease because the mix of activities changes to include more or less time spent in the repetitive activity, as implied by Equation 4.4 (cf. Shull et al., 2001). Second, response rate may increase or decrease because the repetitive activity itself (its topography) changes. The difference in response rates between ratio and interval schedules arises because the schedules reinforce different patterns of responding—that is, different activities. Interval schedules differentially reinforce activities that result in lower rates of key operation, whereas ratio schedules differentially reinforce activities that result in high rates of key operation (Baum, 1981b). Across the low range of reinforcer rates, as reinforcer rate increases across VI schedules (i.e., as the average interval gets shorter), response rate increases and levels off, as Equation 4.4 would predict, but when the VI schedule becomes brief enough, it begins to function like a ratio schedule, and response rate increases up to the same level as for a comparable VR (Baum, 1993a). The increase across the low range of reinforcer rates represents an increase in time spent in low-tempo key pecking, whereas the increase across the high range of reinforcer rates represents an increase in time spent in high-tempo key pecking (Baum, 1981b, 1993a).

Class versus Individual

This difference between the molecular and molar views—the difference between response strength and behavioral allocation—corresponds to the ontological distinction between class and individual (Ghiselin, 1997; Hull, 1988). The molecular view, as laid out by Skinner (1935/1961), relied on the notion of operant classes. A class is defined by specifying a list of properties or rules of membership (e.g., all actions that depress the lever). Classes are abstract in the sense that one can only talk about them, not point to them or measure them. Their abstract nature appears also in the lack of any requirement that they have members or that such members exist (e.g., human beings who can leap over tall buildings in a single bound). Useful classes have members, which, unless they are other classes (a possibility we will ignore here), constitute concrete particulars—concrete in the sense that one can point to them or at least observe them, and particular in the sense that each is just one thing. So, although operant classes are abstract, responses (instances) would be considered concrete (Skinner, 1935/1961).

Besides being members of classes, concrete particulars are individuals (see Ghiselin, 1997, and Hull, 1988, for longer explanations). An individual is a cohesive whole that is situated in space and time—a historical entity. That is, an individual (e.g., B. F. Skinner) has a location, a beginning, and potentially an end. Individuals have no instances (e.g., B. F. Skinner is who he is and has no instances). Individuals cannot be defined except by ostension (i.e., by pointing; e.g., that is my cat there). Individuals have parts (left leg, right leg, liver, and heart), rather than instances. The quote from John Donne at the beginning expresses well the relation of part to whole; as a clod is part of Europe, so any man is a part of mankind. Classes cannot do anything; only individuals can do things (e.g., *cat* cannot walk into the room, whereas *my cat* can).

In particular, whereas individuals can change, classes cannot change. B. F. Skinner changed from boyhood to adulthood, but he was still the same individual, B. F. Skinner. A class remains fixed because it is defined by fixed properties or rules. If the properties or rules change, we only have a new class. The only change associated with a class is in the number of its instances. Were we to discover an individual able to leap tall buildings in a single bound, that class would no longer be empty. Mathematical sets cannot change even in this way, because adding or subtracting elements from a set creates a new set.

Any science that deals with change, whether phylogenetic change, developmental change, or behavioral change, requires entities that can change and yet retain their identity (e.g., *Homo sapiens*, my cat, or my diet), because only such entities provide historical continuity. In other words, because only individuals can change and yet maintain historical continuity, such a science must deal with individuals. Although individual usually means individual organism in everyday discourse, philosophers mean something more general. Organisms exemplify cohesive wholes, but so too do activities or allocations. Just as an organism is made up of a liver, kidneys, brain, and the like, functioning together to produce results in the environment, so too an utterance (e.g., "I need help with this problem") is made up of sounds that function together to produce results in the environment. The various parts of the whole are themselves individuals (e.g., the liver or the uttered word "help"); all individuals are composed of other individuals. This point will be important when we discuss the nesting of activities.

By way of example, we may compare the molecular and molar accounts of differential reinforcement ("shaping"). Skinner (1935/ 1961) recognized that reinforcement of a certain class of responses generates responses that may actually lie outside the reinforced class. He called this process *induction* (see also Segal, 1972). Induction is essential for shaping novel behavior, because the new induced responses may be reinforced. To do this, one defines a new class for reinforcement, one that excludes some of the old members. Reinforcement of this new class leads to induction of further new responses, which allows definition of another new class, and so on, until some target class is reached.

One challenge for this molecular account of shaping is that reinforcement may induce undesirable behavior, sometimes called adjunctive or interim behavior (Staddon & Simmelhag, 1971). The problem is that such behavior interferes with the process of shaping (Breland & Breland, 1961; Segal, 1972) and falls outside the reinforced classes. Consequently, the molecular view treats it as a separate type, distinct from operant behavior and with rules of its own.

The molar view of shaping instead incorporates induced behavior into the account. The process begins, not with a response class, but with an allocation of activities (an individual). Some activities (parts) are reinforced. The allocation changes, the parts reinforced change, and the allocation changes further. Induced activities may enter the allocation at any stage; they become new parts. The end-point of the process (if any) will be a stable allocation maintained by stable reinforcement contingencies.

Although the idea that particular discrete responses are instances of a class remains common (e.g., in textbooks), the molecular view allows at least one other possibility. Glenn, Ellis, and Greenspoon (1992) proposed that the aggregate of particular occurrences be thought of as analogous to a population of organisms. As each individual organism is a part of the population, so each particular discrete unit is a part of a behavioral population, rather than an instance of a class. Thus, one could redefine an operant as a behavioral population, which would be an individual rather than a class. Response rate then would correspond to the size of the population. Such a population would constitute an individual, but different from an activity or allocation, because its parts would be discrete responses (Glenn & Field, 1994). Their proposal illustrates that the molecular view cannot be said to entail the concept of response class in the way that it can be said to entail discrete units.

During the 1960s and 1970s, Skinner's notion of the operant as a class came in for critical discussion (e.g., Schick, 1971; Segal, 1972; Staddon, 1973). The main problem was how to deal with the induction of new behavior. Catania (1973) proposed a solution that resembles the proposal by Glenn et al. (1992). He suggested distinguishing between the descriptive operant and the functional operant—that is, between the operant as specified by class properties and the operant as the pattern of behavior that actually results from reinforcement. Catania's suggestion overlooks, however, that the functional operant constitutes a different ontological kind, one that eludes definition by a list of properties. It overlooks that, in moving from descriptive operant to functional operant, one also moves from class to individual. The functional operant, which Catania represented by drawing frequency distributions, corresponds to a population of responses, but the responses no longer can be seen as instances of a class, because now they are parts of a whole—whatever unspecified responses occur after reinforcement. They could be seen (in the molar view) as parts of an extended behavioral allocation, an individual. That allocation is both engendered by and maintained by the reinforcement it produces. Like Glenn et al., however, Catania based his idea on discrete responses. In their related discussion, Glenn et al. argued:

In the ontological sense, an operant is . . . an entity—a unit, an extant individual. . . . It is composed of a population of behavioral occurrences that are distributed over time, each occurrence having a unique spatiotemporal location. The operant can evolve (as only operants and species can but organisms and responses cannot). (p. 1333)

The main difference between this and the molar view is its implicit reliance on discrete responses ("occurrences"). If one added the point that the occurrences take up time, introducing an analogue to the carrying capacity (limited size of a biological population), and one added that the occurrences were bouts of extended activities, the concept of behavioral population would become almost the same as the concept of allocation. One further concept, implied by the analogy to biological evolution, is the idea that activities are nested, that every activity (allocation) is composed of parts that are other activities.

Species and Activities as Individuals

Another way to understand the concreteness of behavioral allocations, activities, and response rates is by comparison to evolutionary theory. Glenn et al. (1992) were drawing on Ghiselin's (1981, 1997) argument that species are not classes but individuals. That is, the relation of an organism to its species is not the relation of instance to class, but the relation of part to whole. As before, the word *individual* here refers to an integrated entity that may change through time. As before, in contrast to a class, an individual is situated in time and space (i.e., has a beginning and end) and has parts but no instances (e.g., B. F. Skinner). An organism is an individual, of course, but, Ghiselin explains, so too is a species. A species is an individual composed of the organisms that make it up, in the same way as John Donne noted that every man is a part of mankind. All the individual birds in the Galapagos Islands that make up the species *Geospiza fortis* are parts of that whole. Selection may change a species through time, particularly if the environment changes, but the species remains the same individual, just as a person who grows and ages remains the same individual. The existence of a species through time is referred to as its *lineage*. A lineage is an extended temporal entity in much the same way that a pattern of behavior is an extended temporal entity.

Ghiselin's point was at first controversial among biologists, but gradually gained acceptance. Now, even its critics acknowledge that "Only a few biologists and (bio)philosophers have resisted [it]" (Mahner & Bunge, 1997, p. 254).

Like a species, an allocation of behavior—an activity—is an individual. It is an entity with a beginning and an end, integrated by its function; that is, by its effects in the environment. Just as taking away an organism's leg changes its functioning, so taking away part of a behavioral pattern changes its environmental effects. Forget to add baking powder to a cake mix, and the result may be in edible. A particular cake baking, however, is part of a more extended allocation of baking or cooking, including both successful and unsuccessful attempts and all their various outcomes.

This illustrates another parallel between extended activities and species: their similar participation in larger individuals. Common ancestry unites species into more extended individuals at the level of genus. Genera unite into still larger individuals, and so on, right up to phylum and, finally, life (Ghiselin, 1997). Although individuals at these various taxonomic levels may be more or less extended, no matter how large or small they still are individuals. *Homo sapiens*, as a species in the genus *Homo*, is a part of the genus, just as the other species in that genus are parts of it. *Geospiza fortis* is one species of Darwin's finches. It, *Geospiza scandens*, and several other species make up the genus *Geospiza*. In relation to the genus, the species are parts of a whole, not instances of a class.

Nesting of Activities

Activities, like species, are parts of more extended activities. Getting to work each day may be part of working each day. Working each day may be part of holding a job. Holding a job may be part of making a living. Making a living may be part of gaining resources. Gaining resources may end with retirement, and all such parts may make up a whole, which we could call a lifetime lived (Baum, 1995a, 1997; Rachlin, 1994).

The converse holds, too: Activities and species are made up of parts consisting of less extended individuals. A species may be com posed of several populations. A population may be composed of several demes. Demes, populations, and species all are composed of organisms, which are composed of organs, which are composed of cells, and so on. For the purposes of evolutionary theory, one stops at the smallest individual that may evolve—a deme, a population, or a species. An activity like getting to work may be composed of parts like starting the car, driving to the highway, driving on the highway, driving to the campus, hunting for a parking place, and walking to the office. Driving on the highway has parts like adjusting speed, switching lanes, scanning for police cars, and swearing at other drivers. And so on, for each part. Some least extended activity exists for the analysis of behavior, as it does for evolutionary theory, defined by its usefulness and, probably, by its likelihood of evolving. Highly practiced and stereotyped activities like shifting gears in a car change rarely; they attract little interest as targets of modification, whereas driving speed may change significantly and is a frequent target of attempts at modification (e.g., "speed kills").

The Molar View of Everyday Life

We may illustrate the conceptual power of the idea of nested activities with a hypothetical example. Liz is a married woman in her 40s, who lives in a city, works selling retirement plans for a mutual fund company, and has an 18-year-old son who still lives at home. As I suggested in earlier papers (Baum, 1995a, 1997), her life may be divided into four basic activities: personal satisfaction (i.e., health and maintenance), job (i.e., gaining resources), relationships, and family (i.e., reproduction). The left graph in Figure 4.2 shows the pattern of these activities at this point in Liz's life. Leaving out the 9 h she spends sleeping each night, we see that she spends about 35 h (33%) per week in personal satisfaction, 36 h (34%) in gaining resources, 24 h (23%) in making and maintaining relationships, and 10 h (10%) in family activities. The allocation was different 10 or 15 years earlier, when her son was young and she was caring for her husband's children from previous marriages. As an individual, the activity has changed and will change again over the course of Liz's life, but it will remain the same individual, the pattern of Liz's life. All four of the activities shown in the left graph may be analyzed in more detail and seen to be composed of other activities. For example, Liz's family activities consist of occasionally caring for her husband's grandchildren and primarily of caring for her son: feeding him, cleaning up after him, advising him, and interfering in his life sufficiently to make him rebellious.

The middle graph in Figure 4.2 shows the time Liz spends in personal satisfaction decomposed into parts. She spends about 3 h (8%) per week seeing medical practitioners, 10 h (29%) eating, 7 h (20%) in personal hygiene, and 15 h (43%) in recreation. This allocation also is an individual, subject to change, and is nested within or incorporated into the more extended allocation of Liz's life activities. Each of the activities composing the activity of personal satisfaction also is an individual and is itself composed of individuals. The right graph in Figure 4.2 shows Liz's recreational activities broken into parts. She spends about 7 h (47%) per week watching television, 6 h (40%) reading, 0.5 h (3%) watching movies, and 1.5 h (10%) walking for exercise. This allocation of recreational activity constitutes an individual and is part of Liz's personal satisfaction and, because of that, is part of Liz's life activity. Each of the parts of Liz's recreational activities could be further decomposed into parts that also would be individuals (Baum, 1995a, 1997).

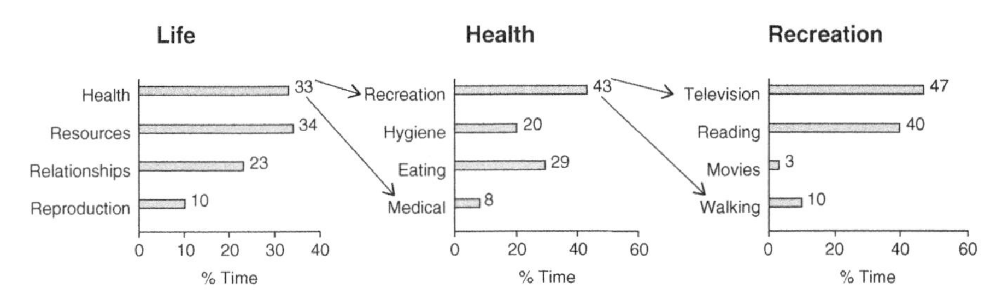

Figure 4.2 One person's hypothetical activity patterns
Note: Left: pattern of life activities, showing time divided among health and maintenance (i.e., personal satisfaction), gaining resources (e.g., job), relationships (e.g., friends), and reproduction (e.g., family). Middle: pattern of health and maintenance nested within the life activities, showing time divided among medical (e.g., visits to practitioners), eating, personal hygiene, and recreation. Right: pattern of recreation nested within health and maintenance, showing time divided among watching television, reading, movies, and walking (i.e., exercise). All of these patterns constitute individuals, because they change without changing their identity.

In contrast, the molecular view invites one to view life as a time line of discrete events, one following another—a behavioral stream (e.g., Schoenfeld & Farmer, 1970). To the molarist, such a characterization, though possible, appears impoverished and to resemble little the way people actually talk about their lives (i.e., inelegant and low on external validity).

Figure 4.2 implies that one might go into any amount of detail about Liz's activities. Where should subdividing stop, and how does one define the parts? Answers would depend on the purpose of the analysis, whether it be therapeutic intervention, basic research, or something else. The issues involved are addressed most directly in the context of laboratory research.

Applications in the Laboratory

As a laboratory example, we may consider activities like key pecking and lever pressing. A pigeon's food peck, when examined in detail, constitutes an individual with parts: forward head motion, eye closing, opening of the beak, head withdrawal, closing of the beak, eye opening (Ploog & Zeigler, 1997; Smith, 1974). It is a stereotyped pattern that researchers almost never seek to change, although other sorts of pecks, containing different parts, exist, such as water-reinforced pecks and exploratory pecks (Jenkins & Moore, 1973; Schwartz & Williams, 1972; Wolin, 1948/1968). A similar, though more varied, list of motions might be made for a rat's lever press. Key pecks or lever presses may be parts of key pecking or lever pressing reinforced, say, on a VI schedule. Key pecking or lever pressing on two different keys or levers may be parts of an allocation of behavior between two sources of reinforcement (Ploog & Zeigler, 1997). We usually measure the responding on one of the keys or levers as a response rate. We measure the allocation as choice or relative response rate.

In contrast, the molecular view sees choice or concurrent performance as consisting of occurrences of two responses, each at a certain rate. The response rates may be compared by calculating some relative measure (proportion or ratio) but such a measure is seen as only a summary or as "derived" (Catania, 1981; Herrnstein, 1961). At least one researcher has suggested that relative measures, as derived, should be viewed with suspicion and that response rate is the only true measure of behavior (Catania, 1981). The limitations of such a view become apparent when we consider a specific example.

Alsop and Elliffe (1988) exposed 6 pigeons to over 30 pairs of concurrent VI schedules, varying both relative and overall rate of reinforcement. I reanalyzed their data by grouping them according to five levels of reinforcer ratio (r): 0.12, 0.25, 1.0, 4.0, and 8.0. Within each group, the obtained reinforcer ratios varied a bit, but the variation from group to group was larger than the variation within a group (although to achieve this, five conditions with aberrant reinforcer ratios out of 186 were omitted—one each for 3 pigeons and two for 1 pigeon). For each reinforcer ratio, overall reinforcer rate varied from about 10 to about 400 reinforcers per hour. Figure 4.3 shows the average results, which were representative of the results for the individual birds.

The top graph in Figure 4.3 shows the total rate of pecking at the two keys as a function of the overall reinforcer rate. The curve represents the least squares fit of Equation 4.4 ($r_O = 9.44$; $k = 94.6$). The only unusual feature of this analysis is that pecks at the two keys were combined, whereas usually Equation 4.4 would apply to pecks at a single key. Recalling that Equation 4.4 describes choice between schedule-reinforced activity and other background activities, we see that choice between key pecking and other activities followed the orderly pattern expected from the matching law. That the various sets of

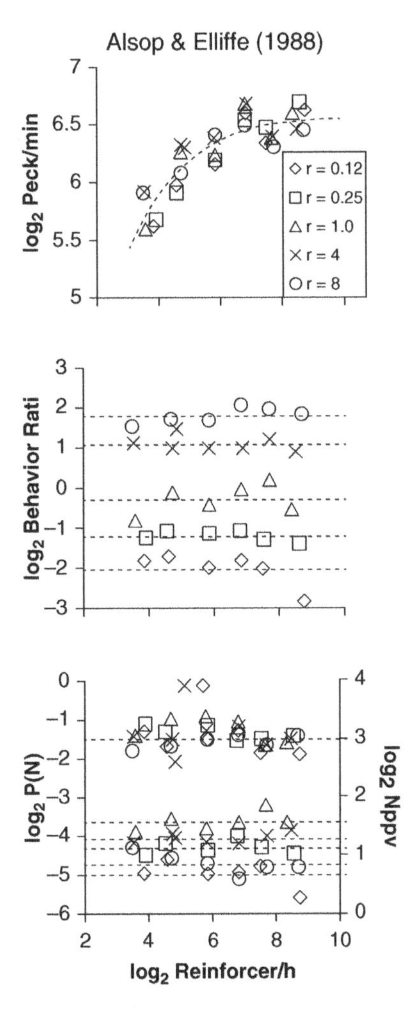

Figure 4.3 A study of concurrent schedules that illustrates the concept of nested patterns
Note: Top: combined rate of key pecking on two concurrent VI schedules as a function of overall rate of reinforcement. The different symbols represent different levels of reinforcer ratio (*r*). The curve represents the least squares fit of Equation 4.4 to all the points. Middle: relative responding at the two alter natives as a pattern nested within the pattern of overall rate of key pecking. The horizontal lines represent the average behavior ratio at each level of reinforcer ratio. Behavior ratio varies, as would be expected, across reinforcer ratio, but is independent of overall rate of reinforcement. Bottom: the fix-and-sample pattern nested within the pattern of relative responding. The bottom five horizontal lines indicate the averages of probability of visiting the nonpreferred alternative, *p*(N), across the levels of reinforcer ratio. The smaller the relative reinforcement for visiting the nonpreferred alternative, the lower the probability of visiting it. Hence, the symbols for *r* equal to 0.12 and 8.0 (circles and diamonds) show the lowest *p*(N), and the symbols for *r* equal to 1.0 (triangles) show the highest *p*(N). The uppermost horizontal line shows the average number of pecks per visit at the nonpreferred alternative (Nppv; right vertical axis). The average duration of a visit to the nonpreferred alternative remained approximately constant at eight pecks, consistent with the fix-and-sample pattern. All variables are transformed to base-2 logarithms.

symbols all overlap each other shows that reinforcer ratio had no effect on this choice pattern.

The middle graph in Figure 4.3 shows that the overall key pecking contained within it another regular pattern. Here the ratio of pecks at the two keys is plotted against overall reinforcer rate. As in the top graph, the different symbols represent data from the different reinforcer ratios from the two keys. A horizontal line, corresponding to the average peck ratio, is drawn through each set of symbols to allow assessment of trend. As overall reinforcer rate varied, each reinforcer ratio maintained a certain peck ratio, which remained approximately invariant across overall reinforcer rate. In other words, regardless of the overall rate of reinforcement, the behavior ratio remained about the same for each reinforcer ratio, in accordance with the generalized matching law (Equation 4.7).

The bottom graph in Figure 4.3 shows that, on still closer examination, a pattern exists within the behavior ratios, the activity that I earlier called fix and sample (Baum et al., 1999). To see whether such a pattern was present, I calculated for each behavior ratio the average number of pecks in a visit to the nonpreferred key and the probability of leaving the preferred key to visit the nonpreferred key (number of visits to the nonpreferred key divided by number of pecks at the preferred key). The fix-and-sample pattern would be revealed by invariant visits to the nonpreferred key combined with variation only in the probability of visiting the nonpreferred key. For convenience, both probability of visiting the nonpreferred key [p(N)] and number of pecks per visit to the nonpreferred key (Nppv) are shown in the same graph; p(N) is represented on the left vertical axis, and Nppv is represented on the right vertical axis. The five lower sets of symbols show that p(N), like the behavior ratio, differed across reinforcer ratios, but remained approximately invariant for each reinforcer ratio as overall reinforcer rate varied. Each reinforcer ratio produced a certain, roughly invariant, probability of a visit to the nonpreferred alternative. The horizontal lines show the averages. As would be expected from the results of Baum et al. (1999), the lowest probabilities of visiting the nonpreferred key occurred for the strongest preferences, generated by the most extreme reinforcer ratios (8 and 0.12). The two intermediate reinforcer ratios (4 and 0.25) produced intermediate p(N), and the 1:1 reinforcer ratio, which produced the weakest preferences, produced the highest probabilities of visiting the nonpreferred side. The higher the relative reinforcement for the nonpreferred key, the higher the frequency of visiting the nonpreferred key. The uppermost sets of symbols show that Nppv, the visit duration to the nonpreferred key, remained invariant with respect to both overall reinforcer rate and reinforcer ratio. Independence from overall reinforcer rate is shown by the adherence of the points to the horizontal line representing the average. That the different symbols all lie on top of one another shows independence from the reinforcer ratio. Visits to the nonpreferred side, regardless of rate or distribution of reinforcement, always lasted about eight pecks (i.e., 23; see right vertical axis), presumably long enough to outlast the 2-s COD. Regardless of the behavior ratio, the same activity of briefly visiting (i.e., sampling) the nonpreferred key held; only the probability of visiting changed to produce the different behavior ratios shown in the middle graph.

Figure 4.3 shows how activities or patterns may be nested within each other. Nested within the pattern of overall responding to the keys (top graph) was a pattern of allocation of the overall responding between the keys, measured by the behavior ratio (middle graph). Nested within the behavior ratio was a pattern of visitation at the two keys, a pattern of fixing on the preferred alternative and briefly sampling the nonpreferred alternative with a frequency depending on the reinforcer ratio (bottom graph). In the molar view, all of these patterns constitute allocations—between pecking and background activities (top graph), between pecking left and pecking right (middle graph), and between fixing and sampling (bottom graph).

Conclusion

An activity, like a species, is an individual, a concrete particular with parts, not a class with instances. Contrary to 19th- and early 20th-century thinking, the concrete particulars of behavior need not be momentary or discrete, but extend through time as parts of behavioral patterns (activities or allocations) over minutes, hours, days, or years. Like species, they only need to have a beginning and potentially an end. This recognition changes the notions of reinforcement and stimulus control, but only moderately. Instead of thinking of reinforcement as a sort of "moment of truth" (e.g., Ferster & Skinner, 1957; Skinner, 1948), defined by contiguity with a momentary response, we may think of reinforcement as a cumulative effect, as selection through time (Skinner, 1981; Staddon, 1973), shaping patterns of behavior (activities) in lineages. Because reinforcement operates on the activity as a whole, we are relieved of any need to imagine that the parts are all separately reinforced. Behavioral chains, for example, need not be held together with imagined conditional reinforcers, because they are reinforced as a whole (Baum, 1973; Rachlin, 1991). Avoidance need not be explained with imagined stimuli and reinforcers (Baum, 1973, 2001). We understand the behavior of a species in relation to the climate and resources available in its evolutionary environment. Similarly, instead of thinking of stimulus control as changing the probability of a response, we may think of discriminative stimuli as setting the context in which certain activities or patterns are reinforced and wax in the time that they occupy. Making these extensions, we increase our ability to explain disparate phenomena, such as variation in asymptotic response rate (k), asymmetrical concurrent performances (Figure 4.1), resistance to change, and the relations among analyses at various levels of generality (Figures 4.2 and 4.3; Hineline, 2001).

Questions remain, of course. If an activity is an individual, how should we think about its coherence? Ghiselin (1997) explains that organisms have a special cohesiveness that species and other taxa lack; an organism functions as an integrated whole. The parts of a species may be less crucial than those of an organism, but a species has coherence because it is defined as a reproductive unit, reproductively isolated from other such units (Mayr, 1970) and because the parts of the species share common ancestry (i.e., are parts of the same lineage). Higher taxa, from genus on up, have coherence only because of common ancestry. In analogy to biological taxa, the parts of an activity share common ancestry—are parts of the same lineage—because they result from the same history of selection (reinforcement). The various parts of the activity we call "holding a job" cohere because they share a common function (gaining resources) and a common history of selection (reinforcement) among variants that functioned better and worse. Activities may become extinct in the same way as species, by a loss of functionality of the whole. The end of an unrewarding marriage (an activity) is the end of an individual, like the extinction of a species. Further evidence of coherence in the parts of an activity may be found in common variation in the face of change in environmental factors (Herrnstein, 1977). Food deprivation, for example, changes time allocation to a host of food-related activities. A more complete answer to the question of coherence awaits further research.

Although the research discussed here suggests advantages to the molar view over the molecular, deciding between the two paradigms depends, not on data, but on satisfactory interpretation of data. No one should doubt that molecular accounts of concurrent performance are possible. The advantages to the molar view lie in its ability to integrate experimental results, in its promotion of quantitative theory, and in its applicability to everyday life. The results of Alsop and Elliffe (1988; Figure 4.3) illustrate the way the molar view both integrates results at various levels of analysis and fits them into a quantitative framework. The hypothetical example of Liz (Figure 4.2) illustrates the power of the molar view to apply to everyday concepts like “holding a job” or “recreation” (see Rachlin, 1994, for further discussion). Taking our cue from another historical science, evolutionary biology, we see that extendedness of allocations or activities in no way excludes them from concreteness. On the contrary, I have argued that the discrete events of the molecular view are abstractions (see Baum, 1997, for further discussion). Although someone committed to the molecular view might disagree, I have argued that on these grounds of plausibility, explanatory power, and elegance the molar view is the superior paradigm.

5

The Molar View of Behavior and Its Usefulness in Behavior Analysis

Foreword

This paper was invited. It aimed to lay out in brief and simple form the molar view of behavior and its conceptual advantages over the traditional molecular view based on discrete responses and contiguity. It was intended for practitioners, applied behavior analysts. It contains a summary of the previous paper and some suggestions for applied behavior analysts.

Abstract

The molar view of behavior contrasts with the older, molecular view. The difference is paradigmatic, not theoretical. No experiment can decide between them, because they interpret all the same phenomena, but in different terms. The molecular view relies on the concepts of discrete, momentary events and contiguity between them, whereas the molar view relies on the concepts of temporally extended patterns of activity and correlations. When dealing with phenomena such as avoidance, rule-governed behavior, and choice, the molar view has the advantage that it requires no appeal to hypothetical constructs. The molecular view always appeals to hypothetical constructs to provide immediate reinforcers and stimuli when none are apparent. As a result, the explanations offered by the molar view are straightforward and concrete, whereas those offered by the molecular view are awkward and implausible. The usefulness of the molar view for applied behavior analysis lies in the flexibility and conceptual power it provides for talking about behavior and contingencies over time.

Keywords: molar view, molecular view, correlation, hypothetical constructs, timeframe

Source: Originally published in *The Behavior Analysis Today*, 4 (2003), pp. 78–81.

Science and Philosophy of Behavior: Selected Papers, First Edition. William M. Baum.

The molar view of behavior is relatively new. Although its origins may be traced back earlier, its first partial articulation was by Baum and Rachlin (1969), in a paper called "Choice as time allocation." It was presented more fully in a paper by Baum (1973), "The correlation-based law of effect." Rachlin (1994) offered a book-length presentation, and Baum (2002) elaborated on his 1973 paper in another paper, "From molecular to molar: A paradigm shift in behavior analysis" and some papers in-between (Baum, 1995a; 1997).

The molar view contrasts with an older view that behavior analysis inherited from nineteenth-century psychology. I call this older view "molecular," because it is based on the notion that explanations of behavior may be constructed by thinking of small discrete units being joined together into larger units, like the joining together of atoms into molecules in chemistry.

The difference between the molecular and molar views of behavior is paradigmatic, not theoretical. No data, no experiment can decide between the two views, because no matter what behavioral phenomenon one chooses, a proponent of either view is able to construct an account of it. The difference between the two lies in the concepts each brings to bear in such an account. The molecular view relies on momentary events and momentary causation, which leads to postulating hypothetical momentary events and causes when none are apparent, whereas the molar view relies on extended activities and extended causation, avoiding postulation of hypothetical constructs.

Replacing the concept of momentary response with the concept of extended activity requires one to become familiar with thinking in more continuous terms—that is, in terms of extended patterns that cannot be seen at a moment in time. A familiar example is the concept of probability. An unbiased coin, when flipped, comes up heads with a probability of .50. What does this mean? On any particular flip, the coin comes up heads or tails; nothing more can be observed. Only for a long series of flips can one observe the probability of .50. If one says that on a particular flip the probability is .50, all one means is that in a long series of such flips about half would show heads. The same is true of response rate. At any particular moment, an activity (lever pressing) is occurring or not. One can only observe the response rate over some substantial time period. A response that occurs 60 times per minute cannot occur 60 times per minute at a moment.

Although Skinner advocated the use of response rate as a dependent variable, he was a molecularist. In his well-known paper on superstition, Skinner (1948) proposed a "snapshot" view of reinforcement, in which delivery of a reinforcer strengthens whatever behavior happens to be occurring at the moment. The molecularity of his approach is perhaps nowhere clearer than in a short piece he wrote called "Farewell, My Lovely!" in which he deplored the absence of cumulative records in the pages of *JEAB* and extolled the virtues of being able to observe "molecular," moment-to-moment changes in behavior (Skinner, 1976). A cumulative recorder, however, is an averaging machine; it only produces smooth curves because the chart moves slowly and the pen moves in small steps. At any particular moment, either a response is occurring or not. The local changes in response rate are changes from one interval to another. If, however, one were to fit a truly continuous curve to a cumulative record, then one might think of momentary rate as the slope of the curve at a particular point. This, however, requires abstracting the continuous function.

In the molecular view, each response is taken as a concrete particular (i.e., the basic observation), and response rate is a "derived" measure (i.e., an abstraction) summarizing behavior over a period of time. The molar view turns this distinction around, making the extended pattern the concrete particular and the momentary response the abstraction. A response rate or activity exists as a pattern through time. Any attempt to infer activity at a moment depends on abstraction, as in the example of the cumulative record. In fact, no behavior can be observed at a moment, because even the simplest unit of behavior—lever press, key peck, button push—takes up time and must unfold from beginning to end before it can be recorded with certainty (for further discussion, see Baum, 1997; 2002). Because every activity takes up time, the concept of behavior at a moment is an abstraction, an inference made after the fact.

Although it has little use for momentary events, the molar view supports analysis in more and less extended time frames (Baum, 1995a; 1997; 2002). That patterns take up time in no way precludes them from being brief. A pigeon's key peck, for example, is an extended pattern that takes a fraction of a second. Analysis may be as local or as extended as suits one's purpose. When trying to change behavior, one should make sure that reinforcers are closely coordinated with the activity one is trying to increase. The molecularist insists reinforcers must immediately follow the responses they are to strengthen; the molarist says reinforcers should coincide closely with the activity to be increased. Such local relations often have powerful effects, sometimes to our grief, when they override more extended relations (Rachlin, 2000). Each additional drink might seem harmless, but in the long run they add up to ruin. The likeliest way to overcome problem drinking is with local reinforcers for abstinent behavior. Thus, the molar view, like the molecular view, says that, in practice, the one who would shape behavior needs to be swift with the reinforcers.

The molecular view has one point in its favor: It coincides with a prejudice toward immediate causes. The notion that the events that affect behavior occur either immediately before or immediately after a response lends simplicity to analysis. One knows just where to look for the antecedents and consequences that control the response. That simplicity, however, comes at a high price: the necessity of inventing immediate antecedents and consequences when none are apparent. Perhaps the best example is explaining avoidance.

To explain avoidance, in which success means that nothing happens following a response, molecularists turn to two-factor theory. Since a reinforcer must follow the avoidance response, even if none is apparent, one has to be invented. Suppose that the stimulus preceding the response becomes a Pavlovian conditional stimulus, eliciting "fear." Then, when the response turns off the stimulus, the reduction in fear reinforces the response. Avoidance responding occurs, however, even if no stimulus precedes or is terminated by the activity (Herrnstein & Hineline, 1966; Herrnstein, 1968). Having already invented the fear-reduction reinforcer, the molecularist now also invents the stimulus. Dinsmoor (2000), for example, argued that response-produced stimuli, paired with a lower frequency of electric shock than their absence, become safety signals. The cost of maintaining the molecular view here is that one must appeal to hypothetical reinforcers and stimuli when none are observable. The result is a theory that cannot be refuted.

The molar view of avoidance is arguably simpler, but requires one to think in terms of temporally extended patterns. Avoidance activity is acquired and maintained because when that activity is present the rate of noxious events is lower than when it is absent. People avoid sensitive topics in conversation to lower the likelihood of embarrassment to themselves and others. People buy insurance to lower the likelihood of financial hardship. Much apparently dysfunctional behavior may be understood as avoidance. If working and failing would be too hard an outcome, one may avoid it by being ill.

Another example of paying a high price to retain a molecular view is in accounting for rule-governed behavior. Rules present a problem for the molecular view because they are invariably associated with behavior that has important consequences in the long run

(Baum, 1994b; 1995b). Since long-delayed effects must be ineffective to the molecularist, if rule-governed behavior is maintained, some immediate (effective) consequences must be found. Why would someone eat vegetables instead of candy when no one else is present to observe? Why would someone save a piece of trash until a trash can appears, when it might have been dropped on the street with impunity? Mallott (2001), in a paper about moral and legal control, provides the molecularist's answer: thoughts and self-punishment. He argues, "For moral control to work, society must have established a special, learned aversive condition—the thought of the wrath of one's God or the thought of the wrath of one's parents. And those thoughts must be aversive, even when no one is looking" (p. 4). Again the molecular view leads directly into the realm of the hypothetical and unverifiable.

The molar view of rule-governed behavior allows that any contingency, no matter how extended, may control behavior, even though more local contingencies may be more powerful than more extended ones (Baum, 1994b; 1995b; Rachlin, 1994; 1995a; 2000). Rules exist, however, because extended contingencies are weak. A rule is a discriminative stimulus produced by one person that induces in another person behavior that is reinforced socially in the short run (and reinforced in some major way in the long run). The behavior may come under the control of the long-term contingency—for example, the relationship between diet and health. Although people often say that then the rule has been "internalized," from the molar point of view, it actually is further externalized, because the control is exerted by a more extended contingency. In looking at rule-governed behavior this way, the molar view introduces no hypothetical events and no new terms.

Perhaps the strongest area of application of the molar view is to choice, the allocation of behavior among alternatives. At any moment, behavior is assigned to only one alternative. Over time, however, one sees a pattern of allocation among alternatives. In the molar view, such a pattern constitutes a concrete particular. The molecular view, focusing on a moment, immediately moves to hypothetical constructs. Each alternative has a certain strength, unobservable at the moment but existing at the moment. The extended pattern of allocation is thought to reveal the relative strengths of the alternatives. If a pigeon pecks twice as often at the left key than the right key, the strength of left pecking is considered twice that of right pecking. If a child spends twice as much time disrupting classroom activities as the child spends doing school work, the strength of disrupting is twice that of remaining on task. In the molar view, no hypothetical strength enters in, because these patterns of allocation are what the science is about.

Even if the molar view seems to allow such phenomena as avoidance, rule-governed behavior, and choice to be understood more readily, the question arises as to whether the molar view has any implications for applied behavior analysis. It makes for the same sort of rule of thumb as the molecular view when one is trying to change behavior: reinforcement must be frequent and quick. Beyond this, however, I think the molar view might have some advantages for applications. First, it offers flexibility in thinking about goals and treatments. No need arises to define some artificial discrete response for reinforcement. One needs only to make sure that reinforcers accompany appropriate activity. For example, in school settings applied behavior analysts already often talk about time on task as a reinforceable activity. The molar view allows this kind of flexible thinking about reinforcement of activities to be extended indefinitely. Second, it frees one to think about time spent instead of response rate. Without artificial discrete responses, activities like reading, playing, grooming, and the like can be measured by timing them. Time spent should be no harder to measure than counting responses and often will be less ambiguous, because one may be able start and stop timing more easily than decide whether exactly the right response occurred. Once applied behavior analysts grow accustomed to the molar way of talking, they will find it more congenial for communicating with one another about behavior and contingencies, because it is more flexible and more concrete.

In conclusion, two points might be made. First, although the molecular view was useful early in the development of behavior analysis, the science has outgrown it, and the molar view supplies the conceptual power required for the new developments. Second, the molar view may be recommended for the flexibility and power that it allows both applied and basic researchers in talking about behavior and contingencies.

6

Molar and Molecular Views of Choice

Foreword

This paper covers some of the same concepts as the preceding one, but more succinctly and with a historical introduction. Historically the molecular view arose out of an attempt to understand the train of thought and the relation between simple ideas and more complex ideas. From the 18th century onwards an atomistic view applied also to behavior. It had critics also who advocated instead for something like the molar view. The main point here is that the molecular view may have been useful in the past, but the sciences have progressed and have revealed irremediable weaknesses that the molar view aims to redress.

When writing this paper, I had not yet come to see the importance of the principle of induction or the necessity of connection to evolutionary theory. Later I argued that understanding behavior in functional terms requires thinking about organisms and behavior as having evolutionary history that determines function and provides the mechanism of induction.

Another missing piece that came later was the recognition of the usefulness and necessity of seeing behavior as process. This ontological category carries with it many of the points made in this paper, because an activity is a process that is a whole consisting of parts that also are processes. Some philosophers argue that an organism is not a thing but a process (Nicholson & Dupré, 2018). Accepting this, we would say that the activities of an organism are parts of its process. Processes are integrated wholes that serve a function, such as survival, maintaining relationships, or gaining resources. In the final analysis function replaces reinforcement.

Abstract

The molar and molecular views of behavior are not different theories or levels of analysis; they are different paradigms. The molecular paradigm views behavior as composed of discrete units (responses) occurring at moments in time and strung together in chains to make up complex performances. The discrete pieces are held together as a result of association by contiguity. The molecular view has a long history both in early thought about reflexes and in associationism, and, although it was helpful to getting a science of behavior started, it has outlived its usefulness. The molar view stems from a conviction that

Science and Philosophy of Behavior: Selected Papers, First Edition. William M. Baum.

behavior is continuous, as argued by John Dewey, Gestalt psychologists, Karl Lashley, and others. The molar paradigm views behavior as inherently extended in time and composed of activities that have integrated parts. In the molar paradigm, activities vary in their scale of organization—i.e. as to whether they are local or extended—and behavior may be controlled sometimes by short-term relations and sometimes by long-term relations. Applied to choice, the molar paradigm rests on two simple principles: (a) All behavior constitutes choice; and (b) All activities take time. Equivalence between choice and behavior occurs because every situation contains more than one alternative activity. The principle that behavior takes time refers not simply to any notion of response duration, but to the necessity that identifying one action or another requires a sample extended in time. The molecular paradigm's momentary responses are inferred from extended samples in retrospect. In this sense, momentary responses constitute abstractions, whereas extended activities constitute concrete particulars. Explanations conceived within the molecular paradigm invariably involve hypothetical constructs, because they require causes to be contiguous with responses. Explanations conceived within the molar paradigm retain direct contact with observable variables.

Keywords: molar, molecular, paradigm, choice, local, extended

Source: Originally published in *Behavioural Processes*, 66 (2004), pp. 349–359. Reproduced with permission of Elsevier. The author thanks Michael Davison for many helpful comments on earlier versions.

Within behavior analysis exist two views of behavior. The older view sees behavior as consisting of discrete units, usually called responses. Since it treats behavior as made up of bits and pieces this way, it is aptly called molecular. The molar view of behavior offers an alternative. It sees behavior as composed of behavioral patterns (hereafter called activities) that, by their very nature, are temporally extended. Activities or patterns differ from discrete responses in that they are integrated wholes, existing not only in space but spanning time. Thus, instead of momentary responses like the lever press or the footstep, the molar view sees the activities of lever pressing and walking. Whereas the molecular view names its discrete units with ordinary nouns, the molar view usually names activities with gerunds.

The Molecular View in Historical Perspective

The molecular view has a long history, because much of its history is the history of associationism. The notion of association of ideas goes back at least to Aristotle (Herrnstein & Boring, 1966). It was developed significantly by John Locke (1632–1704), who considered it not only the means to understand the succession of ideas, but also the means to understand the way that complex ideas are built up out of simpler ones. Locke used expressions like "coalesce," "combination," "connexion," and "tying together" to write about the association of simple ideas into complex ones. James Mill (1773–1836) wrote more explicitly about the building up of complex from simple:

> From a stone I have had, synchronically, the sensation of colour, the sensation of hardness, the sensations of shape, and size, the sensation of weight. When the idea of one of these sensations occurs, the ideas of all of them occur. They exist in my mind synchronically; and their synchronical existence is called the idea of the stone; which, it is thus plain, is not a single idea, but a number of ideas in a particular state of combination (Herrnstein & Boring, 1966, p. 366).

Mill supposed that complex ideas also could connect to make still more complex ideas; for example, he wrote, "Brick is one complex idea, mortar is another complex idea; these ideas, with ideas of position and quantity, compose my idea of a wall" (Herrnstein & Boring, 1966, p. 377).

Few of the associationists discussed action explicitly, the main exception being David Hartley (1705–1757), who argued that the same principles of association that applied to sensations and ideas applied also to motions, and wrote about ideas and motions more or less interchangeably, relying on a vibratory notion of nervous activity to account for both. In supposing that ideas and motions may be conjoined, he anticipated Pavlov's notion of the conditional reflex.

Association formed the backbone of scientific psychology in the nineteenth century. The conception that conscious experience could be analyzed into discrete units that combined with one another to make up the succession and complexity of experience seemed the way to make sense of what otherwise seemed beyond comprehension. When Pavlov began studying reflexes, the concept of association must have occurred to him inevitably as it did to Hartley. Behavior too could be rendered comprehensible by analysis into discrete units that combined with one another to make up the succession and complexity of behavior. Since the reflex was the only behavioral unit that was at all well understood, behaviorists of the early twentieth century incorporated it into their thinking about behavior. Even before that, Thorndike saw the power of the consequences of behavior to change its likelihood to lie in their ability to strengthen the connection between stimuli and responses. He referred to his experiments on the law of effect as "the experimental study of associative processes" (Thorndike, 1911/2000). Like Hartley and Mill before him, he saw complexity as accretion of associations and carefully distinguished between synchronicity and succession:

> We must not mistake for a complex association a series of associations, where one sense-impression leads to an act such as to present a new sense-impression which leads to another act which in its turn leads to a new sense-impression. Of the formation of such series animals are capable to a very high degree. . . . By this power of acquiring a long series animals find their way to distant feeding grounds and back again. But all such cases are examples of the number, not of the complexity, of animal associations (Thorndike, 1911/2000, p. 132).

In this quote, Thorndike explicitly anticipates the notion of a chain of reflexes, written of by early behaviorists such as Hull. From such a notion, the step to the behavioral chains of Keller and Schoenfeld (1950) and Skinner (1953) was a small one.

Thus, the molecular view of behavior was an admirable start for a science. It handled two of the most basic problems, complexity and succession, by conceiving of behavior as composed of discrete units (responses) that could be connected to one another and to stimuli. The problem of complexity was approached by supposing that many simple units could be connected together, and the problem of succession was approached with the notion of chaining—that one behavioral unit could be linked to a stimulus produced by another behavioral unit. The molecular view served in much the same way that corpuscular mechanics served physics as it was getting started. Eventually corpuscular mechanics outlived its usefulness, except as an approximation, and was replaced with more continuous views like field theory, relativity, and quantum mechanics. Behavior analysis is at a point in its development where it too should turn away from a corpuscular view of behavior, away from discrete units toward more continuous concepts.

Criticisms of the Molecular View

From the later nineteenth century, if not earlier, the molecular view had its critics. John Dewey, for example, wrote an article published in 1896 criticizing the idea that the reflex could possibly serve as a sound basis for understanding behavior. He argued against the very idea that stimulus and response could be conceived of as separate entities, maintaining instead that behavior should be understood as entering into coordinations that embrace both behavior and environment. He wrote:

> [T]he reflex arc idea, as commonly employed, is defective in that it assumes sensory stimulus and motor response as distinct psychical existences, while in reality they are always inside a coördination and have their significance purely from the part played in maintaining or reconstituting the coördination (Herrnstein & Boring, 1966, p. 323).

Arguing against viewing behavior as a "series of jerks" and emphasizing instead the "unity of activity," Dewey wrote:

> The 'stimulus,' the excitation of the nerve ending and of the sensory nerve, the central change, are just as much, or just as little, motion as the events taking place in the motor nerve and the muscles. It is one uninterrupted, continuous redistribution of mass in motion. . . .It is redistribution pure and simple; as much so as the burning of a log, or the falling of a house or the movement of the wind. . . . There is just a change in the system of tensions (Herrnstein & Boring, 1966, p. 324).

The coordination, he argued, is one continuous process, "an organization of means with reference to a comprehensive end." This would be true of a "well developed" instinct, such as a hen's sitting on eggs, or a "thoroughly formed" habit, such as walking. Of these, he wrote:

> There is simply a continuously ordered sequence of acts, all adapted in themselves and in the order of their sequence, to reach a certain objective end, the reproduction of the species, the preservation of life, locomotion to a certain place. The end

> has got thoroughly organized into the means. In calling one stimulus, another response we mean nothing more than that such an orderly sequence of acts is taking place. The same sort of statement might be made equally well with reference to the succession of changes in a plant, so far as these are considered with reference to their adaptation to, say, producing seed. It is equally applicable to the series of events in the circulation of the blood, or the sequence of acts occurring in a self-binding reaper (Herrnstein & Boring, 1966, p.325).

Other writers repeated Dewey's criticisms in various forms. The Gestalt psychologists Köhler (1947) and Koffka (1935/1963) argued against the atomism of associationism and urged the continuity of both experience and behavior. E. B. Holt (1915/1965), writing about the Freudian wish, equated the wish with a "course of action" and argued that "intelligent conduct, to say nothing of conscious thought, can never be reduced to reflex arcs and the like; just as a printing-press is not merely wheels and rollers, and still less is it chunks of iron" (p. 50). Those who seek to analyze behavior into reflex arcs, he suggested, have "overlooked the form of organization of these his reflex arcs, has left out of the account that step which assembles wheels and rollers into a printing-press, and that which organizes reflex arcs" (p. 50). In a classic paper, "The Problem of Serial Order in Behavior," Karl Lashley (1951/1961) criticized associative chain theories as inadequate to account for the wholeness of extended sequences of behavior. Although he drew primarily on examples from verbal behavior, he pointed out:

> Temporal integration is not found exclusively in language; the coordination of leg movements in insects, the song of birds, the control of trotting and pacing in a gaited horse, the rat running the maze, the architect designing a house, and the carpenter sawing a board present a problem of sequences of action which cannot be explained in terms of successions of external stimuli (p. 181).

He criticized the idea of associative chains by pointing out that in the enunciation of words a single sound may be followed by a wide variety of sounds, depending on the word being spoken. Similarly, he wrote:

> Words stand in relation to the sentence as letters do to the word; the words themselves have no intrinsic temporal "valence." The word "right," for example, is noun, adjective, adverb, and verb, and has four spellings and at least ten meanings. In such a sentence as "The mill-wright on my right thinks it right that some conventional rite should symbolize the right of every man to write as he pleases," word arrangement is obviously not due to any direct associations of the word "right" itself with other words (p. 183).

Instead an utterance has a wholeness about it that must arise from some sort of overall organization. Lashley wrote:

> There is a series of hierarchies of organization; the order of vocal movements in pronouncing the word, the order of words in the sentence, the order of sentences in the paragraph, the rational order of paragraphs in a discourse. Not only speech, but all skilled acts seem to involve the same problems of serial ordering, even down to the temporal coordination of muscular contractions in such a movement as reaching and grasping (p. 187).

All these criticisms of the molecular view, with its concepts of building complexity by accretion of discrete units and building temporal extension by the linking of discrete units into chains, revolve around two points. Firstly, they say that activities have a continuity or wholeness about them that defies analysis into bits like the pieces of a puzzle. Neither is a printing press a bunch of wheels and rollers, nor is a sentence a bunch of words strung together word by word. Secondly, they say that activities are organized or coordinated with respect to ends or functions. One cannot talk about activity coherently without naming the function it serves. Behavior, they say, is continuous, temporally extended, and organized.

One might think that we face an irresolvable dichotomy between two views of behavior, one atomistic and one holistic. To an extent that is correct, but we need not simply point to the inadequacies of atomism, insist on holism, and let the matter drop there. A third alternative exists that utilizes the ideas that behavior is continuous, temporally extended, and organized, but also allows analysis into parts without reverting to atomism: the concept of levels of organization embodied in the idea of nested temporally extended activities.

The Molar View as an Alternative

In a similar discussion of entities of concern in ecology, Buege (1997) sought to avoid both the atomism of focusing exclusively on organisms and the holism of seeing all nature as fundamentally indivisible. Instead, he suggested that the organism is one level of organization, the species is another, and the ecosystem is yet another, much as Lashley did with his "series of hierarchies of organization." To support his suggestion, Buege distinguished between a class and an individual, much as I did in an earlier paper (Baum, 2002). Buege points out that when people use the word "class," they often mean a *collection* of items. He defines a collection as a grouping of individuals, the potential value of which is equal to the sum of the values of the members grouped. He gives as an example the spare change to be found in someone's pocket at any particular time; its cash value is simply the sum of the coins' values. Skinner's concept of the operant, which he called a class, was probably a collection (Baum, 2002).

In contrast to a collection, Buege defines an "individual" as an entity the potential value of which exceeds the sum of the values of its parts. He gives as an example a baseball team, which constitutes more than just a collection of players. Individuals have at least three properties that collections lack. Firstly, each part stands in the relation of part to whole, as a baseball player is part of a whole team. Secondly, the parts cohere to function as a whole, as the team, not the players, wins and loses games. Thirdly, individuals are subject to the attribute of scale, as a baseball player is an individual at a different scale from the team. Buege notes, "A being's scale is the position in which the being exists in relation to other things" (p. 4). Thus, an organism, a species, and an ecosystem are scaled in that order.

In an earlier paper I proposed that activities are individuals (Baum, 2002). Activities satisfy Buege's three requirements. The parts of any activity are other activities, less extended in time, that function together so as to make the activity effective (to accomplish its ends). Activities have the property of scale because the parts of an activity exist on a smaller scale than it does, and the same activity as a part of another activity exists on a smaller scale than that more extended one. Swinging a tennis racquet, playing a point, playing a game, playing a match, and preparing for the Olympics are activities scaled in that order.

As Buege's (1997) proposal that ecosystems are individuals with scale escapes the atomism/holism dichotomy, so too the molar view of behavior, seeing it as composed of activities with scale, escapes the atomism/holism dichotomy in behavior analysis.

A second feature of the molecular view of behavior, besides its atomism, is its reliance on temporal contiguity as the principle that binds the pieces or discrete units together. Stimuli must be contiguous with responses, and reinforcers must closely follow responses to be effective. This requirement of stimuli and reinforcers immediately contiguous with responses leads the molecular view immediately into positing hypothetical stimuli and reinforcers when it comes to explaining behavior. For example, avoidance, in which no consequences follow effective responses, is explained by resorting to hypothetical reinforcers in the form of fear reduction (e.g., Dinsmoor, 2001). Rule-governed behavior, which depends on long-term consequences, is explained by resorting to thoughts and feelings (Mallott, 2001; see Baum, 2003a, for criticism).

The molar view offers an alternative to contiguity for explanations of behavior. Instead, it suggests that the environment provides correlations that manifest in time, sometimes locally and sometimes on more extended scales. That a rat presses a lever at all may be attributed to occasional episodes of pressing being accompanied by episodes of feeding. A fixed-interval pattern of pressing, however, results from exposure to long-term regularity in presentation of reinforcers. Saving money depends on an extended relation between activity and consequences, whereas spending money on entertainment depends on more local relations entailed in a more local pattern (of socializing perhaps).

Activities contrast with discrete responses in that, rather than being combined like bricks, they comprise integrated parts. Every activity both entails parts that are themselves activities and also is itself a part in a more extended activity. Lever pressing is organized into bouts, which may go together to make up a fixed-interval pattern, which is part of the rat's activities in the experimental chamber. Speaking a sentence is part of conversing, which is part of maintaining a relationship, which is part of living. In other words, less extended activities are nested within more extended activities.

Avoiding Confusion

Activities are episodic; they occur in bouts or visits. This might lead to confusion, because an episode of an activity is a discrete event. Just like a discrete response, it has a beginning and an end, and nothing else happens in-between. Both may have variable duration. So, what's the difference? They differ in two respects: (a) the way that duration enters into measurement; and (b) the relation to reinforcers or rewards.

Firstly, duration is primary for activities, but secondary for discrete responses. Discrete responses vary in duration, particularly if interresponse time (IRT) is included. Researchers often have examined frequency distributions of response and IRT duration. But each response counts for just one response, albeit a response with variable duration. The molecular view treats the duration only as an attribute of the response. In contrast, an episode of an activity counts according to its duration—a brief episode adds less to the measure (time spent) than a long episode does. As an example, in a choice experiment by Baum and Rachlin (1969), in which no discrete responses were defined, time spent on two platforms was measured by running two clocks. Whenever the pigeon was on one side, time on that side accumulated. A 5-s visit to a side added 5 s, and a 30-s visit added 30 s. The episodes accumulated into time spent on the left and right sides. Even if, as subsequent analysis suggests, choice conformed to a pattern of fix and sample

(Baum, 2002), time spent accumulated, except time spent fixing and sampling. As we study IRT distributions, we also may study frequency distributions of episode duration, but the times involved in the two differ in their status, one being a response attribute and the other being the result of switching between activities.

Secondly, discrete responses are thought of as being followed more or less immediately with a reinforcer or reward. If reinforcement is continuous, every response is followed by a reinforcer. If reinforcement is intermittent, only some responses are followed by reinforcers, but each reinforcer is considered to follow a response. Episodes of activity, in contrast, are thought of as accompanied by reinforcers or rewards. A single episode of pressing the left lever or pecking the left key might include several reinforcers (Baum, Schwendiman, & Bell, 1999). In the experiment by Baum and Rachlin, one episode of standing on the left or rich side might include several reinforcers. In everyday life, this is true in spades. An episode of sexual activity includes its rewards, even if some additional ones may follow. The same is true of an episode of reading.

Thus, a discrete response and an episode of an activity differ in ontological terms. They have different properties and play different roles in the conceptual schemes or paradigms of which they are parts.

Distinguishing between the molar and molecular views is complicated also by the long history of usage of the words molar and molecular. For example, Hineline (2001) appears to consider the terms only to anchor the ends of a continuum of scales of analysis, probably referring to various scales of organization, as we have discussed here. Confusion may be avoided if these adjectives are used only to modify nouns like *view*, *thinking*, and *approach*, rather than nouns like analysis, control, or theory. When speaking of analyses or control, one may instead speak of local and extended analyses or short-term and long-term control. Most likely, Hineline's (2001) call for multiscaled analyses is a call for analyses at different scales of temporal extendedness, from more local to more global. Otherwise, his discussion seems in keeping with the present discussion.

The reason to be careful applying the two adjectives to *theory* is that the difference between the molar and molecular views is not theoretical but paradigmatic. The two approaches lead to different theories and different methods, but they are themselves neither theories nor methods, because they are two different paradigms (Kuhn, 1970). A *molecular theory* would mean a theory conceived in molecular terms. For example, a molecularly conceived theory such as Killeen's (1994) behavioral mechanics competes with other molecularly conceived theories such as momentary maximizing (Hinson & Staddon, 1983) in the sense that experiments may be designed that pit them against one another. No such competition occurs between these molecularly conceived theories, however, and molar-conceived theories such as extended optimality (e.g., Baum, 1981b; Rachlin & Burkhart, 1978), because theories conceived in the different paradigms are incommensurate—they are unintelligible to one another, and no experiment could decide between them. Theories conceived in different paradigms differ both in form and in what they attempt to explain. Instead of trying to explain the different cumulative records generated by different schedules of reinforcement, one may try to explain the allocation of behavior between different choice alternatives.

In particular, although short-term relations may exert more control over behavior than long-term relations, such ascendancy in no way contradicts the molar paradigm (Reed, Hildebrandt, DeJongh, & Soh, 2003; Rachlin & Green, 1972). Rather, the molar paradigm proposes the question of short-term versus long-term control as a question for study. What conditions shift control from short-term to long-term and vice versa?

Although no experiment or data can decide between the molar and molecular views, the two paradigms may be tested in other ways. Either view may allow an account of, say, avoidance (Herrnstein, 1969). One will be favored over the other on the basis of plausibility, elegance, and comprehensiveness. No experiment can distinguish between the two-factor theory of avoidance—a molecularly conceived theory—and the shock-frequency-reduction theory—a molar-conceived theory—but one may see the molar view of avoidance as more plausible, more elegant, and more comprehensive (Baum, 2001). The rest of this paper presents arguments that aim to promote the molar view on these grounds, focusing on choice.

Two Simple Principles

Two principles about behavior may be considered fundamental: (a) All behavior constitutes choice; and (b) All activities take time.

The idea that all behavior constitutes choice arises from the recognition that every situation, no matter how restricted, contains more than one behavioral option (Baum, 1974a). Even the most impoverished experimental space, where only lever pressing or key pecking is recorded, still permits the organism to engage in other activities such as grooming, exploration, and resting (Herrnstein, 1970). Choice is ubiquitous, because every situation offers multiple alternatives. In a brief time frame, say a second or less, which might have been taken up entirely by only one alternative, other behavior could have occurred. In a more extended time frame, other behavior will have occurred, resulting in a mix of activities. In an experiment on choice, at any particular moment, a pigeon might peck at the left key or the right key or engage in a number of other activities, but over the course of minutes usually a mix of these different activities occurs. In the activity of living, at any moment I may be spending money or I may be saving money, but over the course of a month I will do some of both.

Recognition of the equivalence of behavior and choice might seem to present no problem for the molecular view. At any moment, one discrete response or another occurs. Discrete responses translate into discrete choices. If over time presses occur at the left lever and the right lever, one may calculate the probability of a press on the left or the relative rate of left presses. These calculations based on extended samples of behavior, however, are regarded as "derived," not primary facets of behavior (Catania, 1981). The argument may be made that even response rate is a derived measure, because one calculates it by picking a time frame and then dividing the number of responses that happen to occur by the duration (Dinsmoor, 2001). In the molecular view, these measures, derived from the concrete occurrences of discrete responses, are abstractions. The lever presses are concrete particulars; the relative rate of left presses—i.e., choice—is an abstraction summarizing something about the presses.

In the molar view, behavior is choice and choice is behavior. Whether local or extended, a relative response rate constitutes an allocation of behavior among alternatives and is a primary feature of behavior. The measure may vary according to the duration of the time frame and may vary from one measurement to another with the same duration, particularly when behavior is in transition from one situation to another, but also even when behavior is stable. Nevertheless it estimates a feature of behavior that is fundamental. A coin comes up either heads or tails on any one flip, for several flips different mixes of heads and tails occur, and for a large number of flips the number of heads may about equal the number of tails; the relative frequency of heads tells something fundamental about the coin flipping. Allocations of behavior in a situation, far from being derived abstractions, are the primary data of the science.

The reason for these assertions in the molar view lies in the second simple principle: that activities take time. Even a lever press or key peck, so brief that the molecular view seeks to treat it as having no duration, still has duration, even if less than a second (Skinner, 1938). This is why measures of response rate never come out infinite. In real life too even singular decisions take time. When I decide to take a vacation in Mexico, neither the decision nor the vacation is instantaneous.

A deeper point, however, is that activities take up time not just in practice but necessarily. Suppose I show you a photograph of a person sitting and holding a book. What may we say the person is doing? Is she reading? The correct answer is that you cannot tell. She might be reading, but she might be pretending to read, or she might be daydreaming. You cannot tell without more observation. You need to see what went before and what comes after. If she is reading, then afterwards she will be able to talk about the book. I might reveal to you that she is actually pretending, because before taking the picture I asked her to, so she is complying with my request—a different activity altogether. To decide whether she is reading or doing something else, you need more than a momentary snapshot; you need a sample of behavior taking some extended time. Indeed, for very extended activities, you need a correspondingly lengthy sample. Is Billy saving? If we just sample for a short time, we may see him depositing money; a longer sample might show us that, in fact, he was frittering his money away. To quote Aristotle, "One swallow does not make a summer, nor does one day; and so too one day, or a short time, does not make a man blessed and happy." [The sense of "make" here is that in the statement, "Three interconnected lines make a triangle."]

The necessity of duration applies to any activity, even to the preparations of the laboratory. If the photograph had contained a rat with its paw on a response lever, and one asked what the rat was doing, the answer would be the same. Does the rat remove its paw without pressing? Did it press the lever just before? Unless one sees what went before and what came after, one cannot tell if the rat is pressing the lever, touching the lever, or exploring. The point becomes even clearer if we ask, when a press occurs, whether the press is part of a fixed-interval pattern.

Thus, no sure identification of momentary behavior is possible without an extended sample that brackets the moment. As a result, the behavior of the moment can be judged with certainty only in retrospect. Once we have an adequate extended observation, we may assert that at that moment the woman in the photograph was complying with my request that she pretend to read or that the rat was pressing the lever. Judgments about momentary behavior necessarily take the form, "At that moment, X was doing Y," where Y is always an extended activity that spans the moment. The extended activity is the matter observed, the concrete particular. The momentary behavior is inferred, an abstraction. Whereas the molecular view treats extended measures as abstractions and tries to treat momentary responses as concrete, the molar view takes the extended activity as concrete and the momentary response as an abstraction.

Significance attaches to this difference because of the molecular view's emphasis on contiguity. Whatever behavior is occurring at the moment of occurrence of a reinforcer is supposed to be strengthened. This requires that if no discrete response is apparent one must be invented. Such invention leads to accounts that are awkward and implausible when dealing with continuous activities. Since a discrete response precedes the reinforcer, some chunk of reading must be thought to precede, say, the garnering of a fact (if that is a reinforcer). In contrast, an episode of an activity comfortably brackets the reinforcer in a manner consistent with the necessity of extension.

Nesting of Activities

Activities are defined by their function. Walking home is an activity, the function of which is to move one through space toward one's home. Let us say it entails walking three blocks in one direction, turning left, and walking five blocks again. Unless these parts occur, the activity fails to perform its function. The parts must be integrated and coordinated for the activity to work.

The parts of walking home, however, are themselves activities, just less extended activities. Walking three blocks along Market Street is an activity, the function of which is to bring me to the corner at which I will turn left. Walking five blocks to the left brings me the rest of the way home. These activities in turn have parts, such as crossing streets and waiting for traffic lights, which must cohere in order to perform their functions.

Thus, every activity both is composed of parts which are less extended activities and is part of some more extended activity. Walking home is part of accomplishing my tasks for the day. Reading has parts such as turning pages and moving eyes. Reading is a part of informing oneself. Pressing a lever has parts such as moving a paw or contracting a muscle. Pressing a lever is part of choosing between two levers. The only exception would be an activity such as living, which may be so extended as only to have parts but to constitute no part of any more extended activity, because no more extended activity exists.

Figure 6.1 shows a hypothetical illustration of nested activities. Activities toward the left are more extended; those toward the right are more local. A pigeon's activities while in an experiment may be broadly divided into three parts: feeding, maintaining the body, and maintaining vigilance. Feeding is composed of pecking and eating. Maintaining the body is composed of resting and grooming. Maintaining vigilance is composed of general activity and sensory scanning. Each of these parts is composed of still less extended activities. If pecking falls into a pattern of fix and sample (Baum et al., 1999; Baum, 2002), then it is composed of fixing on the rich alternative and sampling the lean alternative.

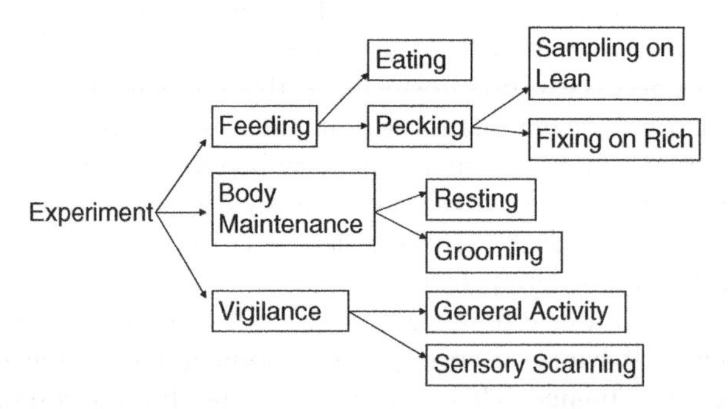

Figure 6.1 The hypothetical nesting of a pigeon's activities in an experiment with concurrent schedules of reinforcement

Molecular Explanations: Failings

The molar and molecular views lead to different types of explanation. Why is the rat pressing the lever? Why is the woman pretending to read? Any complete answer to such a question must make reference to the organism's past—to the rat's prior training and the woman's history with requests in general and my requests in particular. The molecular view frames the answers in terms of discrete stimuli, discrete responses, and immediate reinforcers. The woman's pretending must be broken down into a sequence of pretending responses, some of which must be followed by reinforcers. The molecular view must somehow bring the past into the present by representing the past in stimuli immediately preceding and following responses. The past, however, is invisible. The reinforcers for the woman's compliance are invisible; even my thanks at the end—if they constitute a reinforcer—are too delayed to explain the entire episode, which lasted several minutes. Thus, as with avoidance, the molecular view resorts to hypothetical stimuli and reinforcers to treat the compliance as a chain of discrete stimuli and responses held together by hypothetical reinforcers. In the laboratory, a chain may be arranged by causing responding to produce a discriminative stimulus for some other responding, which produces a reinforcer. The molecular view takes the discriminative stimulus produced to be a conditional reinforcer (analogous to a Pavlovian conditional stimulus), and explains a real-world episode of compliance by imagining that it is a chain similarly held together by discriminative stimuli that double as conditional reinforcers (see Baum, 1973, for discussion). Relying solely on the concepts of discrete responses and contiguity of reinforcers forces such awkward results.

Molar Explanations: Advantages

Both views offer ways of accounting for behavioral episodes like the woman's compliance, but, in keeping with the paradigmatic nature of the difference between them, their accounts differ in elegance, plausibility, and comprehensiveness. As the molar view allows explanation of avoidance by citing the temporally extended dependence of shock rate on response rate (Herrnstein, 1969), so too it allows explanation of the episode of pretending to read by citing the woman's long history of complying with requests of people with whom she interacts socially, her shorter history of interacting with me socially, and her complying with my request still more locally (Baum, 1995a; 2003b). The problem of bringing the past into the present never arises, because her activities past, present, and future are seen as continuous. Her socializing contains as a part her interacting with me, which includes as a part her complying with my request. The reinforcers for socializing have been and continue to be many and varied, accompanying the various parts of socializing with different people in her acquaintance. They take the form both of thanks and reciprocation. Next time she asks me for a favor, I will be likely to deliver consequences that, if remote in time, nevertheless directly stem from her compliance on that occasion. Our complying with one another's requests may be parts of our knowing one another.

Thus, in the molar view, less extended (i.e., more local) activities are explained by their fitting as parts into more extended activities (Rachlin, 1994). For example, the various parts of maintaining a social relationship are less extended activities that may produce their various rewards but may also enable more extended activities in which they participate to produce otherwise unattainable rewards. A conversation may be highly reinforcing, and receiving help from a friend is also reinforcing. The reinforcing value of the whole, however, may exceed the sum of the parts' reinforcers, because some parts may produce rewards that depend on the occurrence of other parts. Helping a friend may produce reinforcers (thanks and obligation) that depend on the likelihood that the friend will help on another occasion in return. Indeed, some parts may produce no reward or even be aversive but still may be essential to the value of the extended activity. Turning down a piece of cake on any particular occasion, a difficult and onerous activity, may be essential to maintaining a diet that has huge reinforcing value in the long run (Rachlin, 1995b). In contrast with the molecular view, this approach to explanation makes no use of hypothetical events or contingencies. The molar view refers to concrete, demonstrable stimuli and consequences that, however, are usually visible only over extended periods of time.

The molecular view resorts to hypothetical causes also when it attempts to explain the occurrence of particular responses on particular occasions. If a lever press is preceded by no observable stimulus, the molecular view explains it as the result of response strength or arousal, hypothetical causes (Skinner, 1938; Killeen, 1994). The molar view, in contrast, requires no account of momentary occurrences, because living, with all its parts, is continuous. Particular episodes of an activity, such as pressing or complying with a request, are occasioned by changes in the environment, such as the passage of time or the occurrence of the request. Nothing analogous to response strength comes in, even when an activity undergoes extinction. When lever pressing no longer produces food, an allocation including substantial lever pressing transforms into one that includes little or none, as other activities increase in frequency. The molar view requires no notion like strength, because in the molar view reinforcement shifts the allocation of activities directly. Competition occurs, not between different types of discrete responses, but between different allocations of activities (Baum, 2002). If one allocation produces more reinforcers in the long run, that allocation will tend to dominate in the long run (Baum, 1981b).

Conclusions

By restricting the terms molar and molecular to refer only to the two different paradigms and by referring to analyses or relations that differ in their temporal scale as varying in extendedness, we gain the ability to contrast the two views without confusing theoretical or methodological considerations with paradigmatic considerations. Although the two paradigms lead to different phenomena and theories, no one should take them to be theories or methods themselves. Recognizing two simple principles, that all behavior entails choice and that all activities take time, illuminates advantages of the molar view. The first advantage is that these principles point up the possibility of conceiving of allocation of time among activities as the fundamental measure of behavior. That idea, in turn, affords a natural way of talking about behavior both in the laboratory and in the real world, without resorting to implausible discrete response units. It offers instead the understanding and explaining of activities as parts of wholes that are more extended activities—the concept of nesting. Finally, it allows explanations that depend on no hypothetical events but only on observable and measurable aspects of behavior and environment.

7

Rethinking Reinforcement: Allocation, Induction, and Contingency

Foreword

The realization that induction can replace reinforcement did not come to me all at once. The seeds were present in 1973 when in "The Correlation-Based Law of Effect" I saw that the molar view of behavior eliminated the need for conditional reinforcement, response strength, two-factor theory, and other hypothetical constructs. My mentor, Richard Herrnstein, once remarked to me—probably around 1970—that a science of behavior could be based either on pushes or pulls. He seemed to think they were equally valid. As research results accumulated, I began to see that pushes had clear advantages—for example, explaining stimulus control and shaping by providing a mechanism: induction. In 1970, Herrnstein pointed out that adaptation is more of a question than an answer. Before long, I began to see that the great lack in understanding behavior was the lack of a mechanism. Indeed, the matching relation by itself was too easily labeled as description, rather than explanation.

During the academic year 1965–1966, I spent a postdoctoral year in the sub-department of animal behavior at Cambridge University studying ethology. Although I accomplished little directly, I came away with the conviction that behavior must be understood in the context of evolutionary theory. When I moved to California in 1999, I started attending meetings of two groups interested in evolution: researchers at University of California, Davis studying cultural evolution and mathematical models of evolution; and philosophers of biology at the California Academy of Sciences working on ontology and epistemology related to evolution. Eventually I concluded that no adequate understanding of behavior is possible without evolutionary theory.

In a series of experiments, Michael Davison and I studied situations in which choice alternatives changed often and randomly, with no signals other than the food deliveries themselves. We saw clearly that the food deliveries functioned just like discriminative stimuli, and we finally produced results that indicated to us that the discriminative function of the food could be seen as its only function. Not only so-called "secondary reinforcers" could be understood solely as discriminative stimuli, but even so-called "primary reinforcers" in general could be understood solely as discriminative stimuli. We suggested that these stimuli serve as "guideposts" (Davison & Baum, 2006, 2010). The conception was bolstered by appeal to evolutionary theory, because in an evolutionary

Science and Philosophy of Behavior: Selected Papers, First Edition. William M. Baum.

context discovery of food, for example, is important only insofar as it predicts survival and reproduction; food is a proxy for fitness. This view solved many problems but left open the question of mechanism.

During the 2000s, I was pondering the question of mechanism and, at some point, revisited the concept of induction. At the same time, I was arguing for evolutionary theory as the appropriate context for behavior analysis. I realized that reinforcers could be seen as phylogenetically important events. By 2007, I was mentioning induction as a mechanism in talks about evolution. I saw that induction could be the mechanism of stimulus control, and that would mean that operant activities were induced by phylogenetically important events, not just adjunctive activities. In 2007 I began giving talks on rethinking reinforcement, and those talks eventually evolved into this paper.

Abstract

The concept of reinforcement is at least incomplete and almost certainly incorrect. An alternative way of organizing our understanding of behavior may be built around three concepts: *allocation, induction,* and *correlation.* Allocation is the measure of behavior and captures the centrality of choice: All behavior entails choice and consists of choice. Allocation changes as a result of induction and correlation. The term induction covers phenomena such as adjunctive, interim, and terminal behavior—behavior induced in a situation by occurrence of food or another *Phylogenetically Important Event* (PIE) in that situation. Induction resembles stimulus control in that no one-to-one relation exists between induced behavior and the inducing event. If one allowed that some stimulus control were the result of phylogeny, then induction and stimulus control would be identical, and a PIE would resemble a discriminative stimulus. Much evidence supports the idea that a PIE induces all PIE-related activities. Research also supports the idea that stimuli correlated with PIEs become PIE-related conditional inducers. Contingencies create correlations between "operant" activity (e.g. lever pressing) and PIEs (e.g. food). Once an activity has become PIE-related, the PIE induces it along with other PIE-related activities. Contingencies also constrain possible performances. These constraints specify feedback functions, which explain phenomena such as the higher response rates on ratio schedules in comparison with interval schedules. Allocations that include a lot of operant activity are "selected" only in the sense that they generate more frequent occurrence of the PIE within the constraints of the situation; contingency and induction do the "selecting."

Keywords: reinforcement, contingency, correlation, induction, inducing stimulus, inducer, allocation, Phylogenetically Important Event, reinstatement

Source: Originally published in *Journal of the Experimental Analysis of Behavior*, 97 (2012), pp. 101–124. Reproduced with permission of John Wiley & Sons. I thank Howard Rachlin, John Staddon, and Jesse Dallery for helpful comments on earlier versions of this paper.

This article aims to lay out a conceptual framework for understanding behavior in relation to environment. I will not attempt to explain every phenomenon known to behavior analysts; that would be impossible. Instead, I will offer this framework as a way to think about those phenomena. It is meant to replace the traditional framework, over 100 years old, in which reinforcers are supposed to strengthen responses or stimulus-response connections, and in which classical conditioning and operant conditioning are considered two distinct processes. Hopefully, knowledgeable readers will find nothing new herein, because the pieces of this conceptual framework were all extant, and I had only to assemble them into a whole. It draws on three concepts: (1) allocation, which is the measure of behavior; (2) induction, which is the process that drives behavior; and (3) contingency, which is the relation that constrains and connects behavioral and environmental events. Since none of these exists at a moment of time, they necessarily imply a molar view of behavior.

As I explained in earlier papers (Baum, 2001; 2002; 2004), the molar and molecular views are not competing theories, they are different paradigms (Kuhn, 1970). The decision between them is not made on the basis of data, but on the basis of plausibility and elegance. No experimental test can decide between them, because they are incommensurable—that is, they differ ontologically. The molecular view is about discrete responses, discrete stimuli, and contiguity between those events. It offers those concepts for theory construction. It was designed to explain short-term, abrupt changes in behavior, such as occur in cumulative records (Ferster & Skinner, 1957; Skinner, 1938). It does poorly when applied to temporally extended phenomena, such as choice, because its theories and explanations almost always resort to hypothetical constructs to deal with spans of time, which makes them implausible and inelegant (Baum, 1989). The molar view is about extended activities, extended contexts, and extended relations. It treats short-term effects as less extended, local phenomena (Baum, 2002; 2010; Baum & Davison, 2004). Since behavior, by its very nature, is necessarily extended in time (Baum, 2004), the theories and explanations constructed in the molar view tend to be simple and straightforward. Any specific molar or molecular theory may be invalidated by experimental test, but no one should think that the paradigm is thereby invalidated; a new theory may always be invented within the paradigm. The molar paradigm surpasses the molecular paradigm by producing theories and explanations that are more plausible and elegant.

The Law of Effect

E. L. Thorndike (2000/1911), when proposing the law of effect, wrote:

> Of several responses made to the same situation, those which are accompanied or closely followed by satisfaction to the animal will, other things being equal, be more firmly connected with the situation, so that, when it recurs, they will be more likely to recur. . . (p. 244).

Early on, it was subject to criticism. J. B. Watson (1930/1970), ridiculing Thorndike's theory, wrote:

> Most of the psychologists. . . believe habit formation is implanted by kind fairies. For example, Thorndike speaks of pleasure stamping in the successful movement and displeasure stamping out the unsuccessful movements (p. 206).

Thus, as early as 1930, Watson was skeptical about the idea that satisfying consequences could strengthen responses. Still, the theory has persisted despite occasional criticism (e.g., Baum, 1973; Staddon, 1973).

Thorndike's theory became the basis for B. F. Skinner's theory of reinforcement. Skinner drew a distinction, however, between observation and theory. In his paper "Are theories of learning necessary?" Skinner (1950) wrote:

> . . . the Law of Effect is no theory. It simply specifies a procedure for altering the probability of a chosen response.

Thus, Skinner maintained that the Law of Effect referred to the observation that when, for example, food is made contingent on pressing a lever, lever pressing increases in frequency. Yet, like Thorndike, he had a theory as to how this came about. In his 1948 paper, "'Superstition' in the pigeon," Skinner restated Thorndike's theory in a new vocabulary:

> To say that a reinforcement is contingent upon a response may mean nothing more than that it follows the response . . . conditioning takes place presumably because of the temporal relation only, expressed in terms of the order and proximity of response and reinforcement.

If we substituted "satisfaction" for "reinforcement" and "accompanied or closely followed" for "order and proximity," we would be back to Thorndike's idea above.

Skinner persisted in his view that order and proximity between response and reinforcer were the basis for reinforcement; it reappeared in *Science and Human Behavior* (Skinner, 1953):

> So far as the organism is concerned, the only important property of the contingency is temporal. The reinforcer simply follows the response. . . We must assume that the presentation of a reinforcer always reinforces something, since it necessarily coincides with some behavior (p. 85).

From the perspective of the present, we know that Skinner's observation about contingency was correct. His theory of order and proximity, however, was incorrect, because a "reinforcer" doesn't "reinforce" whatever it coincides with. That doesn't happen in everyday life; if I happen to be watching television when a pizza delivery arrives, will I be inclined to watch television more? (Although, as we shall see, the pizza delivery matters, in that it adds value to, say, watching football on television.) It doesn't happen in the laboratory, either, as Staddon (1977), among others, have shown.

When Staddon and Simmelhag (1971) repeated Skinner's "superstition" experiment, they observed that many different activities occurred during the interval between food presentations. Significantly, the activities often were not closely followed by food; some emerged after the food and were gone by the time the food occurred. Figure 7.1 shows some data reported by Staddon (1977). The graph on the left shows the effects of presenting food to a hungry pigeon every 12 s. The activities of approaching the window wall of the chamber and wing flapping rose and then disappeared, undermining any notion that they were accidentally reinforced. The graph on the right shows the effects of presenting food to a hungry rat every 30 s. Early in the interval, the rat drinks and runs, but these activities disappeared before they could be closely followed by food. Subsequent research has confirmed these observations many times over, contradicting the notion of strengthening by accidental contiguity (e.g., Palya & Zacny, 1980; Reid, Bacha, & Morán, 1993; Roper, 1978).

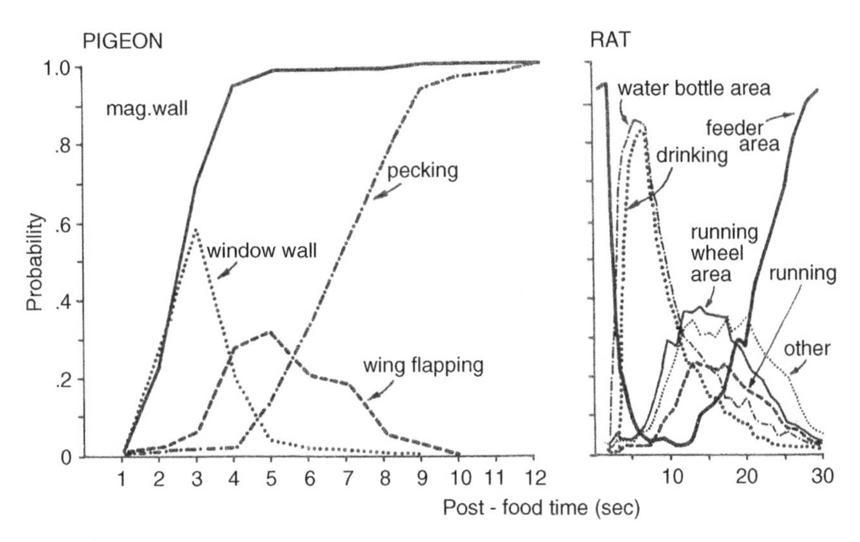

Figure 7.1 Behavior induced by periodic food
Note: Left: Activities of a pigeon presented with food every 12 s. Right: Activities of a rat presented with food every 30 s. Activities that disappeared before food delivery could not be reinforced. Reprinted from Staddon (1977).

Faced with results like those in Figure 7.1, someone theorizing within the molecular paradigm has at least two options. Firstly, contiguity between responses and food might be stretched to suggest that the responses are reinforced weakly at a delay. Secondly, food-response contiguity might be brought to bear by relying on the idea that food might elicit responses. For example, Killeen (1994) theorized that each food delivery elicited *arousal*, a hypothetical construct. He proposed that arousal jumped following food and then dissipated as time passed without food. Build-up of arousal was then supposed to cause the responses shown in Figure 7.1. As we shall see below, the molar paradigm avoids the hypothetical entity by relying instead on the extended relation of induction.

To some researchers, the inadequacy of contiguity-based reinforcement might seem like old news (Baum, 1973). One response to its inadequacy has been to broaden the concept of reinforcement to make it synonymous with optimization or adaptation (e.g., Rachlin, Battalio, Kagel, & Green, 1981; Rachlin & Burkhard, 1978). This redefinition, however, begs the question of mechanism. As Herrnstein (1970) wrote, "The temptation to fall back on common sense and conclude that animals are adaptive, *i.e.*, doing what profits them most, had best be resisted, for adaptation is at best a question, not an answer (p. 243)." Incorrect though it was, the contiguity-based definition at least attempted to answer the question of how approximations to optimal or adaptive behavior might come about. This article offers an answer that avoids the narrowness of the concepts of order, proximity, and strengthening and allows plausible accounts of behavior both in the laboratory and in everyday life. Because it omits the idea of strengthening, one might doubt that this mechanism should be called "reinforcement" at all.

Allocation of Time Among Activities

We see in Figure 7.1 that periodic food changes the allocation of time among activities like wing flapping and pecking or drinking and running. Some activities increase while some activities decrease. No notion of strengthening need enter the picture. Instead, we need only notice the changing allocation.

To be alive is to behave, and we may assume that in an hour's observation time one observes an hour's worth of behavior or that in a month of observation one observes a month's worth of behavior. Thus, we may liken allocation of behavior to cutting up a pie. Figure 7.2 illustrates the idea with a hypothetical example. The chart shows allocation of time in the day of a typical student. About 4 hr are spent attending classes, 4 hr studying, 2.5 hr eating meals, 0.5 hr in bathroom activities, 3 hr in recreation, and 2 hr working. The total amounts to 16 hr; the other 8 hr are spent in sleep. Since a day has only 24 hr, if any of these activities increases, others must decrease, and if any decreases, others must increase. As we shall see later, the finiteness of time is important to understanding the effects of contingencies.

Measuring the time spent in an activity may present challenges. For example, when exactly is a pigeon pecking a key or a rat pressing a lever? Attaching a switch to the key or lever allows one to gauge the time spent by the number of operations of the switch. Luckily, this method is generally reliable; counting switch operations and running a clock whenever the switch is operated give equivalent data (Baum, 1976). Although in the past researchers often thought they were counting discrete responses (pecks and presses), from the present viewpoint they were using switch operations to measure the time spent in continuous activities (pecking and pressing).

Apart from unsystematic variation that might occur across samples, allocation of behavior changes systematically as a result of two sources: induction and contingency. As we see in Figure 7.1, an inducing environmental event like food increases some activities and, because time is finite, necessarily decreases others. For example, food induces pecking in pigeons, electric shock induces aggression in rats and monkeys, and removal of food induces aggression in pigeons. Figure 7.3 shows a hypothetical example. Above (at Time 1) is depicted a pigeon's allocation of time among activities in an experimental situation that includes a mirror and frequent periodic food. The pigeon spends a lot of time eating and resting and might direct some aggressive or reproductive activity toward the mirror. Below (at Time 2) is depicted the result of increasing the interval between food deliveries. Now the pigeon spends a lot of time aggressing toward the mirror, less time eating, less time resting, and little time in reproductive activity.

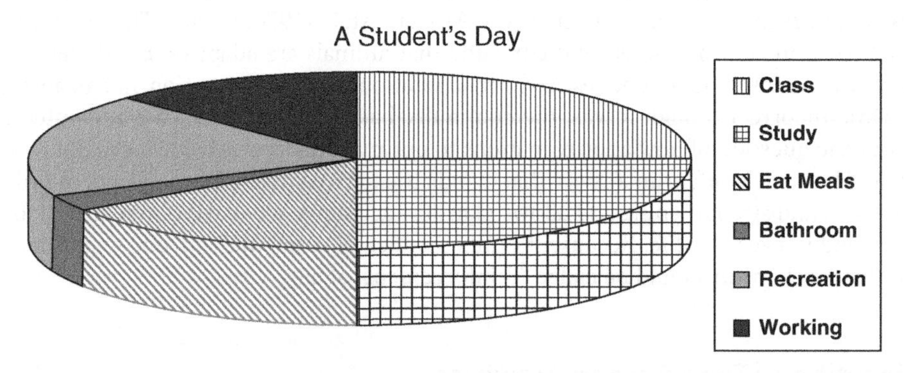

Figure 7.2 A hypothetical illustration of the concept of allocation
Note: The times spent in various activities in a typical day of a typical student add up to 16 h, the other 8 being spent in sleep. If more time is spent in one activity, less time must be spent in others, and vice versa.

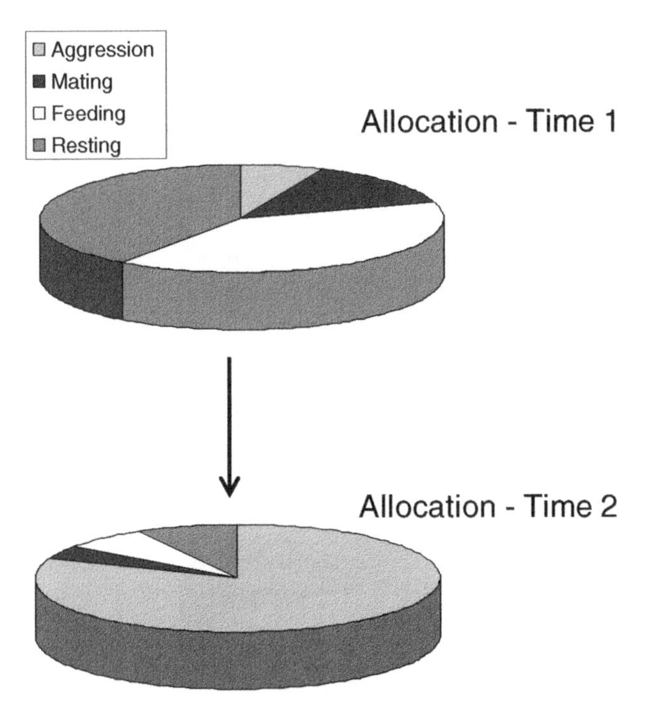

Figure 7.3 A hypothetical example illustrating change of behavior due to induction as change of allocation
Note: At Time 1, a pigeon's activities in an experimental space exhibit Allocation 1 in response to periodic food deliveries. At Time 2, the interval between food deliveries has been lengthened, resulting in an increase in aggression toward a mirror. Other activities decrease necessarily.

A contingency links an environmental event to an activity and results in an increase or decrease in the activity. For example, linking food to lever pressing increases the time spent pressing; linking electric shock to lever pressing usually decreases the time spent pressing. Figure 7.4 illustrates with a hypothetical example. The diagram above shows a child's possible allocation of behavior in a classroom. A lot of time is spent in hitting other children and yelling, but little time on task. Time is spent interacting with the teacher, but with no relation to being on task. The diagram below shows the result after the teacher's attention is made contingent on being on task. Now the child spends a lot of time on task, less time hitting and yelling, and a little more time interacting with the teacher—presumably now in a more friendly way. Because increase in one activity necessitates decrease in other activities, a positive contingency between payoff and one activity implies a negative contingency between payoff and other activities—the more yelling and hitting, the less teacher attention. A way to capture both of these aspects of contingency is to consider the whole allocation to enter into the contingency with payoff (Baum, 2002).

Induction

The concept of induction was introduced by Evalyn Segal in 1972. To define it, she relied on dictionary definitions, "stimulating the occurrence of" and "bringing about," and

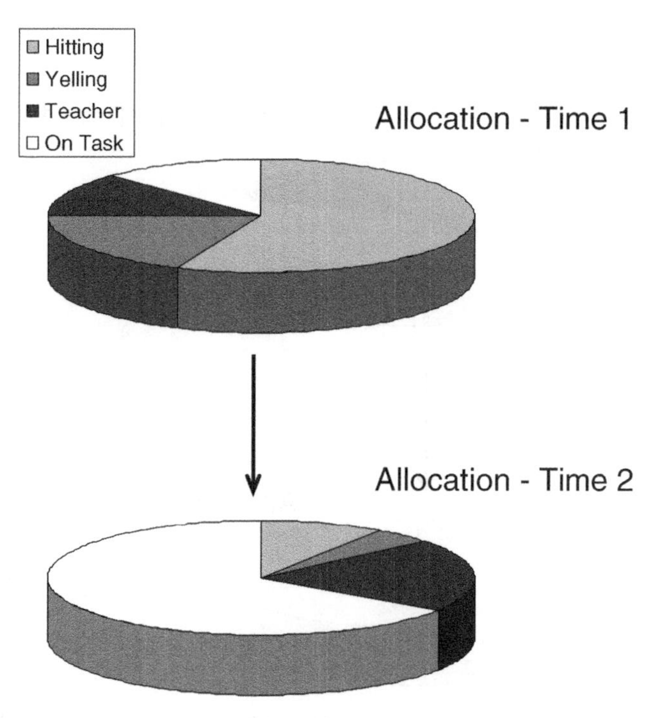

Figure 7.4 A hypothetical example illustrating change of allocation due to correlation
Note: At Time 1, a child's classroom activities exhibit Allocation 1, including high frequencies of disruptive activities. At Time 2, after the teacher's attention has been made contingent upon being on task, Allocation 2 includes more time spent on task and, necessarily, less time spent in disruptive activities.

commented, "It implies a certain indirection in causing something to happen, and so seems apt for talking about operations that may be effective only in conjunction with other factors" (p. 2). In the molar view, that "indirection" is crucial, because the concept of induction means that, in a given context, the mere occurrence of certain events, such as food—i.e., inducers—results in occurrence of or increased time spent in certain activities. All the various activities that have at one time or another been called "interim," "terminal," "facultative," or "adjunctive," like those shown in Figure 7.1, may be grouped together as *induced* activities (Hineline, 1984). They are induced in any situation that includes events like food or electric shock, the same events that have previously been called "reinforcers," "punishers," "unconditional stimuli," or "releasers." Induction is distinct from the narrower idea of elicitation, which assumes a close temporal relation between stimulus and response. One reason for introducing a new, broader term is that inducers need have no close temporal relation to the activities they induce (Figure 7.1). Their occurrence in a particular environment suffices to induce activities, most of which are clearly related to the inducing event in the species' interactions with its environment—food induces food-related activity (e.g., pecking in pigeons), electric shock induces pain-related activity (e.g., aggression and running), and a potential mate induces courtship and copulation.

Induction and Stimulus Control

Induction resembles Skinner's concept of stimulus control. A discriminative stimulus has no one-to-one temporal relation with responding, but rather increases the rate or time spent in the activity in its presence. In present terms, a discriminative stimulus modulates the allocation of activities in its presence; it induces a certain allocation of activities. One may say that a discriminative stimulus sets the occasion or the context for an operant activity. Similarly, an inducer may be said to occasion or create the context for the induced activity. Stimulus control may seem sometimes to be more complicated, say, in a conditional discrimination that requires a certain combination of events, but, no matter how complex the discrimination, the relation to behavior is the same. Indeed, although we usually conceive of stimulus control as the outcome of an individual's life history (ontogeny), if we accept the idea that some instances of stimulus control might exist as a result of phylogeny, then stimulus control and induction would be two terms for the same phenomenon: the effect of context on behavioral allocation. In what follows, I will assume that induction or stimulus control may arise as a result of either phylogeny or life history.

Although Pavlov thought in terms of reflexes and measured only stomach secretion or salivation, with hindsight we may apply the present terms and say that he was studying induction (Hineline, 1984). Zener (1937), repeating Pavlov's procedure with dogs that were unrestrained, observed that during a tone preceding food the dogs would approach the food bowl, wag their tails, and so on—all induced behavior. Pavlov's experiments, in which conditional stimuli such as tones and visual displays were correlated with inducers like food, electric shock, and acid in the mouth, showed that correlating a neutral stimulus with an inducer changes the function of the stimulus. In present terms, it becomes a *conditional inducing stimulus* or a *conditional inducer*, equivalent to a discriminative stimulus. Similarly, activities linked by contingency to inducers become inducer-related activities. For example, Shapiro (1961) found that when a dog's lever pressing produces food, the dog salivates whenever it presses. We will develop this idea further below, when we discuss the effects of contingencies.

Phylogenetically Important Events

Since they gain their power to induce as a result of many generations of natural selection—from phylogeny—I call them *Phylogenetically Important Events* (PIEs; Baum, 2005). A PIE is an event that directly affects survival and reproduction. Some PIEs increase the chances of reproductive success by their presence; others decrease the chances of reproductive success by their presence. On the increase side, examples are food, shelter, and mating opportunities; on the decrease side, examples are predators, parasites, and severe weather. Those individuals for whom these events were unimportant produced fewer surviving offspring than their competitors for whom they were important and are no longer represented in the population.

PIEs are many and varied, depending on the environment in which a species evolved. In humans, they include social events like smiles, frowns, and eye contact, all of which affect fitness because of our long history of living in groups and the importance of group membership on survival and reproduction. PIEs include also dangerous and fearsome events like injuries, heights, snakes, and rapidly approaching large objects ("looming").

In accord with their evolutionary origin, PIEs and their effects depend on species. Breland and Breland (1961) reported the intrusion of induced activities into their attempts to train various species using contingent food. Pigs would root wooden coins as they would root for food; raccoons would clean coins as if they were cleaning food; chickens would scratch at the ground in the space where they were to be fed. When the relevance of phylogeny was widely acknowledged, papers and books began to appear pointing to "biological constraints" (Hinde & Stevenson-Hinde, 1973; Seligman, 1970). In humans, to the reactions to events like food and injury, we may add induction of special responses to known or unknown conspecifics, such as smiling, eyebrow raising, and tongue showing (Eibl-Eiblsfeldt, 1975).

Accounts of reinforcers and punishers that make no reference to evolutionary history fail to explain why these events are effective as they are. For example, Premack (1963; 1965) proposed that reinforcement could be thought of as a contingency in which the availability of a "high-probability" activity depends on the occurrence of a "low-probability" activity. Timberlake and Allison (1974) elaborated this idea by adding that any activity constrained to occur below its level in baseline, when it is freely available, is in "response deprivation" and, made contingent on any activity with a lower level of deprivation, will reinforce that activity. Whatever the validity of these generalizations, they beg the question, "Why are the baseline levels of activities like eating and drinking higher than the baseline levels of running and lever pressing?" Deprivation and physiology are no answers; the question remains, "Why are organisms so constituted that when food (or water, shelter, sex, etc.) is deprived, it becomes a potent inducer?" Linking the greater importance of eating and drinking (food and water) to phylogeny explains why these events are important in an ultimate sense, and the concept of induction explains how they are important to the flexibility of behavior.

Figure 7.5 illustrates that, among all possible events—environmental stimuli and activities—symbolized by the large circle, PIEs are a small subset, and PIE-related stimuli and activities are a larger subset. That the subset of PIE-related stimuli and activities includes the subset of PIEs indicates that PIEs usually are themselves PIE-related, because natural environments and laboratory situations arrange that the occurrence of a PIE like food predicts how and when more food might be available. Food usually predicts more food, and the presence of a predator usually predicts continuing danger.

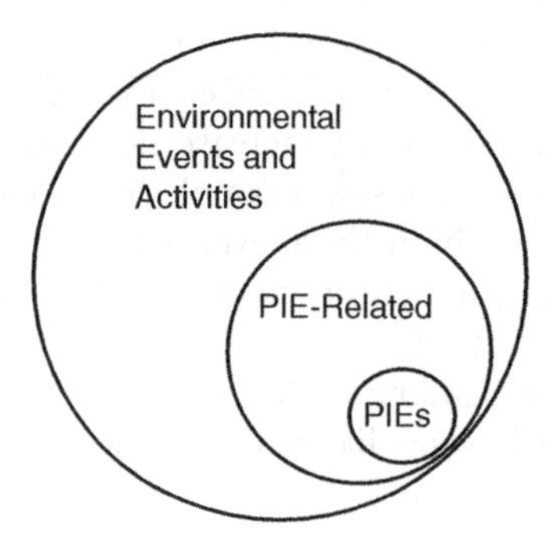

Figure 7.5 Diagram illustrating that Phylogenetically Important Events (PIEs) constitute a subset of PIE-related events, which in turn are a subset of all events

A key concept about induction is that a PIE induces any PIE-related activity. If a PIE like food is made contingent on an activity like lever pressing, the activity (lever pressing) becomes PIE-related (food-related). The experimental basis for this concept may be found in experiments that show PIEs to function as discriminative stimuli. Many such experiments exist, and I will illustrate with three examples showing food contingent on an activity induces that activity.

Bullock and Smith (1953) trained rats for 10 sessions. Each session consisted of 40 food pellets, each produced by a lever press (i.e., Fixed Ratio 1), followed by 1 hr of extinction (lever presses ineffective). Figure 7.6 shows the average number of presses during the hour of extinction across the 10 sessions. The decrease resembles the decrease in "errors" that occurs during the formation of a discrimination. That is how Bullock and Smith interpreted their results: The positive stimulus (S^D) was press-plus-pellet, and the negative stimulus (S^Δ) was press-plus-no-pellet. The food was the critical element controlling responding. Thus, the results support the idea that food functions as a discriminative stimulus.

The second illustrative finding, reported by Reid (1958), is referred to as "reinstatement," and has been studied extensively (e.g., Ostlund & Balleine, 2007). Figure 7.7 shows the result in a schematic form. Reid studied rats, pigeons, and students. First he trained responding by following each response with food (rats and pigeons) or a token (students). Once the responding was well-established, he subjected it on Day 1 to 30 min of extinction. This was repeated on Day 2 and Day 3. Then, on Day 3, when responding had disappeared, he delivered a bit of food (rats and pigeons) or a token (students). The free delivery was immediately followed by a burst of responding (circled in Figure 7.7). Thus, the freely delivered food or token reinstated the responding, functioning just like a discriminative stimulus.

As a third illustration, we have the data that Skinner himself presented in his 1948 "superstition" paper. Figure 7.8 reproduces the cumulative record that Skinner claimed

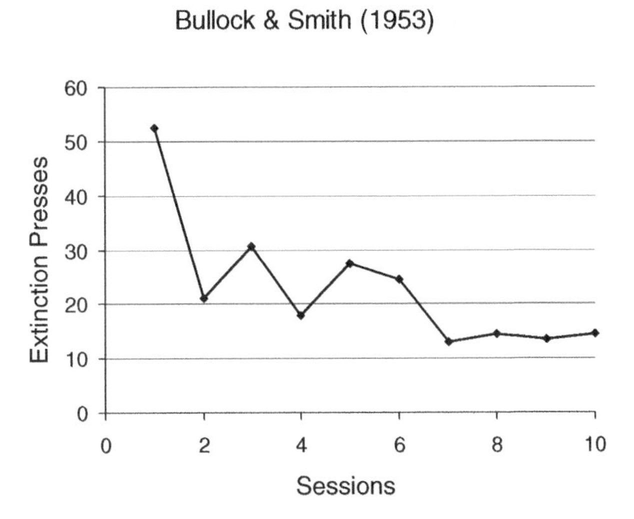

Figure 7.6 Results of an experiment in which food itself served as a discriminative stimulus
Note: The number of responses made during an hour of extinction following 40 response-produced food deliveries decreased across the ten sessions of the experiment. These responses would be analogous to "errors." The data were published in a paper by Bullock and Smith (1953).

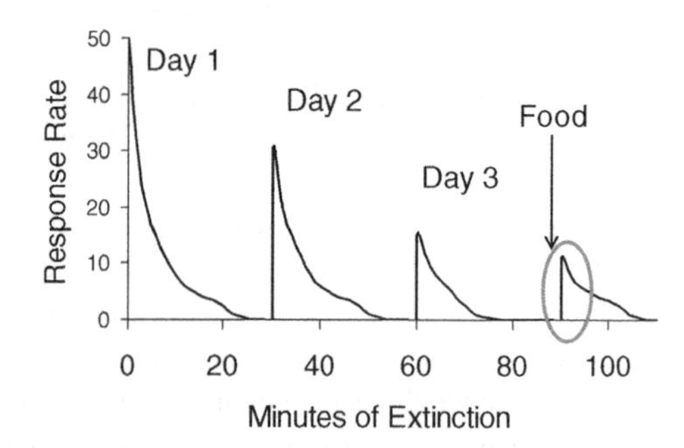

Figure 7.7 Cartoon of Reid's (1958) results with reinstatement of responding following extinction upon a single response-independent presentation of the originally contingent event (food for rats and pigeons; token for students)
Note: A burst of responding (circled) followed the presentation, suggesting that the contingent event functioned as a discriminative stimulus or an inducer.

to show "reconditioning" of a response of hopping from side to side by adventitious reinforcement. Looking closely at the first two occurrences of food (circled arrows), however, we see that no response occurred just prior to the first occurrence and possibly just prior to the second occurrence. The hopping followed the food; it did not precede it. Skinner's result shows reinstatement, not reinforcement. In other words, the record shows induction by periodic food similar to that shown in Figure 7.1.

All three examples show that food induces the "operant" or "instrumental" activity that is correlated with it. The PIE induces PIE-related activity, and these examples show that PIE-related activity includes operant activity that has been related to the PIE as a result of contingency. Contrary to the common view of reinforcers—that their primary function is to strengthen the response on which they are contingent, but that they have a secondary role as discriminative stimuli—the present view asserts that the stimulus function is all there is. No notion of "strengthening" or "reinforcement" enters the account.

One way to understand the inducing power of a PIE like food is to recognize that in nature and in the laboratory food is usually a signal that more food is forthcoming in that

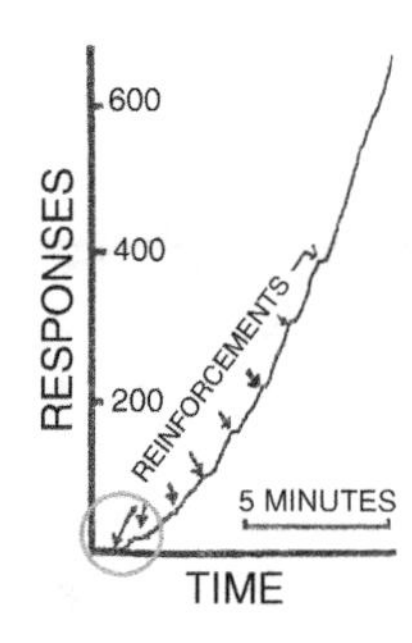

Figure 7.8 Cumulative record that Skinner (1948) presented as evidence of "reconditioning" of a response of hopping from side to side
Note: Food was delivered once a minute independently of the pigeon's behavior. Food deliveries (labeled "reinforcements") are indicated by arrows. No response immediately preceded the first and possibly the second food deliveries (circled), indicating that the result is actually an example of reinstatement.

environment for that activity. A pigeon foraging for grass seeds, on finding a seed, is likely to find more seeds if it continues searching in that area ("focal search;" Timberlake & Lucas, 1989). A person searching on the internet for a date, on finding a hopeful prospect on a web site is likely to find more such if he or she continues searching that site. For this reason, Michael Davison and I suggested that the metaphor of strengthening might be replaced by the metaphor of guidance (Davison & Baum, 2006).

An article by Gardner and Gardner (1988) argued vigorously in favor of a larger role for induction in explaining the origins of behavior, even suggesting that induction is the principal determinant of behavior. They drew this conclusion from studies of infant chimpanzees raised by humans. Induction seemed to account for a lot of the chimpanzees' behavior. But induction alone cannot account for selection among activities, because we must also account for the occurrence of the PIEs that induce the various activities. That requires us to consider the role of contingency, because contingencies govern the occurrence of PIEs. The Gardners' account provided no explanation of the flexibility of individuals' behavior, because they overlooked the importance of contingency within behavioral evolution (Baum, 1988).

Analogy to Natural Selection

As mentioned earlier, optimality by itself is not an explanation, but rather requires explanation by mechanism. The evolution of a population by natural selection offers an example. Darwinian evolution requires three ingredients: (1) variation; (2) recurrence; and (3) selection (e.g., Dawkins, 1989a; Mayr, 1970). When these are all present, evolution occurs; they are necessary and sufficient. Phenotypes within a population vary in their reproductive success—i.e., in the number of surviving offspring they produce. Phenotypes tend to recur from generation to generation—i.e., offspring tend to resemble their parents—because parents transmit genetic material to their offspring. Selection occurs because the carrying capacity of the environment sets a limit to the size of the population, with the result that many offspring fail to survive, and the phenotype with more surviving offspring increases in the population.

The evolution of behavior within a lifetime (i.e., shaping), seen as a Darwinian process, also requires variation, recurrence, and selection (Staddon & Simmelhag, 1971). Variation occurs within and between activities. Since every activity is composed of parts that are themselves activities, variation within an activity is still variation across activities, but on a smaller time scale (Baum, 2002; 2004). PIEs and PIE-related stimuli induce various activities within any situation. Thus, induction is the mechanism of recurrence, causing activities to persist through time analogously to reproduction causing phenotypes to persist in a population through time. Selection occurs because of the finiteness of time, which is the analog to the limit on population size set by the carrying capacity of the environment. When PIEs induce various activities, those activities compete, because any increase in one necessitates a decrease in others. As an example, when a pigeon is trained to peck at keys that occasionally produce food, a pigeon's key pecking is induced by food, but pecking competes for time with background activities such as grooming and resting (Baum, 2002). If the left key produces food five times more often than the right key, then the combination left-peck-plus-food induces more left-key pecking than right-key pecking—the basis of the matching relation (Herrnstein, 1961; Davison & McCarthy, 1988). Due to the upper limit on pecking set by its competition with background activities, pecking at the left key competes for time with pecking at the right key (Baum, 2002).

If we calculate the ratio of pecks left to pecks right and compare it with the ratio of food left to food right, the result may deviate from strict matching between peck ratio and food ratio across the two keys (Baum, 1974a; 1979), but the results generally approximate matching, and simple mathematical proof shows matching to be optimal (Baum, 1981b).

Thus, evolution in populations of organisms and behavioral evolution, though both may be seen as Darwinian optimizing processes, proceed by definite mechanisms. Approximation to optimality is the result, but the mechanisms explain that result.

To understand behavioral evolution fully, we need not only to understand the competing effects of PIEs, but we need to account for the occurrence of the PIEs—how they are produced by behavior. We need to include the effects of contingencies.

Contingency

The notion of reinforcement was always known to be incomplete. By itself, it included no explanation of the provenance of the behavior to be strengthened; behavior has to occur before it can be reinforced. Segal (1972) thought that perhaps induction might explain the provenance of operant behavior. Perhaps contingency selects behavior originally induced. Before we pursue this idea, we need to be clear about the meaning of "contingency."

In the quotes we saw earlier, Skinner defined contingency as contiguity—a reinforcer needed only to follow a response. Contrary to this view based on "order and proximity," a contingency is not a temporal relation. Rescorla (1968; 1988) may have been the first to point out that contiguity alone cannot suffice to specify a contingency, because contingency requires a comparison between at least two different occasions. Figure 7.9 illustrates the point with a 2-by-2 table, in which the columns show the presence and absence of Event 1, and the rows show the presence and absence of Event 2. Event 1 could be a tone or key pecking; Event 2 could be food or electric shock. The upper left cell represents the conjunction of the two events (i.e., contiguity); the lower right cell represents the conjunction of the absence of the two events. These two conjunctions (checkmarks) must both have high probability for a contingency to exist. If, for example, the probability of Event 2 were the same, regardless of the presence or absence of Event 1 (top left and right cells), then no contingency would exist. Thus, the presence of a contingency requires a comparison across two temporally separated occasions (Event 1 present and Event 1 absent). A correlation between rate of an activity and food rate would entail more than two such comparisons—for example, noting various food rates across various peck rates (the basis of a feedback function; Baum, 1973, 1989). Contrary to the idea that contingency requires only temporal conjunction, Figure 7.9 shows that accidental contingencies should be rare, because an accidental contingency would require at least two accidental conjunctions.

	E1 Present	E1 Absent
E2 Present	√	
E2 Absent		√

Figure 7.9 Why a contingency or correlation is not simply a temporal relation
Note: The 2-by-2 table shows the conjunctions possible of the presence and absence of two events, E1 and E2. A positive contingency holds between E1 and E2 only if two conjunctions occur with high probability at different times: the presence of both and the absence of both (indicated by checks). The conjunction of the two alone (contiguity) cannot suffice.

It is not that temporal relations are entirely irrelevant, but just that they are relevant in a different way from what traditional reinforcement would require. Whereas Skinner's formulation assigned to contiguity a direct role, the conception of contingency illustrated in Figure 7.9 suggests instead that the effect of contiguity is indirect. For example, inserting unsignalled delays into a contingency usually reduces the rate of the activity (e.g., key pecking) producing the PIE (e.g., food). Such delays must affect the tightness of the correlation. A measure of correlation such as the correlation coefficient (r) would decrease as average delay increased. Delays affect the clarity of a contingency like that in Figure 7.9 and affect the variance in a correlation (for further discussion, see Baum, 1973).

Avoidance

No phenomenon better illustrates the inadequacy of the molecular view of behavior than avoidance, because the whole point of avoidance is that when the response occurs, nothing follows. For example, Sidman (1953) trained rats in a procedure in which lever pressing postponed electric shocks that, in the absence of pressing, occurred at a regular rate. The higher the rate of pressing, the lower the rate of shock, and if pressing occurred at a high enough rate, the rate of shock was reduced close to zero. To try to explain the lever pressing, molecular theories resort to unseen "fear" elicited by unseen stimuli, "fear" reduction resulting from each lever press—so-called "two-factor" theory (Anger, 1963; Dinsmoor, 2001). In contrast, a molar theory of avoidance makes no reference to hidden variables, but relies on the measurable reduction in shock frequency resulting from the lever pressing (Herrnstein, 1969; Sidman, 1953; 1966). Herrnstein and Hineline (1966) tested the molar theory directly by training rats in a procedure in which pressing reduced the frequency of shock but could not reduce it to zero. A molecular explanation requires positing unseen fear reduction that depends only on the reduction in shock rate, implicitly conceding the molar explanation (Herrnstein, 1969).

In our present terms, the negative contingency between shock rate and rate of pressing results in pressing becoming a shock-related activity. Shocks then induce lever pressing. In the Herrnstein-Hineline procedure, this induction is clear, because the shock never disappears altogether, but in Sidman's procedure, we may guess that the experimental context—chamber, lever, etc.—also induces lever pressing, as discriminative stimuli or conditional inducers. Much evidence supports these ideas. Extra response-independent shocks increase pressing (Sidman, 1966; cf. reinstatement discussed earlier); extinction of Sidman avoidance is prolonged when shock rate is reduced to zero, but becomes faster on repetition (Sidman, 1966; cf. the Bullock & Smith study discussed earlier); extinction of Herrnstein-Hineline avoidance in which pressing no longer reduces shock rate depends on the amount of shock-rate reduction during training (Herrnstein & Hineline, 1966); sessions of Sidman avoidance typically begin with a "warm-up" period, during which shocks are delivered until lever pressing starts—i.e., until the shock induces lever pressing (Hineline, 1977).

Figure 7.10 illustrates why shock becomes a discriminative stimulus or a conditional inducer in avoidance procedures. This two-by-two table shows the effects of positive and negative contingencies (columns) paired with fitness-enhancing and fitness-reducing PIEs (rows). Natural selection ensures that dangerous (fitness-reducing) PIEs (e.g., injury, illness, or predators) induce defensive activities (e.g., hiding, freezing, or fleeing) that remove or mitigate the danger. (Those individuals in the population that failed to behave so reliably produced fewer surviving offspring.) In avoidance, since shock or injury is

Phylogenetically Important Events and Contingencies Determine Induction of Activities

	Positive Correlation or Contingency	Negative Correlation or Contingency
Fitness-Enhancing PIE (e.g., Food or Mate)	TARGET ACTIVITY (e.g., key pecking; AKA Positive Reinforcement)	OTHER ACTIVITY (e.g., off-key pecking; AKA Negative Punishment)
Fitness-Reducing PIE (e.g., Injury or Predator)	OTHER ACTIVITY (e.g., off-key pecking; AKA Positive Punishment)	TARGET ACTIVITY (e.g., key pecking; AKA negative reinforcement)

Figure 7.10 Different Correlations or Contingencies induce either the target activity or other-than-target activities, depending on whether the Phylogenetically Important Event (PIE) involved usually enhances or reduces fitness (reproductive success) by its presence

fitness-reducing, any activity that would avoid it will be induced by it. After avoidance training, the operant activity (e.g., lever pressing) becomes a (conditional) defensive, fitness-maintaining activity, and is induced along with other shock-related activities; the shock itself and the (dangerous) operant chamber do the inducing. This lower-right cell in the table corresponds to relations typically called "negative reinforcement."

Natural selection ensures also that fitness-enhancing PIEs (e.g., prey, shelter, or a mate) induce fitness-enhancing activities (e.g., feeding, sheltering, or courtship). (Individuals in the population that behaved so reliably left more surviving offspring.) In the upper left cell in Figure 7.10, a fitness-enhancing PIE (e.g., food) stands in a positive relation to a target (operant) activity, and training results in the target activity's induction along with other fitness-enhancing, PIE-related activities. This cell corresponds to relations typically called "positive reinforcement."

When lever pressing is food-related or when lever pressing is shock-related, the food or shock functions as a discriminative stimulus, inducing lever pressing. The food predicts more food; the shock predicts more shock. A positive correlation between food rate and press rate creates the condition for pressing to become a food-related activity. A negative correlation between shock rate and press rate creates the condition for pressing to become a shock-related activity. More precisely, the correlations create the conditions for food and shock to induce allocations including substantial amounts of time spent pressing.

In the other two cells of Figure 7.10, the target activity would either produce a fitness-reducing PIE—the cell typically called "positive punishment"—or prevent a fitness-enhancing PIE—the cell typically called "negative punishment." Either way, the target activity would reduce fitness and would be blocked from joining the other PIE-related activities. Instead, other activities, incompatible with the target activity, that would maintain fitness, are induced. Experiments with positive punishment set up a conflict, because the target activity usually produces both a fitness-enhancing PIE and a fitness-reducing PIE. For example, when operant activity (e.g., pigeons' key pecking) produces both food and electric shock, the activity usually decreases below its level in the absence of shock (Azrin & Holz, 1966; Rachlin & Herrnstein, 1969). The food induces pecking, but other activities (e.g., pecking off the key) are negatively correlated with shock rate and are induced by the shock. The result is a compromise allocation including less key pecking. We will come to a similar conclusion about negative punishment, in which target activity (e.g., pecking) cancels a fitness-enhancing PIE (food delivery), when we discuss negative automaintence. The idea that punishment induces alternative activities to the punished activity is further supported by the observation that if the situation includes another activity that is positively correlated with food and produces no shock, that activity dominates (Azrin & Holz, 1966).

Effects of Contingency

Contingency links an activity to an inducing event and changes the time allocation among activities by increasing time spent in the linked activity. The increase in time spent in an activity—say, lever pressing—when food is made contingent on it was in the past attributed to reinforcement, but our examples suggest instead that it is due to induction. The increase in pressing results from the combination of contingency and induction, because the contingency turns the pressing into a food-related (PIE-related) activity, as shown in the experiments by Bullock and Smith (Figure 7.6) and by Reid (Figure 7.7). We may summarize these ideas as follows:

1. Phylogenetically Important Events (PIEs) are unconditional inducers.
2. A stimulus correlated with a PIE becomes a conditional inducer.
3. An activity positively correlated with a fitness-enhancing PIE becomes a PIE-related conditional induced activity—usually called "operant" or "instrumental" activity.
4. An activity negatively correlated with a fitness-reducing PIE becomes a PIE-related conditional induced activity—often called "operant avoidance."
5. A PIE induces operant activity related to it.
6. A conditional inducer induces operant activity related to the PIE to which the conditional inducer is related.

The effects of contingency need include no notion of strengthening or reinforcement. Consider the example of lever pressing maintained by electric shock. Figure 7.11 shows a sample cumulative record of a squirrel monkey's lever pressing from Malagodi, Gardner, Ward, and Magyar (1981). The scallops resemble those produced by a fixed-interval schedule of food, starting with a pause and then accelerating to a higher response rate up to the end of the interval. Yet, in this record the last press in the interval produces an electric shock. The result requires, however, that the monkey be previously trained to press this same lever to avoid the same electric shock. Although in the avoidance training shock acted as an aversive stimulus or punisher, in Figure 7.11 it acts, paradoxically, to maintain the lever pressing, as if it were a reinforcer. The paradox is resolved when we recognize that the shock, as a PIE, induces the lever pressing because the prior avoidance training made lever pressing a shock-related activity. No notion of reinforcement enters in—only the effect of the contingency. The shock induces the lever pressing, the pressing produces the shock, the shock induces the pressing, and so on, in a loop.

The diagrams in Figure 7.12 show how the contingency closes a loop. The diagram on the left illustrates induction alone. The environment E produces a stimulus S (shock), and the organism O produces the induced activities B. The diagram on the right illustrates the effect of contingency. Now the environment E links the behavior B to the stimulus S, resulting in an induction-contingency loop.

The reader may already have realized that the same situation and the same diagram apply to experiments with other PIEs, such as food. For example, Figure 7.1 illustrates that, among other activities, food induces pecking in pigeons. When a contingency arranges also that pecking produces food, we have the same sort of loop as shown in

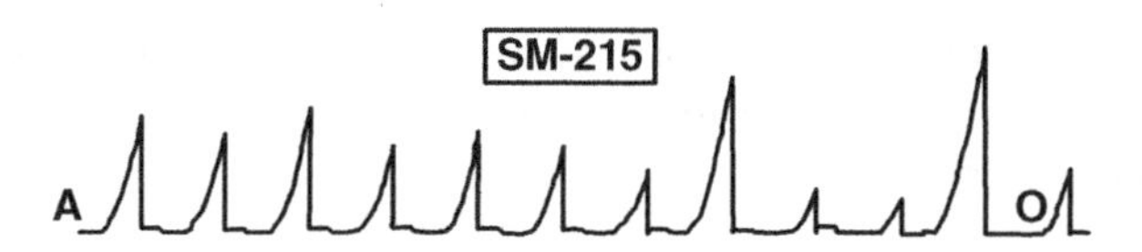

Figure 7.11 Cumulative record of a squirrel monkey pressing a lever that produced electric shock at the end of the fixed interval

Note: Even though the monkey had previously been trained to avoid the very same shock, it now continues to press when pressing produces the shock. This seemingly paradoxical result is explained by the molar view of behavior as an example of induction. Reprinted from Malagodi, Gardner, Ward, and Magyar (1981).

Figure 7.12, except that the stimulus S is food and the behavior B is pecking. Food induces the activity, and the (operant) activity produces the food, which induces the activity, and so on, as shown by the results of Bullock and Smith (Figure 7.6) and Reid's reinstatement effect (Figure 7.7).

Figure 7.13 shows results from an experiment that illustrates the effects of positive and negative contingencies (Baum, 1981a). Pigeons, trained to peck a response key that occasionally produced food, were exposed to daily sessions in which the payoff schedule began with a positive correlation between pecking and food rate. It was a sort of variable-interval schedule, but required only that a peck occur anywhere in a programmed interval for food to be delivered at the end of that interval (technically, a conjunctive variable-time 10-s fixed-ratio 1 schedule). At an unpredictable point in the session, the correlation switched to negative; now, a peck during a programmed interval canceled the food delivery at the end of the interval. Finally, at an unpredictable point, the correlation reverted to positive. Figure 7.13 shows four representative cumulative records of four pigeons from this experiment. Initially, pecking occurred at a moderate rate, then decreased when the correlation switched to negative (first vertical line), and then increased again when the correlation reverted to positive (second vertical line). The effectiveness of the negative contingency varied across the pigeons; it suppressed key pecking completely in Pigeon 57, and relatively little in Pigeon 61. Following our present line, these results suggest that when the correlation was positive, the food induced key pecking at a higher rate than when the correlation was negative. In accord with the results shown in Figure 7.1, however, we expect that the food continues to induce pecking. As we expect from the upper-right cell of Figure 7.10, in the face of a negative contingency, pigeons typically peck off the key, sometimes right next to it. The negative correlation between food and pecking on the key constitutes a positive correlation between food and pecking off the key; the allocation between the two activities shifts in the three phases shown in Figure 7.13.

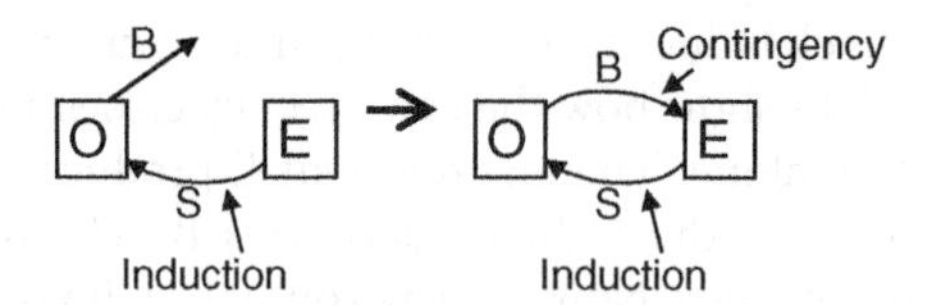

Figure 7.12 How contingency completes a loop in which an operant activity (B) produces an inducing event (S), which in turn induces more of the activity

Note: O stands for organism. E stands for environment. Left: induction alone. Right: the contingency closes the loop. Induction occurs because the operant activity is or becomes related to the inducer (S; a PIE).

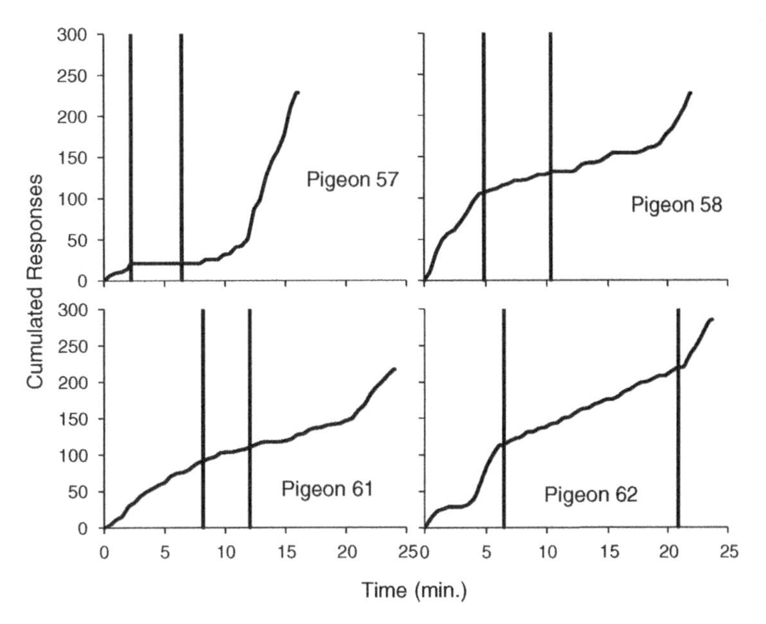

Figure 7.13 Results from an experiment showing discrimination of correlation
Note: Four cumulative records of complete sessions from four pigeons are shown. At the beginning of the session, the correlation between pecking and feeding is positive; if a peck occurred anywhere in the scheduled time interval, food was delivered at the end of the interval. Following the first vertical line, the correlation switched to negative; a peck during the interval canceled food at the end of the interval. Peck rate fell. Following the second vertical line, the correlation reverted to positive, and peck rate rose again.

The results in the middle phase of the records in Figure 7.13 resemble negative automaintenance (Williams & Williams, 1969). In autoshaping, a pigeon is repeatedly presented with a brief light on a key, which is followed by food. Sooner or later, the light comes to induce key pecking, and pecking at the lit key becomes persistent; autoshaping becomes automaintenance. In negative automaintenance, pecks at the key cancel the delivery of food. This negative contingency causes a reduction in key pecking, as in Figure 7.13, even if it doesn't eliminate it altogether (Sanabria, Sitomer, & Killeen, 2006). Presumably, a compromise occurs between pecking on the key and pecking off the key—the allocation of time between pecking on the key and pecking off the key shifts back and forth. Possibly, the key light paired with food induces pecking on the key, whereas the food induces pecking off the key.

A Test

To test the molar view of contingency or correlation, we may apply it to explaining a puzzle. W. K. Estes (1943; 1948) reported two experiments in which he pre-trained rats in two conditions: (a) a tone was paired with food with no lever present; and (b) lever pressing was trained by making food contingent on pressing. The order of the two conditions made no difference to the results. The key point was that the tone and lever never occurred together. Estes tested the effects of the tone by presenting it while the lever pressing was undergoing extinction (no food). Figure 7.14 shows a typical result. Pressing decreases across 5-min intervals, but each time the tone occurred, pressing increased.

This finding presented a problem for Estes's molecular view—that is, for a theory relying on contiguity between discrete events. The difficulty was that, because the tone and the lever had never occurred together, no associative mechanism based on contiguity could explain the effect of the tone on the pressing. Moreover, the food pellet couldn't mediate between the tone and the lever, because the food had never preceded the pressing, only followed it. Estes "explanation" was to make up what he called a "conditioned anticipatory state" (CAS). He proposed that pairing the tone with food resulted in the tone's eliciting the CAS and that the CAS then affected lever pressing during the extinction test. This idea, however, begs the question, "Why would the CAS increase pressing?" It too would never have preceded pressing. The "explanation" only makes matters worse, because, whereas the puzzle started with accounting for the effect of the tone, now it has shifted to an unobserved hypothetical entity, the CAS. The muddle illustrates well how the molecular view of behavior leads to hypothetical entities and magical thinking. The molecular view can explain Estes's result, but only by positing unseen mediators (Trapold & Overmier, 1972). (See Baum, 2002, for a fuller discussion.)

The molar paradigm (e.g., Baum, 2002; 2004) offers a straightforward explanation of Estes's result. First, pairing the tone with the food makes the tone an inducing stimulus (a conditional inducer). This means that the tone will induce any food-related activity. Second, making the food contingent on lever pressing—i.e., correlating pressing with food—makes pressing a food-related activity. Thus, when the tone is played in the presence of the lever, it induces lever pressing. When we escape from a focus on contiguity, the result seems obvious.

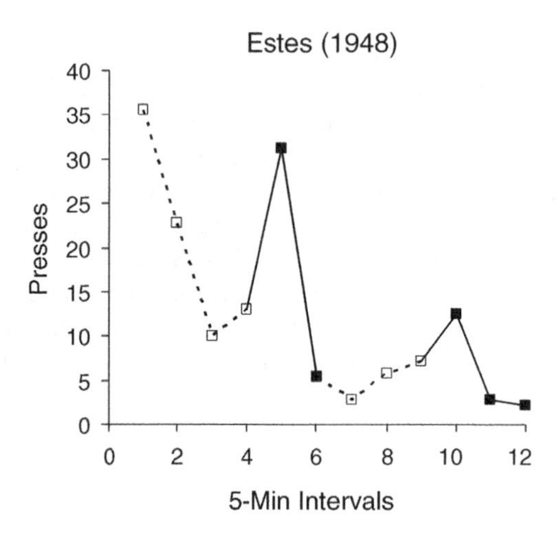

Figure 7.14 Results from an experiment in which lever pressing undergoing extinction was enhanced by presentation of a tone that had previously been paired with the food that had been used in training the lever pressing

Note: Because the tone and lever had never occurred together before, the molecular view had difficulty explaining the effect of the tone on the pressing, but the molar view explains it as induction of pressing as a food-related activity. The data were published by Estes (1948).

Constraint and Connection

The effects of contingency go beyond just "closing the loop." A contingency has two effects: (a) it creates or causes a correlation between the operant activity and the contingent event—the equivalent of what an economist would call a *constraint*; (b) as it occurs repeatedly, it soon causes the operant activity to become related to the contingent (inducing) event—it serves to *connect* the two.

Contingency ensures the increase in the operant activity, but it also constrains the increase by constraining the possible outcomes (allocations). Figure 7.15 illustrates these effects. The vertical axis shows frequency of an inducing event represented as time spent in a fitness-enhancing activity, such as eating or drinking. The horizontal axis shows time spent in an induced or operant activity, such as lever pressing or wheel running. Figure 7.15 diagrams situations studied by Premack (1971) and by Allison, Miller, and Wozny (1979); it resembles also a diagram by Staddon (1983; Figure 7.2). In a baseline condition with no contingency, a lot of time is spent in the inducing activity and relatively little time is spent in the to-be-operant activity. The point labeled "baseline" indicates this allocation. When a contingency is introduced, represented by the solid diagonal line, which indicates a contingency such as in a ratio schedule, a given duration of the operant activity allows a given duration of the inducing activity (or a certain amount of the PIE). The contingency constrains the possible allocations between time in the contingent activity (eating or drinking) and time in the operant activity (pressing or running). Whatever allocation occurs must lie on the line; allocations off the line are no longer

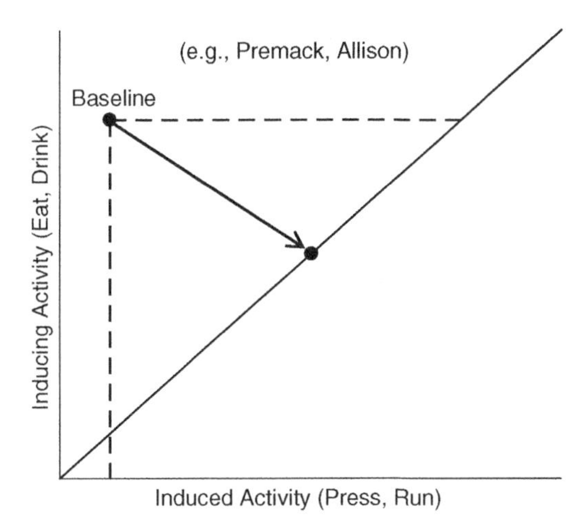

Figure 7.15 Effects of contingency
Note: In a baseline condition with no contingency, little of the to-be-operant activity (e.g., lever pressing or wheel running; the to-be-induced activity) occurs, while a lot of the to-be-contingent PIE-related activity (e.g., eating or drinking; the inducing activity) occurs. This is the baseline allocation (upper left point and broken lines). After the PIE (food or water) is made contingent on the operant activity, whatever allocation occurs must lie on the solid line. The arrow points to the new allocation. Typically, the new allocation includes a large increase in the operant activity over baseline. Thus, contingency has two effects: a) constraining possible allocations; and b) making the operant behavior PIE-related so it is induced in a large quantity.

possible. The point on the line illustrates the usual result: an allocation that includes more operant activity than in baseline and less of the PIE than in baseline. The increase in operant activity would have been called a reinforcement effect in the past, but here it appears as an outcome of the combination of induction with the constraint of the contingency line [also known as the *feedback function* (Baum, 1973; 1989; 1992)]. The decrease in the PIE might have been called punishment of eating or drinking, but here it too is an outcome of the constraint imposed by the contingency (Premack, 1965).

Another way to express the effects in Figure 7.15 might be to say that, by constraining the allocations possible, the contingency "adds value" to the induced operant activity. Once the operant activity becomes PIE-related, the operant activity and the PIE become parts of a "package." For example, pecking and eating become parts of an allocation or package that might be called "feeding," and the food or eating may be said to lend value to the package of feeding. If lever pressing is required for a rat to run in a wheel, pressing becomes part of running (Belke, 1997; Belke & Belliveau, 2001). If lever pressing is required for a mouse to obtain nest material, pressing becomes part of nest building (Roper, 1973). Figure 7.15 indicates that this doesn't occur without some cost, too, because the operant activity doesn't increase enough to bring the level of eating to its level in baseline; some eating is sacrificed in favor of a lower level of the operant activity. In an everyday example, suppose Tom and his friends watch football every Sunday and usually eat pizza while watching. Referring to Figure 7.9, we may say that watching football (E1) is correlated with eating pizza (E2), provided that the highest frequencies are the conjunctions in the checked boxes, and the conjunctions in the empty boxes are relatively rare. Rather than appeal to reinforcement or strengthening, we may say that watching football induces eating pizza or that the two activities form a package with higher value than either by itself. We don't need to bring in the concept of "value" at all, because all it means is that the PIE induces PIE-related activities, but this interpretation of "adding value" serves to connect the present discussion with behavioral economics. Another possible connection would be to look at demand as induction of activity (e.g., work or shopping) by the good demanded and elasticity as change in induction with change in the payoff schedule or feedback function. However, a full quantitative account of the allocations that actually occur is beyond the scope of this article. [See Allison (1983), Rachlin, Green, Kagel, and Battalio (1976), Rachlin, Battalio, Kagel, and Green (1981), and Baum (1981b) for examples.] Part of a theory may be had by quantifying contingency with the concept of a feedback function, because feedback functions allow more precise specification of the varieties of contingency.

Correlations and Feedback Functions

Every contingency or correlation specifies a feedback function (Baum, 1973; 1989; 1992). A feedback function describes the dependence of outcome rate (say, rate of food delivery) and operant activity (say, rate of pecking). The constraint line in Figure 7.15 specifies a feedback function for a ratio schedule, in which the rate of the contingent inducer is directly proportional to the rate of the induced activity. The feedback function for a variable-interval schedule is more complicated, because it must approach the programmed outcome rate as an asymptote. Figure 7.16 shows an example. The upper curve has three properties: (a) it passes through the origin, because if no operant activity occurs, no food can be delivered; (b) it approaches an asymptote; and (c) as operant activity approaches zero, the slope of the curve approaches the reciprocal of the parameter *a* in the equation shown (*1/a*; Baum, 1992). It constrains possible performance, because whatever the rate of operant activity, the outcome (inducer) rate must lie on the curve. The lower curve suggests a frequency distribution of operant activity on various occasions. Although average operant rate and average inducer rate would correspond to a point on the feedback curve, the distribution indicates that operant activity should vary around a mode and would not necessarily approach the maximum activity rate possible.

Figure 7.17 shows some unpublished data, gathered by John Van Syckel, a student at the University of New Hampshire. Several sessions of variable-interval responding by a rat were analyzed by examining successive 2-min time samples and counting the number of lever presses and food deliveries in each one. All time samples were combined according the number of presses, and the press rate and food rate calculated for each number of presses. These are represented by the plotted points. The same equation as in Figure 7.15 is fitted to the points; the parameter *a* was close to 1.0. The frequency distribution of press rates is shown by the curve without symbols (right-hand vertical axis). Some 2-min samples contained no presses, a few contained only one (0.5 press/min), and a mode appears at about 5 press/min. Press rates above about 15 press/min (30 presses in 2 min) were rare. Thus, the combination of induction of lever pressing by occasional food together with the feedback function results in the stable performance shown.

Although the parameter *a* was close to 1.0 in Figure 7.17, it usually exceeds 1.0, ranging as high as 10 or more when other such data sets are fitted to the same equation (Baum, 1992). This presents a puzzle, because one might suppose that as response rate dropped to extremely low levels, a food delivery would have set up prior to each response, and each response would deliver food, with the result that the feedback function would approximate that for a fixed-ratio 1 schedule near the origin. A possible explanation as to why the slope (*1/a*) continues to fall short of 1.0 near the origin might be that when food deliveries become rare, a sort of reinstatement effect occurs, and each delivery induces a

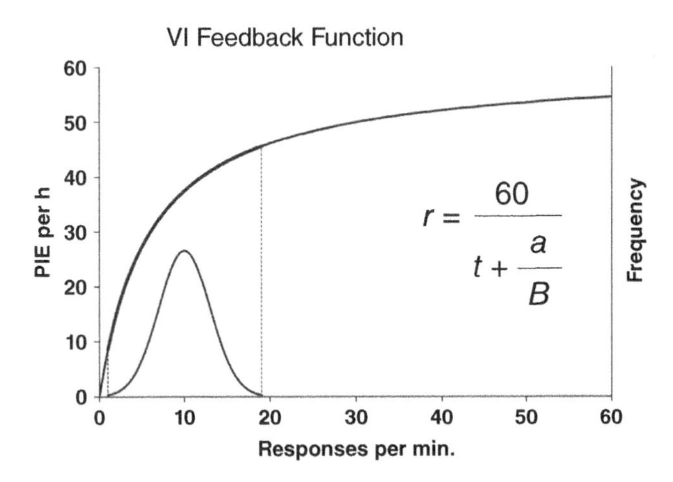

Figure 7.16 Example of a feedback function for a variable-interval schedule
Note: The upper curve, the feedback function, passes through the origin and approaches an asymptote (60 PIEs per h; a VI 60s). Its equation appears at the lower right. The average interval *t* equals 1.0. The parameter *a* equals 6.0. Response rate is expected to vary from time to time, as shown by the distribution below the feedback function (frequency is represented on the right-hand vertical axis).

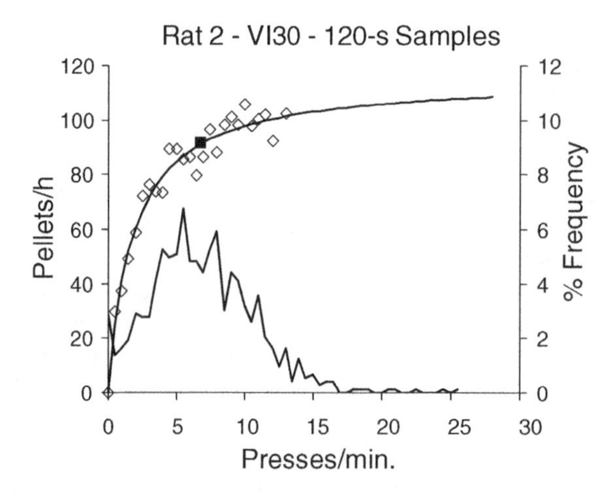

Figure 7.17 Example of an empirical feedback function for a variable-interval schedule
Note: Successive 120-s time windows were evaluated for number of lever presses and number of food deliveries. The food rate and press rate were calculated for each number of presses per 120 s. The unfilled diamonds show the food rates. The feedback equation from Figure 7.15 was fitted to these points ($t = 0.52$; $a = 0.945$). The frequency distribution (right-hand vertical axis) below shows the percent frequencies of the various response rates. The filled square shows the average response rate.

burst of pressing. Since no delivery has had a chance to set up, the burst usually fails to produce any food and thus insures several presses for each food delivery.

The relevance of feedback functions has been challenged by two sorts of experiment. In one type, a short-term contingency is pitted against a long-term contingency, and the short-term contingency is shown to govern performance to a greater extent than the long-term contingency (e.g., Thomas, 1981; Vaughan & Miller, 1984). In these experiments, the key pecking or lever pressing produces food in the short-term contingency, but cancels food in the long-term contingency. Responding is sub-optimal, because the food rate would be higher if no responding occurred. That responding occurs in these experiments supports the present view. The positively correlated food induces the food-related activity (pressing or pecking); the other, negatively correlated food, when it occurs, would simply contribute to inducing the operant activity.

These experiments may be compared to observations such as negative automaintenance (Williams & Williams, 1969) or what Herrnstein and Loveland (1972) called "food avoidance." Even though key pecking cancels the food delivery, still the pigeon pecks, because the food, when it occurs, induces pecking (unconditionally; Figure 7.1). Sanabria, Sitomer, and Killeen (2006) showed, however, that under some conditions the negative contingency is highly effective, reducing the rate of pecking to low levels despite the continued occurrence of food. As discussed earlier in connection with Figure 7.13, the negative contingency between key pecking and food implies also a positive contingency between other activities and food, particularly pecking off the key. Similar considerations explain contrafreeloading—the observation that a pigeon or rat will peck at a key or press a lever even if a supply of the same food produced by the key or lever is freely available (Neuringer, 1969; 1970). The pecking or pressing apparently is induced by the food produced, even though other food is available. Even with no pecking key available, Palya and

Zacny (1980) found that untrained pigeons fed at a certain time of day would peck just about anywhere (any "spot") around that time of day.

The other challenge to the relevance of feedback functions is presented by experiments in which the feedback function is changed with no concomitant change in behavior. For example, Ettinger, Reid, and Staddon (1987) studied rats' lever pressing that produced food on interlocking schedules—schedules in which both time and pressing interchangeably advance the schedule toward the availability of food. At low response rates, the schedule resembles an interval schedule, because the schedule advances mainly with the passage of time, whereas at high response rates, the schedule resembles a ratio schedule, because the schedule advances mainly due to responding. Ettinger et al. varied the interlocking schedule and found that average response rate decreased in a linear fashion with increasing food rate, but that variation in the schedule had no effect on this relation. They concluded that the feedback function was irrelevant. The schedules they chose, however, were all functionally equivalent to fixed-ratio schedules, and the rats responded exactly as they would on fixed-ratio schedules. In fixed-ratio performance, induced behavior other than lever-pressing tends to be confined to a period immediately following food, as in Figure 7.1; once pressing begins, it tends to proceed uninterrupted until food again occurs. The post-food period increases as the ratio increases, thus tending to conserve the relative proportions of pressing and other induced activities. The slope of the linear feedback function determines the post-food period, but response rate on ratio schedules is otherwise insensitive to it (Baum, 1993a; Felton & Lyon, 1966). The decrease in average response rate observed by Ettinger et al. occurred for the same reason it occurs with ratio schedules: because the post-food period, even though smaller for smaller ratios, was an increasing proportion of the inter-food interval as the effective ratio decreased. Thus, the conclusion that feedback functions are irrelevant was unwarranted.

When we try to understand basic phenomena, such as the difference between ratio and interval schedules, feedback functions prove indispensible. The two different feedback functions—linear for ratio schedules (Figure 7.15) and curvilinear for interval schedules (Figures 7.16 and 7.17)—explain the difference in response rate on the two types of schedule (Baum, 1981b). On ratio schedules, the contingency loop shown in Figure 7.12 includes only positive feedback and drives response rate toward the maximum possible under the constraints of the situation. This maximal response rate is necessarily insensitive to variation in ratio or food rate. On interval schedules, where food rate levels off, food induces a more moderate response rate (Figure 7.17). The difference in feedback function is important for understanding both these laboratory situations and also everyday situations in which people deal with contingencies like ratio schedules, in which their own behavior alone matters to production, versus contingencies, like interval schedules, in which other factors, out of their control, partially determine production.

Explanatory Power

The conceptual power of this framework—allocation, induction and contingency—far exceeds that of the contiguity-based concept of reinforcement. Whether we redefine the term and call it "reinforcement," or whether we call the process something else ("inducement"?), it explains a large range of phenomena. We have seen that it explains standard results such as operant and respondent conditioning, operant-respondent interactions, and adjunctive behavior. We saw that it explains avoidance and shock-maintained responding. It can explain the effects of non-contingent events (e.g., "non-contingent reinforcement," an oxymoron). Let us conclude with a few examples from the study of choice, the study of stimulus control, and observations of everyday life.

Dynamics of Choice

In a typical choice experiment, a pigeon pecks at two response keys, each of which occasionally and unpredictably operates a food dispenser when pecked, based on the passage of time (variable-interval schedules). Choice or preference is measured as the logarithm of the ratio of pecks at one key to pecks at the other. For looking at putative reinforcement, we examine log ratio of pecks at the just-productive key (left or right) to pecks at the other key (right or left). With the right procedures, analyzing preference following food allows us to separate any strengthening effect from the inducing effect of food (e.g., Cowie, Davison, & Elliffe, 2011). For example, a striking result shows that food and stimuli predicting food induce activity other than the just-productive activity (Krägeloh, Davison, and Elliffe, 2005; Davison & Baum, 2006; 2010; Boutros, Davison, & Elliffe, 2011).

A full understanding of choice requires analysis at various time scales, from extremely small—focusing on local relations—to extremely large—focusing on extended relations (Baum, 2010). Much research on choice concentrated on extended relations by aggregating data across many sessions (Baum, 1979; Herrnstein, 1961). Some research has examined choice in a more local time frame by aggregating data from one food delivery to another (Aparicio & Baum, 2006; 2009; Baum & Davison, 2004; Davison & Baum, 2000; Rodewald, Hughes, & Pitts, 2010). A still more local analysis examines changes in choice within inter-food intervals, from immediately following food delivery until the next food delivery. A common result is a pulse of preference immediately following food in favor of the just-productive alternative (e.g., Aparicio & Baum, 2009; Davison & Baum, 2006; 2010). After a high initial preference, choice typically falls toward indifference, although it may never actually reach indifference. These preference pulses seem to reflect local induction of responding at the just-productive alternative. A still more local analysis looks at the switches between alternatives or the alternating visits at the two alternatives (Aparicio & Baum, 2006, 2009; Baum & Davison, 2004). A long visit to the just-productive alternative follows food produced by responding at that alternative, and this too appears to result from local induction, assuming, for example, that pecking left plus food induces pecking at the left key whereas pecking right plus food induces pecking at the right key. Since a previous paper showed that the preference pulses are derivable from the visits, the long visit may reflect induction more directly (Baum, 2010).

Differential Outcomes Effect

In the differential-outcomes effect, a discrimination is enhanced by arranging that the response alternatives produce different outcomes (Urcuioli, 2005). In a discrete-trials procedure, for example, rats are presented on some trials with one auditory stimulus (S1) and on other trials with another auditory stimulus (S2). In the presence of S1, a press on the left lever (A1) produces food (X1), whereas in the presence of S2, a press on the right lever (A2) produces liquid sucrose (X2). The discrimination that forms—pressing the left lever in the presence of S1 and pressing the right lever in the presence of S2—is greater than when both actions A1 and A2 produce the same outcome. The result is a challenge for the molecular paradigm because, focusing on contiguity, it sees the effect of the different outcomes as an effect of a future event on present behavior. The "solution" to this problem has been to bring the future into the present by positing unseen "representations" or "expectancies" generated by S1 and S2 and supposing that those invisible events determine A1 and A2. In the present molar view, we may see the result as an effect of differential induction by the different outcomes (Ostlund & Balleine, 2007). The combination of S1 and X1 makes S1 a conditional inducer of X1-related activities. The contingency between A1 and X1 makes A1 an X1-related activity. As a result, S1 induces A1. Similarly, as a result of the contingency between S2 and X2 and the contingency between A2 and X2, S2 induces A2, as an X2-related activity. In contrast, when both A1 and A2 produce the same outcome X, the only basis for discrimination is the relation of S1, A1, and X in contrast with the relation of S2, A2, and X. A1 is induced by the combination S1-plus-X, and A2 is induced by the combination S2-plus-X. Since S1 and S2 are both conditional inducers related to X, S1 would tend to induce A2 to some extent, and S2 would tend to induce A1 to some extent, thereby reducing discrimination. These relations occur, not just on any one trial, but as a pattern across many trials. Since no need arises to posit any invisible events, the explanation exceeds the molecular explanation on the grounds of plausibility and elegance.

Cultural Transmission

The present view offers a good account of cultural transmission. Looking on cultural evolution as a Darwinian process, we see that the recurrence of a practice from generation to generation occurs partly by imitation and partly by rules (Baum, 1995b; 2000; 2005; Boyd & Richerson, 1985; Richerson & Boyd, 2005). A pigeon in a chamber where food occasionally occurs sees another pigeon pecking at a Ping-Pong ball and subsequently pecks at a similar ball (Epstein, 1984). The model pigeon's pecking induces similar behavior in the observer pigeon. When a child imitates a parent, the parent's behavior induces behavior in the child, and if the child's behavior resembles the parent's, we say the child imitated the parent. As we grow up, other adults serve as models, particularly those with the trappings of success—teachers, coaches, celebrities, and so forth—and their behavior induces similar behavior in their imitators (Boyd & Richerson, 1985; Richerson & Boyd, 2005). Crucial to transmission, however, are the contingencies into which the induced behavior enters following imitation. When it occurs, is it correlated with resources, approval, or opportunities for mating? If so, it will now be induced by those PIEs, no longer requiring a model. Rules, which are discriminative stimuli generated by a speaker, induce behavior in a listener, and the induced behavior, if it persists, enters into correlations with both short-term PIEs offered by the speaker and, sometimes, long-term PIEs in long-term relations with the (now no longer) rule-governed behavior (Baum, 1995b; 2000; 2005).

Working for Wages

Finally, let us consider the everyday example of working for wages. Zack has a job as a dishwasher in a restaurant and works every week from Tuesday to Saturday for 8 hr a day. At the end of his shift on Saturday, he receives an envelope full of money. Explaining his persistence in this activity presents a problem for the molecular paradigm, because the money on Saturday occurs at such a long delay after the work on Tuesday or even the work on Saturday. No one gives Zack money on any other occasion—say, after each dish washed. The molecular paradigm's "solution" to this has been to posit hidden reinforcers that strengthen the behavior frequently during his work. Rather than having to resort to such invisible causes, we may turn to the molar view and assert, with common sense, that the money itself maintains the dishwashing. We may look upon the situation as similar to a rat pressing a lever, but scaled up, so to speak. The money induces dishwashing, not just in one week, but as a pattern of work across many weeks. We explain Zack's first week on the job a bit differently, because he brings to the situation a pattern of showing up daily from having attended school, which is induced by the promises of his boss. The pattern of work must also be seen in the larger context of Zack's life activities, because the money is a means to ends, not an end in itself; it permits other activities, such as eating and socializing (Baum, 1995a; 2002).

Omissions and Inadequacies

As I said at the outset, to cover all the phenomena known to behavior analysts in this one article would be impossible. The article aims to set out a conceptual framework that draws those phenomena together in a way not achieved by the traditional framework.

Perhaps the most significant omission here is a full treatment of the effects of context or what is traditionally called stimulus control. In the molar paradigm, context needs to be treated similarly to behavior—as extended in time and occurring on different time scales. In a simple discrimination, the inducing stimuli—say, a red light and a green light—may be seen in a relatively short time span. In matching to sample, however, where a sample precedes the exposure of the choice alternatives, the time frame is larger, because it has to include the sample plus the correct alternative (i.e., parts). More complex arrangements require still longer time spans. For example, experiments with relational contexts inspired by Relational Frame Theory require long time spans to specify the complex context for a simple activity like a child's pointing at a card (e.g., Berens & Hayes, 2007). In everyday life, the contexts for behavior may extend over weeks, months, or years. A farmer might find over a period of several years that planting corn is more profitable than planting alfalfa, even though planting alfalfa is more profitable in some years. I might require many interactions with another person over a long time before I conclude he or she is trustworthy.

I have said less than I would have liked about the traditional metaphor of *strength* as it is used in talking about response strength, bond strength, or associative strength. One virtue of the molar paradigm is that it permits accounts of behavior that omit the strength metaphor, substituting allocation and guidance instead (Davison & Baum, 2006; 2010). The concept of extinction of behavior has been closely tied to the notion of response strength. A future paper will address the topic of extinction within the molar paradigm.

I left out verbal behavior. One may find some treatment of it in my book, *Understanding Behaviorism* (Baum, 2005). It is a big topic that will need to be addressed in the future, but the molar paradigm lends itself to looking at verbal behavior in a variety of time-frames. The utterance stands as a useful concept, but variable parts occur within similar utterances that have different effects on listeners (e.g., "Let me go" versus "Never let me go.") and are induced by different contexts. Relatively local parts like a plural ending or a possessive suffix also are induced by different contexts. Utterances, however, are parts of more extended activities, such as conversations. Usually conversations are parts of still more extended activities, like building a relationship or persuading someone to part with money.

I have said little in this paper about rule-governed behavior, which is particularly important to understanding culture. I discussed it in earlier papers (Baum, 1995b, 2000).

The molar paradigm is fruitful for discussing rules in relation to short-term and long-term relations for groups and for individuals within groups.

The list of omissions is long, no doubt, but the reader who is interested in pursuing this conceptual framework will find many ways to apply it.

Conclusions

Taken together, the concepts of allocation, induction, and correlation (or contingency) account for the broad range of phenomena that have been recognized by behavior analysis since the 1950s. They provide a more elegant and plausible account than the older notions of reinforcement, strengthening, and elicitation. They render those concepts superfluous.

Induction not only solves the problem of provenance; induction accounts for the stimulus effects of PIEs and PIE-related stimuli. Induction also explains the effects of contingencies, because a contingency establishes a correlation that makes the operant activity into a PIE-related activity. Taken together, contingency and induction explain allocation or choice. They also explain change in allocation—that is, changes in behavior. None of this requires the concept of reinforcement as strengthening.

The popular formulation of operant behavior as "Antecedent-Behavior-Consequence" (A-B-C) may be reconciled with this conclusion. When "consequences" increase an activity, they don't strengthen it, but they induce more of the same activity, just as "antecedents" do. Indeed, the "consequence" is the antecedent of the behavior that follows, as shown in Figures 7.6 and 7.7. When "consequences" decrease an activity, they don't weaken it, but induce other behavior incompatible with it (Figure 7.10).

The formulation sketched here stems from a molar view of behavior that focuses on time spent in activities instead of discrete responses (e.g., Baum, 2002; 2004; Baum & Rachlin, 1969). Criticism of the molecular view, which is based on discrete responses and temporal contiguity as central concepts, may be found in earlier papers (Baum, 1973; 2001; 2002; 2004). The present discussion shows the superiority of the molar view in helping behavior analysts to rethink their central concepts and to better explain behavioral phenomena both in the laboratory and in everyday life.

8

Driven by Consequences: The Multiscale Molar View of Choice

Foreword

Although this paper overlaps with some others in this collection, it is the only one that presents a molar perspective on self-control and impulsiveness—long-term control and short-term control. It begins by reviewing choice in standard stationary conditions as a lead-up to self-control. It contrasts the molar perspective with the common molecular-inspired conception based on delay and discrete events. Experiments on self-control overwhelmingly are conceived in the molecular view, and most study humans by offering series of binary choices asking the subjects to guess how they would choose between two offers, usually an immediate amount and a delayed larger amount—for example, $100 now versus $200 in a month. All of this research depends on people being able to accurately predict how they would choose if these alternatives were actual instead of hypothetical. This assumption raises a problem because research indicates that people are in fact poor at predicting or accounting for their own behavior (e.g. Nisbett & Wilson, 1977). Comparisons across groups—e.g. smokers versus non-smokers—however, have demonstrated differences as to what people say they would choose. In this paper I suggest that therapeutic intervention to discourage bad habits and encourage good habits would be facilitated by adopting a molar perspective. The key is to arrange conditions that bring behavior into contact with long-term relations by increasing the timeframe within which outcomes are assessed.

Abstract

In the Multiscale Molar View of behavior, all behavior is seen as choice and is measured as time allocation. Because time is limited, activities compete for the limited time available. When Phylogenetically Important Events (PIE) that ultimately affect fitness and ontogenetic proxies of these PIEs occur as consequences of an activity, they drive time spent in that activity. Time allocation is studied in the laboratory with concurrent payoff schedules, in which two or more schedules operate simultaneously. The Generalized Matching Law describes choice in relation to relative consequences. It has been verified for food and other PIEs and for pairs of variable-interval schedules and variable-interval schedules paired with variable-ratio schedules. Because behavior produces consequences in the environment and those consequences in turn affect behavior, the environmental feedback functions and behavioral functional relations may be characterized as a feedback system. When different activities produce different consequences, choice depends also on the substitutability of the consequences. When consequences are perfectly substitutable, exclusive preference may occur, but when they are imperfectly substitutable, partial preferences may occur. Choice may become a dilemma pitting impulsivity against self-control when consequences are not stationary with respect to time. Evaluated in short timeframes, an activity may be strongly induced by its consequences, but evaluated in long timeframes, its consequences may be extremely negative; such an activity (e.g. using cocaine or lying) is a bad habit. A good habit (e.g. tooth brushing or helping others) presents the opposite conflict: bad consequences in short timeframes and positive consequences in long timeframes. Research on choice between good and bad habits may reveal factors that increase time spent in good habits relative to time spent in bad habits. The Multiscale Molar View helps to clarify various complexities that underlie choice viewed as time allocation.

Keywords: molar view, choice, Phylogenetically Important Event, feedback function, substitutability, partial preferences, impulsivity, self-control, bad habit, good habit

Source: Originally published in *Managerial and Decision Economics*, 37 (2016), pp. 239–248. Reproduced with permission of John Wiley & Sons. The author thanks Howard Rachlin for many helpful comments.

All behavior entails choice. Whatever an organism's situation, more than one activity is always possible. Even in the laboratory, one cannot create a situation so impoverished that only one activity is possible (Herrnstein, 1970). Moreover, in any significant period of time, several activities occur actually. Thus, choice may be understood as an allocation of behavior among several or many activities. This approach to studying behavior is the Multiscale Molar View of behavior (Baum, 1973; 1981b; 1989; 2002; 2004; 2012a; 2013).

The Multiscale Molar View contrasts with the traditional view, which takes behavior to consist of discrete responses that are strengthened by an immediately following reinforcer. The traditional view, though useful in its time, is implausible and unwieldy, not only when discussing everyday behavior, but even when studying behavior in the laboratory (Baum, 2012a; 2013). For example, how would one identify a discrete response for an activity like watching television or a rat's licking, biting, and chewing of a response lever (Baum, 1976)? How would one plausibly explain that a person works for a salary if reinforcers must be immediate? Instead, the Multiscale Molar View takes behavior to consist of temporally extended activities that are selected by their covariance with temporally extended consequences and induced or driven by those consequences (Baum, 2012a; 2013). In this view, choice—the allocation of behavior among activities—is fundamental.

A simplification that has proven successful takes the allocation of behavior to be the allocation of time among activities (Baum, 1973; 2010; 2012b; Baum & Rachlin, 1969). For example, one might represent a person's allocation of time among life activities as in Figure 8.1, which shows time allocation among four major activities across several weekdays. An adult with a family spends time in activities that maintain health (e.g., exercise, eating, and sleeping), activities that gain resources (e.g., working and shopping), activities

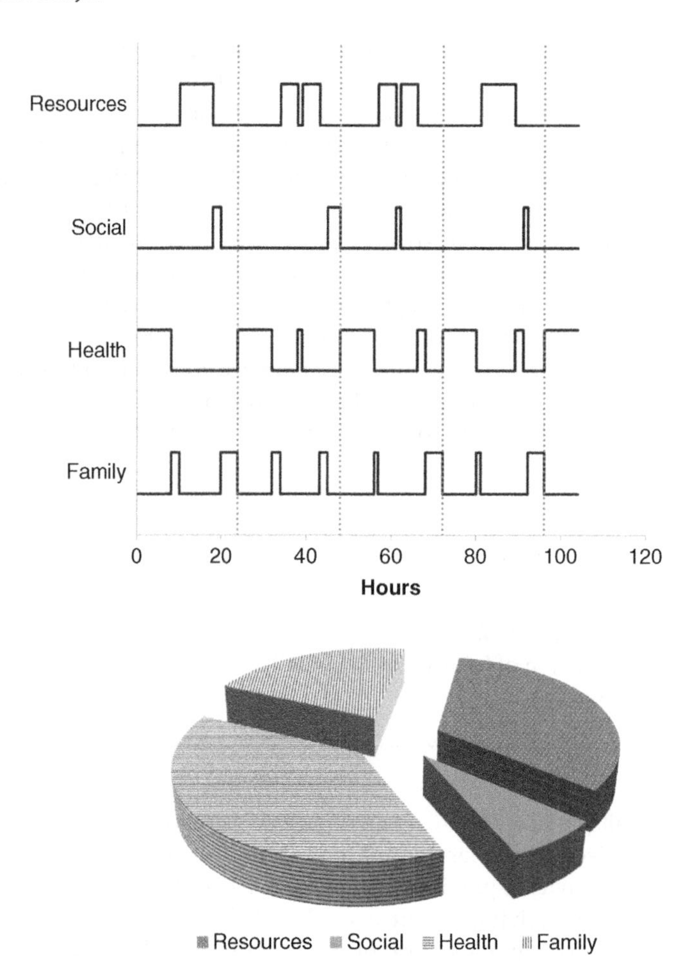

Figure 8.1 Hypothetical time allocation among 4 activities: Gaining Resources, Socializing, Maintaining Health, and Engaging with Family
Note: Top: time lines show episodes (high portions) of each activity. Dotted lines mark every 24 hours. Bottom: summary of time spent during the period of observation.

that maintain relationships with others (e.g., socializing and chatting), and activities that promote reproduction (e.g., sexual and other interacting with one's spouse and caring for children). The top panel in Figure 8.1 shows the activities occurring in episodes through time. The bottom panel shows a summary of time spent in the activities during the span shown in the top panel.

Since time is limited, because a day contains only 24 hours or a period of measurement is of a definite length, time allocation like that in Figure 8.1 implies that activities compete for time. If one activity increases, others must decrease. In everyday life, the competition leads to tension among activities and to dynamics in which adjustments occur across spans of time. For example, time management becomes an important skill, and "work-life balance" becomes an important issue. A songbird's time allocates dynamically among foraging for prey and nest material, protecting a territory and a mate, feeding nestlings, and avoiding predators. Every creature lives on a time budget (e.g., Barnard, 1980).

Activities compete because of the consequences they produce. The songbird must spend time foraging because foraging produces food, and it must spend time in vigilance because vigilance avoids predators. Likewise, a human must spend time working because working produces resources and must spend time in other activities such as maintaining relationships because relationships avoid isolation and instability. The consequences and antecedents that drive or induce behavior do so ultimately as a result of evolutionary history; they are Phylogenetically Important Events (PIE; Baum, 2005; 2012a). PIEs that tend to enhance reproductive success, like resources, mates, and shelter, may be called "good." PIEs that tend to diminish reproductive success, like predators, illness, and injury, may be called "bad." Both good and bad PIEs induce activities specific to them. Good PIEs induce behavior that makes them likely to remain or occur; bad PIEs induce behavior that makes them unlikely to remain or occur. The influences of culture and individual experience that increase activities that make good PIEs more likely and bad PIEs less likely depend ultimately on evolutionary history with respect to those PIEs. A neutral object or event becomes a proxy for a PIE when it covaries with the PIE, as, for example, money covaries with resources. The covariance selects money from other environmental objects and events, and once money becomes a proxy, it induces activities like shopping or working much as the resources themselves would. Whether we consider food or predators, a mate or an injury, the relevant activities are induced by the PIEs themselves or their proxies (money, alarm, flirtation, weapons, etc.), and the activities compete for time. [For further discussion, see Baum (2005; 2012a).]

The Matching Law

In the laboratory, choice as behavioral allocation has been studied extensively in a variety of species, including rats, monkeys, pigeons, and humans [see Baum (1979) and Davison & McCarthy (1988) for reviews]. Figure 8.2 shows the results of a typical experiment. A pigeon was exposed to several situations, each presenting two keys continuously at which the pigeon could peck, each for enough daily sessions until no further systematic change in allocation could be seen. Pecking at the keys occasionally produced food,

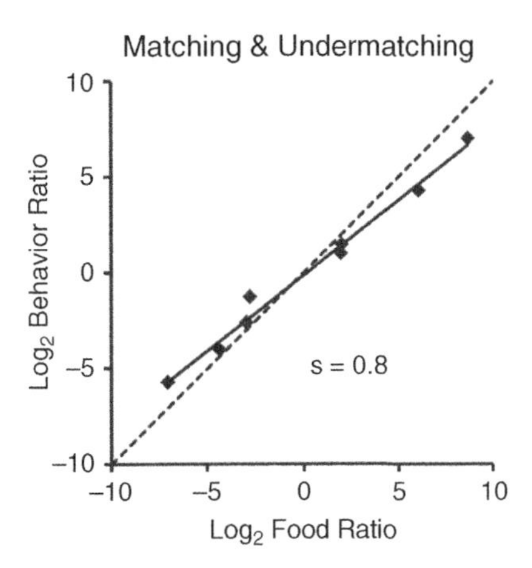

Figure 8.2 The generalized matching law

Note: Points show results for one pigeon in a typical experiment with pairs of VI schedules. The solid line shows the least-squares regression line, with slope equal to 0.8. The broken line shows the locus of strict matching.

according to irregular time-based schedules (variable-interval schedules), and the food rate differed between the two keys, sometimes by as much as 100:1 or more. Each point shows the stable allocation of pecks between keys as a function of the allocation of food between the keys.

The regression line in Figure 8.2 has the equation (Baum, 1974a):

$$log\frac{B_1}{B_2} = slog\frac{r_1}{r_2} + logb, \tag{8.1}$$

where B_1 and B_2 are the times spent pecking at keys 1 and 2 measured as numbers of pecks, r_1 and r_2 are the food rates delivered by pecking at keys 1 and 2, s is sensitivity to variation in the food ratio, and b measures any bias due to factors other than food rate. Equation 8.1 is known as the Matching Law. When s and b both equal 1.0, perfect matching of behavior ratio to food ratio occurs. Often, however, sensitivity to food ratio falls short of 1.0, as in Figure 8.2, where s equals 0.8. Although the behavior ratio tracks the food ratio across situations, it often falls a bit short of equaling the food ratio, a result known as *undermatching* (Baum, 1974a).

Research on the Matching Law has examined consequences other than food, varying deprivation, qualitatively and quantitatively different consequences across keys, penalties for switching between keys, and frequency of changing food ratios.

Although most of the research on the Matching Law concerns just two alternative activities, some has studied allocation among 3 or more alternatives (e.g., Aparicio & Cabrera, 2001; Jensen, 2014; Jensen & Neuringer, 2009; Schneider & Davison, 2005). One may generalize Equation 8.1 to any number of alternatives n as follows. The general arithmetic version of Equation 8.1 is a power function:

$$\frac{B_i}{B_j} = \frac{b_i\, r_i^{s_i}}{b_j\, r_j^{s_j}}, \tag{8.2}$$

where b_i/b_j replaces b, r_i and r_j are rates of consequences not necessarily food, s_i and s_j are possibly unequal, and i and j denote two alternatives out of n alternatives.

For any alternative i, we may multiply together its ratios with respect to itself and all of the alternatives j:

$$\prod_{j=1}^{n}\frac{B_i}{B_j} = \prod_{j=1}^{n}\frac{b_i r_i^{s_i}}{b_j r_j^{s_j}}$$

Taking the logarithm of this equation, we arrive at a working equation:

$$logB_i - \frac{1}{n}\Sigma logB_j = s_i logr_i - \frac{1}{n}\Sigma s_j logr_j - \frac{1}{n}\Sigma logb_j \tag{8.3}$$

The Matching Law succeeds as a description of behavioral allocation, but where does it come from? A number of derivations from more basic principles have been proposed, but the law has also been suggested to be basic itself (e.g., McDowell, 1986; Gallistel et al., 2007).

A possibility that might reconcile all speculation about the origins of the Matching Law would be that the power functions which comprise it are the basic relations that underlie it. If r represents the rate of a phylogenetically important event (e.g., food, mate, shelter), and a PIE induces the activity that produces it ("operant" activity), then the function governing the induction of any activity B_i that produces r_i might be:

$$B_i = b_i r_i^{s_i} \tag{8.4}$$

This possibility was suggested early (Baum & Rachlin, 1969; Killeen, 1971; Rachlin, 1971) and has received some empirical support recently (Baum & Davison, 2014).

If correct, Equation 8.4 would fit well with the recognition that behavior and environment together constitute a feedback system.

The Behavior-Environment Feedback System

Behavior produces effects in the environment, and those changes to the environment in turn affect behavior. If the world is arranged so an activity (e.g., shopping or foraging) produces a good (e.g., food or prey), the good produced also induces the activity—break either of these relations, and the activity is no longer maintained. This interlocking is characteristic of a feedback system (Baum, 1973; 1981b; 1989; 2012a).

Figure 8.3 diagrams the behavior-environment feedback system in a rudimentary way. It shows the system in the most general terms. The activity *B* governs *r* according to a feedback function, *f*(*B*), which is a characteristic of the environment, and *r* feeds back to the organism to induce the activity according to a functional relation, *g*(*r'*), perhaps like Equation 8.4. The criterion *C* sets limits to *r* and depends on the other activities that compete with *B*; *r'* represents this constrained effective rate. The time spent in the activity (*B*) is the output from the organism. Depending on the feedback function and other activities present, *B* may stabilize, maintaining equilibrium.

Up to now, the discussion of the Matching Law focused on situations in which the alternatives constituted variable-interval schedules, in which time has to pass before the activity can produce food, thereby setting an upper limit to *r*. The curves in Figure 8.4 show some examples of VI feedback functions. An approximate equation for these curves (Baum, 1992) is:

$$r = \frac{1}{t + \frac{a}{B}}, \tag{8.5}$$

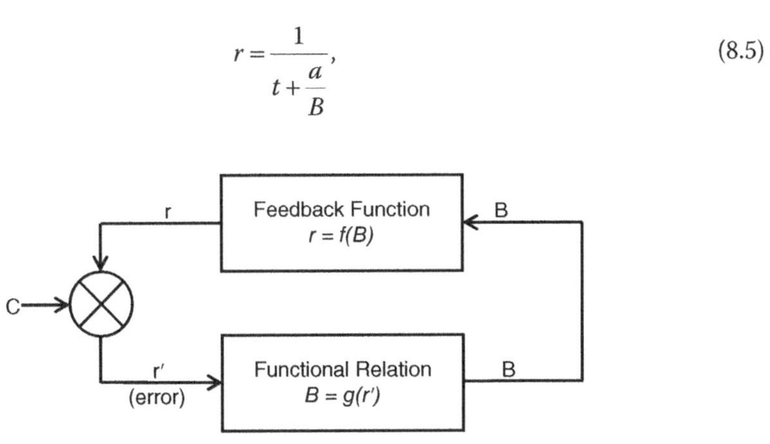

Figure 8.3 The behavior-environment feedback system
Note: The upper half shows activity *B* affecting the environment to produce feedback *r* (e.g., reinforcer rate). The criterion *C* indicates the competition with other activities that limits effective *r* to *r'*. The lower half indicates the organism's contribution: a functional relation with *r* as input and activity *B* as output.

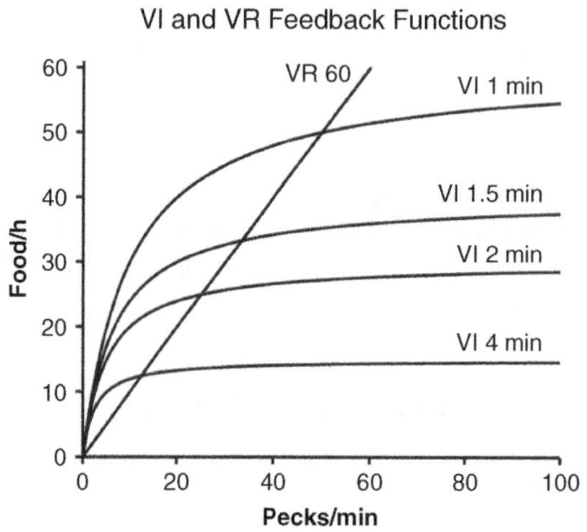

Figure 8.4 Typical feedback functions
Note: The curves show feedback functions for a pigeon pecking a response key paying off according to a VI schedule. The straight line illustrates the feedback function for a VR 60 schedule: direct proportionality. VI schedules model situations in which payoff rate is limited by some uncontrollable factor (e.g., time). VR schedules model situations in which labor alone is effective.

where t is the average interval required to set up food, and a is a constant reflecting length of bouts of the activity. As B increases, r approaches an upper limit of $1/t$.

Plotted in the coordinates of Figure 8.4, the power function in Equation 8.4, with s less than or equal to 1.0, would pass through any number of these VI feedback functions. (If s is greater than 1.0, it will still pass through some, depending on t, b, and s.) Thus, if the n alternatives in Equation 8.3 are all VI schedules or entail negatively accelerated feedback functions like those in Figure 8.4, then the Matching Law applies.

Ratio schedules

The situation is different for ratio schedules, which model feedback relations in which only labor counts—for example, gambling or hunting. A ratio includes no time dependency, and specifies that r is directly proportional to B. The feedback function for a ratio schedule is:

$$r = \frac{B}{v}, \tag{8.6}$$

where v is the average time spent in activity B required for payoff. The straight line in Figure 8.4 shows a feedback function for a variable-ratio (VR) schedule—a situation in which payoff requires an uncertain amount of time spent in B.

If a fixed VR schedule is paired with various VI schedules, matching still is possible, and experiments show that it occurs, but with a constant bias in favor of the VR schedule (Baum & Aparicio, 1999; Herrnstein & Heyman, 1979; Heyman & Herrnstein, 1986). An exception occurs if the VI schedule is richer than the VR schedule, because preference tends then to favor the VI exclusively (Baum & Aparicio, 1999).

If a fixed VI schedule is paired with various VR schedules, optimality predicts substantial undermatching, but experiments so far support matching instead (Baum, 1981b; Herrnstein & Heyman, 1979; Heyman & Herrnstein, 1986).

Since B equals rv in a VR schedule, we may substitute for B in Equation 8.2 or 8.3 and discover that if the alternatives are both ratio schedules the ratio r_i/r_j is constant—i.e., unaffected by B_i/B_j. If v_i equals v_j, no preference can be predicted. To predict the outcome if v_i and v_j are unequal, suppose that the ratio on the right side of Equation 8.2 is greater than 1.0—i.e., Alternative i is richer, which means v_i is smaller than v_j, and r_i/B_i is greater than r_j/B_j. Whatever B_i/B_j equals, the only corrective action that will tend toward the relation in Equation 8.2 is for B_i to increase and B_j to decrease. The inequality never goes away, and eventually all time is allocated to B_i and none to B_j. The exclusive preference for B_i satisfies Equation 8.2 trivially, because both sides become indeterminate. By extension, Equation 8.3 cannot be met if the alternatives are all VR schedules, because the richest schedule will attract all the time (Herrnstein & Loveland, 1975). The prediction of exclusive preference, however, depends on all the alternatives producing the same outcome.

Substitutability and partial preferences

Up to now, all the results we have considered occurred in experiments in which the alternative activities produced the same (identical) outcome—generally, the same opportunity to eat. Identical goods guarantee perfect substitutability, but in the world outside the laboratory, the products of our activities are often imperfectly substitutable. When will I give up some peanut butter for some jam? Some clothes for some money? Some money for some love?

Some experiments have studied choice between qualitatively different outcomes. Miller (1976), for example, studied pigeons' behavioral allocation between pairs of VI schedules that produced different grains. The results conformed to Equation 8.1 with the difference only contributing to bias (b not equal to 1.0), suggesting unequal b_i and b_j in Equation 8.2. Hollard and Davison (1971) obtained similar results studying VI schedules that paid off with food and electrical brain stimulation. When only bias is affected, the two qualitatively different outcomes would be completely substitutable.

Some studies have been done of concurrent VR schedules with qualitatively different outcomes (e.g., Belke, Pierce, & Duncan, 2006; Green & Freed, 1993; Green & Rachlin, 1991). In research on foraging, however, numerous experiments on dietary choice have been done (e.g., Krebs & Davies, 1993). Foraging may be thought of as equivalent to a ratio schedule, because the more time is spent, the more prey are obtained. Optimal diet theory in its simplest form took calories as a currency and ignored other nutrients that might affect preference for various prey. As a result, early optimal diet theory predicted exclusive preference for a more calorie-rich prey item when a forager is given a choice. Instead, researchers found partial preferences, implying that the different prey could not be measured on a single currency and were imperfectly substitutable.

Doubtless, different prey items contain different nutrients, any of which might be crucial to a health-maintaining diet. If we think of all the different nutrients required, we may conceive of them as complements, at least when any of them is scarce (Rapport, 1980; 1981). If a forager requires calcium, and the usual prey are deficient in calcium, the forager will switch when a rare calcium-rich prey item appears, and may for a time prefer such prey exclusively. Switching among prey produces partial preferences. Partial preferences occur also with concurrent VR schedules in the laboratory, if the different schedules produce imperfectly substitutable outcomes. Preference in a situation like that might tend to be optimal, just as performance on concurrent VI schedules tends to be optimal, but the two situations generate switching for different reasons. In concurrent VI schedules, switching occurs because time spent with one schedule results in higher likelihood of reward from the other schedule (Rachlin, Green, & Tormey, 1988). In concurrent VR schedules, switching would depend on non-substitutable outcomes. For example, rats' choice between food and water cannot be exclusive, although Rachlin and Krasnoff (1983) found evidence that when water is easily available, drinking may substitute to a degree for eating. Belke, Pierce, and Duncan (2006) found evidence of substitutability between sucrose and wheel running in rats. A version of Equation 8.2 describes substitutability and complementarity, with the exponent indicating degree of substitutability or complementarity (Green & Freed, 1993; Green & Rachlin, 1991, Equation 4).

Impulsivity, Self-Control, and Time Allocation

Up to this point, the discussion has assumed that the inducing consequences of an activity are stationary with respect to time. Apart from increasing variability with smaller sample sizes, whether we measure a pigeon's time allocation between concurrent VI schedules for 20 minutes or 3 hours makes no difference to the choice relation. A person's time allocation between work and family might remain the same whether measured for a month or a year. Some activities, however, change consequences depending on the time frame within which they are evaluated. In the short term, eating candy, chips, and soda may be strongly induced, because these junk foods stimulate receptors for sweet, salt, and fat, but in the long term consuming junk food has bad consequences, because a diet heavy with junk food leads to health problems and early death. Similarly, activities like smoking, drinking alcohol, injecting heroin, spending money, lying, cheating, and crime may be strongly induced in the short term and have bad consequences in the long term. Conversely, putting off visiting the dentist may be strongly induced in the short term, because it takes time and may be uncomfortable and painful, but in the long term, visiting the dentist helps maintain health and prolong life. The same pattern holds for activities like saving money, paying taxes, cooperating with other people, helping strangers, and using public transit: In the short term, their consequences induce avoidance, but in the long term they have good consequences. In all these examples, when the activity shifts consequences with timeframe, the better alternative is the one that produces better long-term consequences, but the long-term consequences are weakly inducing in competition with short-term consequences that are strongly inducing. Money in the bank is weakly inducing in competition with money in the hand, public transit is weakly inducing in competition with taking one's own car, and the benefits of sobriety are weakly inducing in competition with a drink available immediately.

Activities like eating junk food and smoking may be called "bad habits," and activities like visiting the dentist and saving money may be called "good habits." Figure 8.5 illustrates how a bad habit affects quality of life and changes consequences with time frame. The top graph indicates the decrease in quality of life to expect with the passage of time from the inception of the bad habit when the activity occurs at a low, medium, or high rate. Low time allocation to an activity like drinking alcohol has little deleterious effect on quality of life—and might even enhance it—but a medium allocation (say, heavy drinking

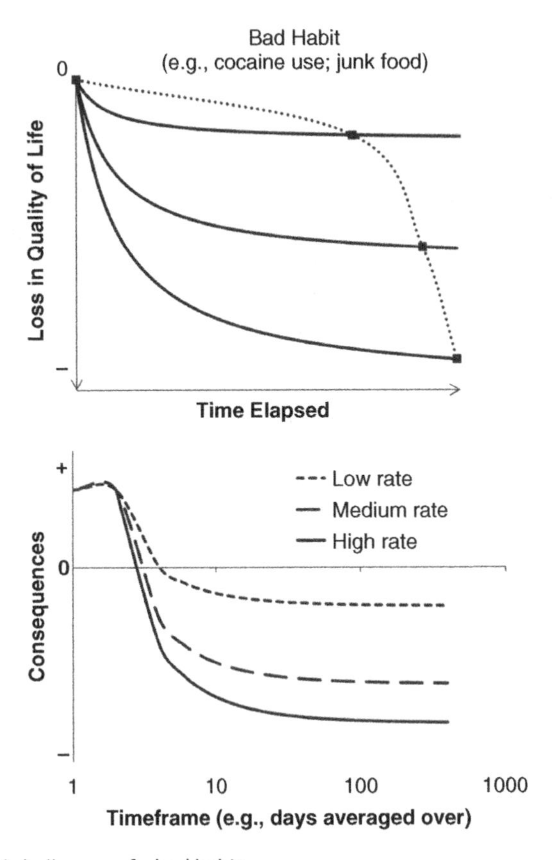

Figure 8.5 Effects and challenges of a bad habit
Note: Top: Loss in quality of life (downward into negative y-axis) of three different rates of time spent in a bad habit as time goes by. The upper curve shows loss due to a low rate, the middle curve shows loss due to a medium rate, and the lower curve shows the large loss due to a high rate. The dotted curve connects points at which loss reaches 90 % of asymptotic loss, and suggests that loss accelerates with rate of time spent in the bad habit. Bottom: Conflict of timeframes results from a shift from positive (good) consequences of the bad habit in short timeframes to negative (bad) consequences of the bad habit in long timeframes. Evaluated over a day or two, snorting cocaine might have positive consequences, but evaluated over many days, its bad consequences reduce quality of life.

on weekends) reduces quality of life (possibly hurting health, job performance, and losing friends), and a high allocation (daily drunkenness) lowers quality of life hugely by effects like losing one's job, spouse, friends, house, and health. A person who is addicted to alcohol, heroin, pornography, or gambling engages in the bad habit at a high rate and suffers the loss. The dotted line suggests the long-term relation between activity rate or time allocation and loss in quality of life; loss accelerates with rate.

The lower graph in Figure 8.5 illustrates the dependence of the consequences of a bad habit on the timeframe in which it is evaluated. The vertical axis goes in the opposite direction to that of the upper graph. Consequences on the vertical axis range from positive to negative (good to bad). The horizontal axis indicates the time frame over which the consequences are calculated; it is in days and is logarithmic to consider timeframes on the order of a day up to timeframes of months or years. Eating junk food or snorting cocaine has high positive consequences for a matter of hours, but when pursued repeatedly over a longer timeframe, its consequences shift to negative—less negative for a low rate, more negative for a medium or high rate.

A bad habit presents the problem that short-term consequences conflict with long-term consequences. Because of genetics and environmental effects (i.e., life history), some people's behavior fails to come into contact with the long-term effects on quality of life, and those people suffer. They may be trained, however, in the long-term contingencies—alcoholics and other addicts sometimes can learn to abstain, spendthrifts can learn to save, gamblers can quit, and criminals can go straight.

Figure 8.6 illustrates how a good habit affects quality of life and changes consequences with time frame. The top graph indicates how the benefit to quality of life increases with time elapsed since the inception of the good habit—less for a low rate or time allocation, more for a medium rate or allocation, and most for a high rate or allocation. Caring for one's teeth or eating fruits and vegetables, if infrequent, may slightly increase health and

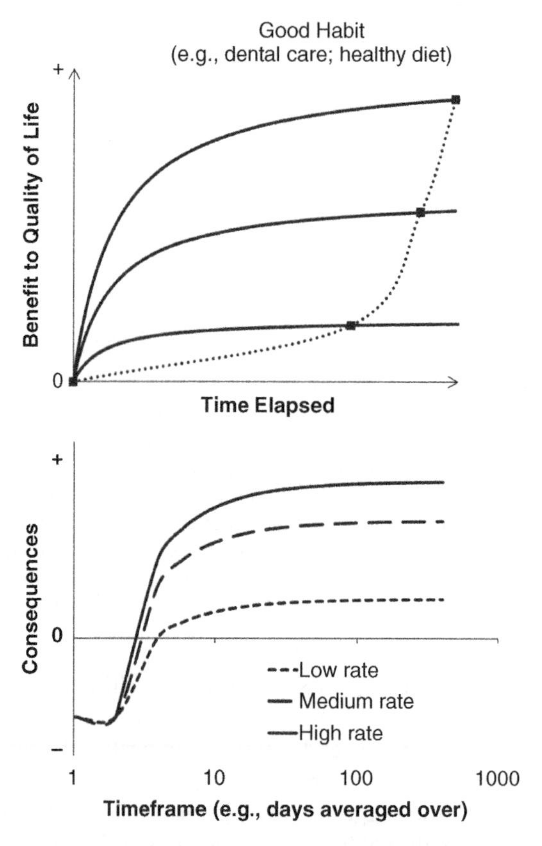

Figure 8.6 Effects and challenges of a good habit
Note: Top: Benefit to quality of life (upward into positive y-axis) of three different rates of time spent in a good habit as time goes by. The lower curve shows benefit of a low rate, the middle curve shows benefit of a medium rate, and the upper curve shows the high benefit of a high rate. The dotted curve connects points at which benefit reaches 90 % of asymptotic benefit, and suggests that benefit accelerates with rate of time spent in the good habit. Bottom: Conflict of timeframes results from a shift from negative consequences of the good habit in short timeframes to positive consequences of the good habit in long timeframes. Evaluated over a day or two, the bad consequences of going to the dentist might induce procrastination, but evaluated over many days, visiting the dentist has good consequences in benefits to health.

well-being, but if engaged in more often, they increase quality of life more. The dotted line suggests the relation between benefit in quality of life and rate or time allocated to the good habit; benefit accelerates with rate.

The lower graph in Figure 8.6 illustrates the way a good habit's consequences change with the time frame over which they are evaluated. The axes are the same as in the lower graph in Figure 8.5. The good habit typically has negative (bad) consequences when evaluated over a time frame on the scale of a day, but evaluated on scales of days, months, or years, its consequences shift to positive (good)—less positive if its rate is low and most positive if its rate is high.

A good habit presents the opposite problem to a bad habit. Short-term consequences conflict with long-term consequences, but in the opposite direction. Saving money or helping a stranger is bad in the short term, but enhances one's quality of life in the long term, because one may have money when it is needed or because one may live in an environment where people help one another when in need. As with bad habits, some people's behavior, because of genetics or life history, may fail to come into contact with the long-term contingencies between the good habit and its benefits. As with bad habits, those people may be trained in the long-term benefits. If they missed it in preschool, they can sometimes be taught to cooperate and plan ahead as adults.

This way of thinking about good and bad habits is not usual (but see Rachlin, 1995a; 2000; 2002; Rachlin & Locey, 2011). Impulsivity (i.e., bad habits) and self-control (i.e., good habits) are most often thought of as depending on temporal discounting or probability discounting. Experiments on discounting typically give a subject choices between two outcomes, one constant and usually immediate (e.g., 10 dollars now) and one that varies from choice to choice in either delay or probability (e.g., 100 dollars in a week or 100 dollars with a probability of .5). As the delay or probability varies across choices, a delay or probability is found for which the subject is indifferent between the two amounts. The immediate amount is taken as the measure of the delayed or probabilistic amount. This procedure is repeated for several different immediate amounts, and indifference points may be analyzed as a function of delay or odds-against transformation of probability (e.g., Green & Myerson, 2013; Green, Myerson, Oliveira, & Chang, 2014).

Experiments on discounting afford a way to measure impulsivity by equating it to degree of discounting. They have been popular because they allow one to study the environmental factors that favor impulsivity over self-control. Addicts, for example, often discount more steeply than non-addicts (see Odum, 2011, for overview). One great benefit the experiments may offer would be to understand the environmental factors or therapeutic procedures that would shift control of behavior from short-term timeframes to long-term. Figure 8.3 indicates that the consequences of an operant activity induce the activity that produces them (i.e., r feeds back to the activity B), but the diagram specifies nothing about timeframe. It applies to the addict's injecting heroin and the effects of the heroin inducing continued injecting as much as to eating a good diet and the good health inducing continued healthful eating. Future research might suggest ways to shift the addict's behavior from short-term control by short-term benefits to long-term control by long-term benefits, and some research has already addressed the problem (Locey & Rachlin, 2012, 2013; Mazur & Logue, 1978; Rachlin, 1995b).

Although experiments on discounting may offer a measure of impulsivity or self-control, how they relate to the real-world extended behavioral patterns of bad habits and good habits remains unclear. The outcomes in experiments on discounting are invariably discrete events like receiving 10 dollars or 100 dollars, because only discrete events can be unambiguously delayed or probabilistic. Real-world outcomes, however, rarely consist of discrete events. Good health cannot arrive suddenly after a delay nor can it usually be lost suddenly one day. Instead, real-world outcomes are extended conditions like sobriety, contributing to the welfare of others, having a good marriage, having a growing bank account, and enjoying the admiration of others. None of these can reasonably be considered delayed or probabilistic, because they extend through time. A bad habit or a good habit, though extended in time, might be thought of as a pattern of many choices that were discrete, but the extended habit pattern also has extended consequences, as suggested by the feedback system represented in Figure 8.3. Discounting experiments seem to have little to do with such temporally extended relations (Rachlin, 1995a; 2002). How one might model extended outcomes with discounting experiments remains to be seen (see Heyman, 2009 and Rachlin, 2000 for extended discussions).

Conclusion

The Multiscale Molar View taken in this paper affords accounts of behavior that are simple, elegant, and plausible. These natural-science accounts are actually or potentially quantitative, as illustrated by the Matching Law, and they omit anti-scientific concepts like free will and agency (Baum, 1995c; 2005). Additionally, the multiscale molar view ties the study of behavior directly to evolutionary theory by means of phylogenetically important events and the activities they induce. The concept of reinforcement takes a different form, because instead of strengthening activities they follow, phylogenetically important events induce behavior on which they are contingent (Baum, 2012a). The present paper focused on choice, but because all behavior entails choice, the treatment of choice is the treatment of behavior in general. Taking choice as the allocation of time among competing activities offers a general framework for understanding the behavior of humans and other animals, including choice between qualitatively different outcomes and self-control.

9

Reinforcement

Foreword

When Hal Miller invited me to write this encyclopedia entry, I think he expected something unorthodox. He was not surprised, then, by this brief summary of the molar view of behavior. At least he did not balk. Eventually, I expect that the word "reinforcement" will drop out of the vocabulary of behavior analysis, tied as it is to the molecular view, discrete responses, and the hypothetical construct "strength." Eventually, when behaviorists escape from focusing on behavior of the moment, they will see changes in behavior as adaptation only and see no need for hidden dispositions.

Abstract

The concept of reinforcement is crucial to understanding behavior of humans and other animals. It refers to the increases and decreases in behavior that occur as a result of contingencies between different activities and different phylogenetically important events. Historically, it was equated with strength. As a result of increased understanding, reinforcement is now a bit of a misnomer, because changes in behavior do not reflect changes in strength, but only changes in frequency or time spent in activities.

Keywords: reinforcer, punisher, activity, phylogenetically important event, induction, impulsivity, self-control

Source: Originally published in H.L. Miller (ed.), *The SAGE Encyclopedia of Theory in Psychology*, pp. 795–798. Thousand Oaks: Sage Publications, Inc., 2016. Reproduced with permission of Sage Publications, Inc.

Reinforcers and Punishers

Some life events are good: mating and raising offspring; gaining food, water, shelter, and other resources; exercising, sleeping, reading, meditating, and other activities that maintain health; in a social species like ours, making and maintaining relationships with friends, family, spouse, and others. When possible, humans and other creatures behave in ways that make these events more likely and avoid behaving in ways that make them less likely. They seek mates, work or hunt for resources, and write or telephone friends and family. Because the connection between working and gaining resources, for example, supports working, good events like gaining resources are called *reinforcers*. In general, when an activity occurs frequently because of its ability to make a good life event more likely, that activity is called an *operant* activity and is said to be *reinforced* by the good event it makes more likely, and the increment in the activity as a result of its connection to the reinforcer is called *reinforcement*.

Some life events are bad: encountering a predator, fighting with a competitor, falling from a height, and other forms of injury; excessive cold or heat; snakes and plants that may be toxic; bacteria, worms, and other parasites that cause illness; rejection by a mate or friend. When possible, humans and other creatures behave in ways that make these events less likely and avoid behaving in ways that make them more likely. They hide or flee from predators, threaten competitors without actually fighting, build shelters, and choose what to eat. Because the connection between dangerous behavior and injury discourages dangerous behavior, bad events like injury are called *punishers*. In general, when an activity occurs infrequently because of its ability to make a bad life event more likely, that operant activity is said to be *punished* by the bad event it makes more likely, and the decrement in the activity as a result of its connection to the punisher is called *punishment*.

Contingencies

Reinforcers and punishers affect behavior because of their connections to it. They are often referred to collectively as "consequences." Gaining resources is a consequence of working, and injury is a consequence of fighting. Such a connection is called a *contingency*. Gaining resources is contingent on working, and injury is contingent on fighting, in the sense that the more one works or fights, the more likely is gaining resources or being injured. A contingency in which more of an activity makes a consequence more likely is a *positive* contingency. In contrast, a contingency in which more of an activity makes a consequence less likely is a *negative* contingency. The key point about contingencies is that the likelihood of the consequence, measured as its rate, amount, or intensity, covaries with the time spent in the activity. In a positive contingency, like foraging in relation to gaining food, the more time spent foraging the more food gained. In a negative contingency, like digging a burrow in relation to exposure to cold, the more digging the less exposure.

When a positive contingency between an activity and a good event causes the activity to increase, the increase is called *positive reinforcement*. Foraging is positively reinforced by obtaining food. When a negative contingency causes an activity to increase because the activity avoids a bad event, the increase is called *negative reinforcement*. Lying about how nice a dress looks on a friend is negatively reinforced by avoidance of hurting the friend's feelings. When a positive contingency between an activity and a bad event causes a decrease in the activity, the decrease is called *positive punishment*. Lying is positively punished by social rejection. When a negative contingency causes an activity to decrease because it prevents a good event, the decrease is called *negative punishment*. Breaking curfews is negatively punished by removal of privileges ("grounding").

When an activity increases or decreases in reinforcement or punishment, other activities increase or decrease also. More time spent digging a burrow means less time spent otherwise, perhaps foraging. Since time is limited to 24 hours in a day or 365 days in a year, increases in one activity necessarily require decreases in some other activities. More time spent working usually means less time in leisure activities. Avoidance of fighting means more time spent in alternatives to fighting like friendly social behavior. Since trade-offs between activities occur, such that increase in one activity necessarily decreases another activity, reinforcement of one activity means punishment of the other and vice versa. For example, increased pay or prestige might increase (reinforce) time spent working, but if leisure activities decrease, then the increased pay or prestige punishes spending time in leisure activities. If the threat of going to prison punishes stealing cars, then it also negatively reinforces more acceptable activities that remove the threat.

Sometimes time spent in an activity determines consequences all by itself, as in gambling, piecework wages, or hunting, but these pure contingencies are rare. Usually other factors besides behavior limit the possible rate of good or bad events. No matter how many times you check the mailbox, mail comes only six times per week. When you are in a conversation, you limit your time spent talking to let the other person talk. In the laboratory, the pure contingency is studied as a *ratio schedule,* in which a certain amount of an activity is required for a certain amount of a good event, such as access to food. The contingency with an additional factor is studied as an *interval schedule,* in which a timer has to time out before a certain amount of activity can produce a good event; the timer limits the rate of the good event. Researchers study other complications to contingencies also, with the aim of understanding how consequences determine increases and decreases in activities.

Phylogeny and Phylogenetically Important Events

Contingencies with good events and bad events affect behavior as a result of evolutionary history or phylogeny. Good events like gaining resources, making friends, and mating improve fitness—that is, make reproductive success more likely. Bad events like injury, illness, and encountering predators decrease fitness—that is, make reproductive success less likely. As a result, natural selection favors those individuals in a population that behave in ways that make good events more likely and bad events less likely, because those individuals are more likely to contribute offspring to subsequent generations than individuals that behave less effectively. Individuals that evade predators well live to reproduce, and so do their offspring. Individuals that forage well are healthy enough to reproduce, and so will their offspring be. Thus, abilities like running when a predator is near and digging a burrow to run to are selected, as are abilities like finding and cracking seeds or stalking and chasing animal prey.

Good and bad events that affect reproductive success are important to phylogeny. They are *phylogenetically important events* (PIEs). Any event that affects health, gaining resources, making and maintaining relationships, or mating and rearing offspring is a PIE.

Situations in which good events or bad events are likely to occur *induce* behavior selected as appropriate to the good or bad events—that is, when those situations arise, all activities appropriate to the likely PIE increase in frequency. When a predator is present, fleeing or hiding becomes likely; the presence of the predator induces fleeing or hiding. When a plant is shedding seeds, a bird searches beneath the plant; the shedding plant induces searching and pecking. The sight of an antelope induces stalking and chasing in a cheetah. A situation in which a PIE is likely induces activities that, because of phylogeny, are appropriate to it—that is, make the good PIE more likely or the bad PIE less likely. Anyone who has fed a pet dog knows that at feeding time the dog engages in a mix of activities: tail wagging, barking, dancing about, and approaching the person with the food. I. P. Pavlov, famous for his studies of salivation in response to a tone that anticipated food, measured only that specific part of a dog's induced activities in advance of feeding.

Besides activities that are induced by PIE-likely situations as a result of normal development and are characteristic of most members of a species, additional activities may be induced as a result of an individual's personal history; these induced activities are idiosyncratic. Historically, the adjective "operant" has been reserved for these idiosyncratic activities, even though no clear line divides them from the species-characteristic activities. A PIE-likely situation induces a mix of activities, partly species-characteristic and partly idiosyncratic. In the laboratory, rats may be trained to press a lever using a positive contingency between lever pressing and food or to press a lever using a negative contingency between lever pressing and electric foot-shock (simulating injury). When the PIE is food, the rats press the lever in a variety of ways, pawing, licking, and biting it, as they would food. When the PIE is shock, training lever pressing is more difficult, because the experimental situation induces activities like freezing that interfere, and the rats do not treat the lever as food. Similarly, pigeons may be trained to peck a key using a positive contingency between key pecking and food or using a negative contingency between key pecking and electric shock. When food is the PIE, the pigeons peck at the key as they would at a seed, with an open beak, and training key pecking to avoid shock is difficult because of interfering activities induced by the shock-likely situation.

Human behavior in the everyday world also includes mixes of induced activities. In the presence of a potential mate, with sex as the PIE, a heterosexual man behaves in the ways we call courtship or "dating"—he lavishes attention on the woman, feeds her in restaurants, gives her gifts, flirts, and is sexually aroused. In a situation where eating is likely, we typically sit down and salivate. Humans who are hunter-gatherers hunt and forage when resources are low; all the activities that constitute hunting and foraging are induced by the low-resource situation. When agriculture was invented, the onset of the growing season induced all the activities that constitute cultivation and planting. The presence of a powerful person—someone who controls access to PIEs—induces activities that we call "subordination," whereas a friend induces activities that we call "affection" and "exchange."

Idiosyncratic or operant activities become PIE-appropriate as a result of entering into contingencies with PIEs. When a rat's lever pressing enters into a positive contingency with food or a negative contingency with shock, the contingency occurs in a certain situation—for example, the experimental chamber. Lever pressing is induced in that food-likely or shock-likely situation. Lever pressing joins those activities that in phylogeny made feeding more likely or injury less likely. As an everyday example, a dancer's idiosyncratic style derives from her personal history with contingencies between her dancing and approval (a PIE for humans), and her dance style has become applause-appropriate, with the result that applause-likely situations like being on stage before an audience induce her idiosyncratic dance style. Thus, the increases and decreases in activities that are called "reinforcement" and "punishment" arise because every reinforcer and every punisher is either a PIE or a PIE-likely situation and every PIE-likely situation induces all activities appropriate to the PIE.

Not only may induced behavior be idiosyncratic, but the PIE-likely situations that induce activities may also be idiosyncratic. One pigeon may be trained to peck at a green key but not at a red key, whereas another pigeon may be trained to peck at a red key but not at a green key. In everyday life, inducing situations also may be idiosyncratic. Mushrooms may induce disgust in one person and delight in another. Men and women differ in the situations that arouse them sexually. A speaker's casual remark may induce laughing in one listener and aggression in another, depending on the listeners' personal history. Situations become PIE-likely when they are correlated with PIEs or other PIE-related situations. If a person eats lobster and becomes ill, eating lobster becomes a dangerous situation to be avoided. If a person eats strawberries and finds them good, eating strawberries becomes a good situation to be approached. Other foods are not affected, only lobster and strawberries. When one situation induces certain activities and another situation does not, that difference is technically known as *discrimination*. Scientific usage is that behavior discriminates, not the organism.

Additional flexibility in behavior occurs because one temporally extended situation may entail other, more local, situations. In the laboratory, for example, a pigeon may be trained in a contingency in which pecking at a red key sometimes turns the key green, and pecking at the green key enters into a more local contingency with receiving food. In nature, hunting entails searching for prey, but sighting a prey item induces a new mix of activities. For humans, the classic example is the contingency between working and money. Having money is a situation that entails many other situations. It enables, for example, possessing a washing machine, which entails clean clothes, which entails a healthy body, avoidance of illness, and social acceptance. Historically, when a contingency exists between an activity (e.g., working) and an extended situation that is not immediately PIE-likely but entails other situations that are PIE-likely (e.g., having money), the extended situation has been called a *secondary* reinforcer or punisher, to distinguish it from situations in which PIEs are more immediately likely, which are called *primary* reinforcers and punishers. The difference is a matter of degree, however, rather than a sharp division.

Mixed Consequences, Impulsivity, and Self-Control

Typically, behavior has mixed consequences. At the least, when one activity increases, opportunities to engage in some other activities decrease. Enjoying the company of one friend may mean giving up some time spent with another friend. Sometimes the mixed consequences are of more concern. Drugs like alcohol, nicotine, cocaine, and heroin, which either mimic the effects of good PIEs or ameliorate the effects of bad PIEs, easily become addictive because of short-term effects that compete with bad long-term effects. The alcoholic who enjoys drinking excessively also performs poorly at work and in social situations. All addictions have this character of being enjoyable in the short term and disastrous in the long term. In the short term, the inebriating effects of drinking avoid PIEs like feeling ill or social timidity, but in the long term, drunkenness is a situation in which much worse PIEs like serious illness, loss of resources, and social rejection are likely. Such conflicts are also common in framing public policy—the difficulty taxpayers and politicians have in paying for infrastructure or disease prevention in the short term, rather than have bridge collapses and epidemics in the long term.

These short-term versus long-term conflicts are conflicts of contingencies. The same activity—drinking or spending money—enters into contingencies both with short-term inducers and with long-term huge inducers. The problem is that the short-term contingency tends to be highly effective, even more effective than the long-term contingency with PIEs that usually have major effects on reproductive success. When activities are induced by the short-term contingency, the activities are called *impulsivity*—for example, a person on a diet accepting an offer of ice cream or a person on a budget buying an unneeded dress at the mall. When activities are induced instead by the more extended contingency, the activities are called *self-control*—for example, a person on a diet declining an offer of ice cream or a person on a budget visiting the mall with limited cash and no credit card. The tendency for long-term situations, like health, adequate resources, and relationships, to fall short of the inducing power they would have if they were more immediate is called *discounting*. Researchers study ways to decrease discounting and make long-term contingencies more effective in their competition with short-term contingencies. Two strategies that help are *commitment* and *rules*. Commitment strategies limit the ability of a short-term contingency to induce dysfunctional activities by intervening physically—for example, leaving credit cards at home when visiting the mall or placing an alarm clock across the room from one's bed. A rule is a salient signal correlated with the long-term contingency that tends to induce activities appropriate to the long-term contingency. Rules are usually verbal, but may be nonverbal too. For example, a person on a diet may post pictures on her refrigerator of a hippopotamus and a woman in a bikini or an alcoholic may repeat to himself, "I love my wife and children and must decline all offers of alcohol." When commitment is impossible, rules may help.

In summary, "reinforcement" is a bit of a misnomer, because reinforcers and punishers are phylogenetically important events or PIE-likely situations that induce activities related to them by phylogeny or life history.

Cross-references: Behaviorism, Biological Constraints, Classical Conditioning, Law of Effect, I. P. Pavlov, Self-control, B. F. Skinner

Further Reading

Baum, W. M. (2012). Rethinking Reinforcement: Allocation, Induction, and Contingency. *Journal of the Experimental Analysis of Behavior, 97*, 101–124.

Baum, W. M. (2017). *Understanding behaviorism: Behavior, culture, and evolution*, 3rd ed. Malden, MA: Wiley Blackwell Publishing.

Rachlin, H. (1994). *Behavior and mind: The roots of modern psychology*. Oxford University Press: Oxford, England.

Rachlin, H. (2009). *The science of self-control*. Cambridge, MA: Harvard University Press.

10

Avoidance, Induction, and the Illusion of Reinforcement

Foreword

Segal's (1972) concept of induction did not just apply to fitness-enhancing inducers (phylogenetically important events; PIE), but also to fitness-threatening inducers (PIEs). In the world of nature, fitness-threatening PIEs include predators, illness, toxic organisms, and injury. Injury induces avoidance, defense, and other mitigating activities. In the laboratory, electric shock stands in as a surrogate for injury, and experiments on avoidance almost always employ electric shock.

The molecular view based on discrete responses and immediate consequences never afforded a plausible explanation of avoidance. Adherents invoked two-factor theory, which proposed hidden "fear" and a hidden consequence of "fear-reduction" following a response that avoided shock. Teaching students about behavior, I had a hard time keeping a straight face when describing this theory, because it was so fantastic, so based on imaginary events that it transcended the bounds of legitimate science.

In a landmark paper in 1969, called "Method and Theory in the Study of Avoidance," Herrnstein offered an account of avoidance based on reduction of shock frequency—a molar view based on extended variables and extended relations. Sidman (1962) had suggested that avoidance is maintained, not by "fear" reduction, but by reduction in shock rate, and Herrnstein built on this idea. The higher the rate of avoidance activity, the lower the shock rate.

Sidman (1953) had produced the largest and most comprehensive data set on avoidance. I requested the data many years after, and he still had them and sent them to me. Not until 2019 did I get around to analyzing them, and this paper is the result. The paper offers analysis of Sidman's data and a few other sets, with a theory that explains avoidance in molar terms and without any imaginary stimuli. Shock-rate reduction turned out to be only half the story; induction by shock made the other half.

In the 2012 paper, "Rethinking Reinforcement: Allocation, Induction, and Contingency," I argued that the concept of reinforcement and its attendant mysterious concept of "response strength" might be replaced by induction as a mechanism of behavior maintenance and modification. Most of the evidence I offered to support this argument could still be interpreted within the molecular framework pretty easily. With the present paper, I offered a plausible quantitative account of avoidance—something that reinforcement

Science and Philosophy of Behavior: Selected Papers, First Edition. William M. Baum.

theory had never done and could never do. With this result, I can now argue that induction is indeed a superior concept, because it explains behavioral phenomena both more plausibly and more comprehensively.

Abstract

Environmental events that impact reproductive success may be called *phylogenetically important events* (PIEs). Some promote reproductive success, like mates and food; others threaten reproductive success, like predators and injury. Beneficial PIEs induce activities that enhance them, and detrimental PIEs induce activities that mitigate or avoid them. Free-operant avoidance relies on electric shock as a proxy for injury, a PIE. One theory takes avoidance behavior to be reinforced by its reducing shock rate. A more complete explanation is that avoidance both reduces shock rate and is induced by the PIEs it usually prevents. Shocks received act in concert with shock-rate reduction, in a feedback system. Four parametric data sets were analyzed to show that avoidance is induced by received shock rate according to power functions. Avoidance is not reinforced at all; avoidance is induced by its failures. Induction explains not only avoidance itself, but also phenomena unique to avoidance, like warmup and effects of unavoidable shock. Induction explains behavior more generally than reinforcement, because induction explains not only food-maintained operant and non-operant behavior, but also shock-maintained behavior, including avoidance. Reinforcement fails to explain behavior when reinforcement is defined as strengthening by consequences. Induction erases the distinction between consequences and antecedents.

Keywords: avoidance, induction, reinforcement, covariance, molar view

Source: Originally published in *Journal of the Experimental Analysis of Behavior,* 114 (2020), pp. 116–141. Reproduced with permission of John Wiley & Sons.

The law of effect, as stated by Thorndike (1911/2012) and, later, by Skinner (1938, 1948), asserted that, to strengthen a response, a satisfier or a reinforcer had to follow the response closely in time—had to be contiguous with the response. This contiguity-based law of effect seemed to apply readily to consequences like food or water, and even to punishers like injury or illness that should weaken the response. It seemed to fail, however, with avoidance, because an avoidance response prevents an occurrence; successful avoidance is followed by nothing.

Attempting to overcome the seemingly paradoxical nature of avoidance, Solomon and Wynne (1954) proposed two-factor theory, derived from studies of signaled avoidance. In signaled avoidance, a tone or light precedes the onset of electric shock, and a subject moves or otherwise responds to prevent the shock. The theory held that the signal causes "fear," and the avoidance response reduces this fear as an immediate consequence. Critics pointed out that neither the fear nor its reduction was observable, and hence the theory was difficult or impossible to test (Herrnstein, 1969). Herrnstein reviewed the empirical evidence and theoretical vacuousness of two-factor theory. For example, in signaled avoidance, avoidance is still acquired and maintained even if the response does not terminate the signal but only avoids the shock (Kamin, 1956, 1957). If the signal was required to cause "fear," its failure to terminate ought to have prevented avoidance from being maintained. Since the signal apparently functions only as a discriminative stimulus inducing the avoidance activity, two-factor theorists had to posit unobservable stimuli in addition to the unobservable fear. They had to appeal to proprioceptive and kinesthetic stimuli generated by the response itself, thus rendering the theory untestable and vacuous, because if avoidance activity reinforces itself, then avoidance activity occurs because avoidance activity occurs. The theory gives only the appearance of an explanation, not the simplifying and predictive power an explanation should possess.

Dinsmoor (2001) attempted to rescue two-factor theory by recasting it in terms of "aversiveness" and "safety." He maintained that stimuli associated with shock acquire aversiveness and that stimuli associated with the absence of shock signal safety. He argued that the avoidance response is associated with relative safety and absence of the response is associated with relative aversiveness, and hence the avoidance response effectively reinforces itself. Such a theory, based on hypothetical, unobservable, entities is untestable. To quote myself (Baum, 2001):

> . . . two-factor theory cannot explain avoidance without resorting to hypothetical entities (Baum, 1973, 1989). The hypothetical entity in Dinsmoor's (2001) attempted explanation is the reinforcement. His appeals to "aversiveness" and "safety" are no more defensible than was Mowrer's (1960) appeal to "fear." We know that under certain circumstances creatures will behave so as to avoid electric shock. To say that the reason for this is the "aversiveness" of the shock is to add nothing to the account. It is exactly like saying that objects are heavy because they possess letharge, are hot because they possess caloric, or burn because they possess phlogiston. Dinsmoor needs this imaginary essence only because he needs something to transfer from shock to signal. (pp. 339–340).

Add to the vacuousness of the imaginary essence that the stimulus to which it supposedly transfers is unobservable, and "safety" theory becomes completely untestable.

If we set aside the subjective connotations of "fear" and "safety," which seem to imply a role for rats' private events, and consider what objective factor these terms might refer to, they seem just to refer to the actuality that the likelihood of shock (i.e., shock rate) is lower following avoidance activity than preceding avoidance activity. In other words, these terms just adumbrate the avoidance contingencies. In effect they merely state that avoidance activity occurs because avoidance occurs.

Free-operant avoidance, initiated by Sidman (1953), attempted to challenge two-factor theory on empirical grounds. If two-factor theory were to be testable, it would assume an observable stimulus, and Sidman's procedure included no signal to cause fear. Sidman trained rats in procedures in which shocks occurred at regular intervals (S-S interval) in the absence of avoidance lever pressing, and a lever press resulted in postponing shock for an interval (R-S interval). For example, if no pressing occurred, shocks might occur every 5 s, but each press might postpone shock for 10 s. If the rat pressed the lever often enough, the shock rate might approach zero. Since the procedure included no overt signal, and two-factor theory requires a signal, the observed avoidance seemed to invalidate two-factor theory. Anger (1963), however, attempted to save two-factor theory by pointing to the constancy of the S-S and R-S intervals. He argued that these regular intervals offer temporal stimuli that may substitute for the signal, and referred to these as "conditioned aversive temporal stimuli." Sidman (1966), however, showed that avoidance often occurs without any temporal discrimination, and thus that temporal stimuli at least are not necessary for maintained free-operant avoidance. Herrnstein and Hineline (1966) studied avoidance in a procedure that precluded any signals, overt or temporal. In the absence of lever pressing, shocks occurred at a high rate at variable intervals, and a lever press scheduled a shock after a variable interval that was, on average, longer than the variable S-S interval. Thus, the only effect of lever pressing was to lower the shock rate. Herrnstein and Hineline found avoidance in this procedure that excluded any possibility of temporal discrimination, thus invalidating both Anger's theory and two-factor theory in general.

Instead of two-factor theory, Sidman (1962, 1966) proposed that the reinforcer maintaining free-operant avoidance is the reduction of shock rate caused by the responding. This proposal not only offered an alternative to two-factor theory, but opened the door to an alternative law of effect based on temporally extended variables and relations (Baum, 1973, 2002, 2018a). Just as positive reinforcement could be seen as due to positive correlation between responding and reinforcer rate, so perhaps negative reinforcement could be seen as due to negative correlation between responding and shock rate (Herrnstein, 1969).

A problem remains, however, because Sidman's (1962) proposal depends not on shock rate, but on shock-rate reduction. A true analogy between positive and negative reinforcement would consider shock rate itself, because shocks are concrete events the way food deliveries are concrete events. The explanation of avoidance should refer to the negative correlation between responding and shock rate, but this is not enough. In Sidman's (1953, 1966) study, pressing reduced shock rate, but never all the way to zero. Some shocks continued to occur even when the R-S was 50, 90, or 150 s. If the shock is eliminated altogether, however, avoidance activity eventually ceases (extinguishes; Boren & Sidman, 1957). Even Dinsmoor's account requires a role for received shocks, because "safety" only exists in contrast to danger. Were it not for danger (i.e., received shocks), self-reinforced avoidance activity would continue forever when the shock is altogether eliminated. The shocks received must play a crucial role in maintaining avoidance. The question arises, "How could shocks maintain avoidance?"

Taking the broader perspective offered by evolutionary theory, we may view electric shock as a proxy for injury, and injury belongs to a category of events that must be avoided for an organism to survive and reproduce. Such events belong to a larger category: environmental events that affect reproductive success, are therefore important in phylogeny, and which therefore may be called *phylogenetically important events* (PIE; Baum, 2012a, 2018b). Some PIEs enhance the likelihood of reproductive success, like a potential mate, food, and shelter, but some PIEs threaten reproductive success, like predators, illness, and injury. The good PIEs induce activities that enhance their effects, like courting a mate, capturing prey, and constructing a nest. The bad PIEs induce activities that mitigate their effects, like hiding from predators, avoiding toxic foods, and avoiding injury. These basic induced activities are the result of natural selection (Baum, 2018a, 2018b).

Segal (1972) introduced the concept of *induction*. It may be contrasted with elicitation. Whereas elicitation relies on a narrow time frame and a one-to-one correspondence between stimulus and response, induction is inherently a temporally extended relation, in which the occurrence of a PIE in a context results in an increase in time spent in the induced activities in that context. Induction includes elicitation, but also more temporally extended phenomena. For example, when bits of food are occasionally delivered to a hungry rat in an experimental chamber, the time spent in appetitive activities like drinking and chewing increases. These non-operant activities sometimes are called "adjunctive"

(Falk, 1971, 1977). When electric shocks are occasionally delivered to a rat's feet in an experimental chamber, the time spent in defensive activities like avoidance and aggression increases (e.g., Pear, Hemingway, & Keiser, 1978; Pear, Moody, & Persinger, 1972). Segal included some examples of induction of operant activity, and extending induction to operant activity in general required only a small step (Baum, 2012a).

Induction may also be compared to stimulus control. A discriminative stimulus is said to "modulate" or "control" operant activity, increasing some activities and decreasing others. Its effects are extended in time in the same way that the effects of occasional food or shock are extended in time. As an earlier paper explained, a discriminative stimulus is the same as a conditional inducer that induces the activity it is said to "control" (Baum, 2012a). If a rat is trained to press a lever when a light is on and not when the light is off, the light induces pressing and its absence does not. Induction supplies the mechanism by which a discriminative stimulus affects behavior. Although lever pressing is a conditionally induced (i.e., "operant") activity, and the inducing effect of a light is conditional on training, presumably electric shock should be considered an unconditional inducer.

When a PIE like food is made dependent on an activity like lever pressing, food rate covaries with press rate. If the covariance is positive, the activity becomes operant activity—that is, a conditionally-induced activity, because the PIE now induces it along with whatever other non-operant activities the PIE may induce (Baum, 2012a, 2018a, 2018b). We know that this is so because of observations like reinstatement, in which non-contingent presentation of the PIE reinstates (induces) previously extinguished operant activity (Reid, 1958). Operant activity like lever pressing is induced by food (a PIE) and produces the food, thus closing a loop that maintains the operant activity as long as it continues to covary with the PIE.

A parallel may be drawn between an operant activity induced by a good PIE in positive covariance with the activity and an operant avoidance activity induced by a bad PIE in negative covariance with the avoidance activity. If an organism is trained to avoid injury (shock) by pressing a lever or jumping over a barrier, then whenever the shock occurs (a "failure"), the shock induces the pressing or jumping. In signaled avoidance, the signal induces the avoidance response, but in free-operant avoidance, only the occasional occurrence of the shock induces the activity. If the shock is no longer forthcoming or the bad PIE no longer covaries with the activity, avoidance responding eventually ceases (Boren & Sidman, 1957; Herrnstein & Hineline, 1966; Powell & Peck, 1969). If avoidance training alternates with discontinuation of shock, a discrimination forms between presence and absence of shock (Boren & Sidman, 1957), just as a discrimination forms between presence and absence of food when training with food alternates with discontinuation of food (Baum, 2012c; Bullock & Smith, 1953). One could say that the shocks play the role of a discriminative stimulus, meaning that the shocks induce the avoidance activity.

This paper aims to show that shock rate, coupled with shock-rate reduction, maintains free-operant avoidance. Accordingly, I reanalyze Sidman's (1953) data and a few other substantial parametric data sets I could find. I show that they are well-fitted by the same sort of power functions that fit operant activities maintained by food (e.g., Baum, 2015; Baum & Aparicio, 2020; Baum & Davison, 2014b; Baum & Grace, 2020). The function form is:

$$B = cr^{s} \tag{10.1}$$

where B is the rate of operant activity, s is sensitivity of the activity to variation in r, which here is PIE rate (food or shock), and c is a coefficient that equalizes units on the two sides of the equation. I will refer to Equation 10.1 as *power-function induction.*

Power-function induction fits well with matching theory, as I will discuss below (Baum & Aparicio, 2020; Baum & Grace, 2020). One question about power functions arises because a power function has no asymptote, no upper limit. Should we suppose that response rate can increase without limit? An activity like lever pressing must have a physical upper limit, because a rat can only jiggle a lever just so fast (Baum, 1976). Researchers studying avoidance often report that high-rate bursts of pressing immediately follow delivered shocks (e.g., de Villiers, 1972, 1974; Herrnstein & Hineline, 1966). These bursts may show how high press rate could go in principle, if a rat did nothing else but press at the highest rate possible. Apart from the post-shock bursts, and even including them, the press rates observed in experiments on avoidance fall far below the maximum possible.

Analyses

Sidman (1953)

The procedure, often referred to as "Sidman avoidance," depends on two intervals: the shock-shock interval (S-S) and the response-shock interval (R-S). The avoidance activity for rats and monkeys is usually pressing a lever, but other experiments have used locomotion, jumping, pressing a treadle, and operating a key. If the avoidance activity fails to occur, shocks are delivered at a rate set by the S-S; for example, if the S-S equals 10 s, and no lever press occurs for more than 10 s following a shock, a shock is delivered at 10 s since the last shock. Each press on the lever postpones the shock for the R-S; for example, if the R-S equals 20 s, as long as the organism presses the lever again before 20 s elapses, no shocks are delivered. If the organism receives a shock, the S-S begins timing. Thus, in Sidman's procedure, if the rat pressed the lever often enough, no shocks would occur.

Figure 10.1 illustrates the sort of negative covariance that Sidman's procedure maintains. The vertical axis represents PIE (shock) rate, and the horizontal axis represents response (press) rate. The 3 decreasing curves show the covariances given by S-S equal to 10 s and R-S equal to 5, 10, and 20 s. An example of positive covariance for variable-interval (VI) 12 s (gray increasing curve) is included for comparison. Such relations, positive or negative, are called "feedback functions," because they may be viewed as part of a behavior-environment feedback system (Baum, 1973, 1981b, 1989, 2017c, 2018a, 2018b).

A feedback function defines a constraint. It specifies what combinations of response rate and PIE rate are possible in the situation, given the programming. For example, the gray VI feedback function in Figure 10.1 shows how response rate determines food rate. Every response rate maintains a certain food rate. Similarly, the three avoidance feedback functions show press rate determining shock rate. Depending on the programming of the avoidance schedule, each press rate determines a certain shock rate. The dotted vertical line intersects the feedback functions at different levels, illustrating that the same press rate results in different shock rates according to the different feedback functions. The dotted horizontal line illustrates the converse: that the same maintained shock rate requires different press rates according to the different feedback functions.

The three avoidance feedback functions shown in Figure 10.1 are illustrative only. A discussion of the exact shape of the avoidance feedback function appears in the appendix.

The three avoidance feedback functions in Figure 10.1 illustrate two key features of Sidman avoidance. As response rate increases, received shock rate declines and eventually approaches zero as an asymptote. The zero asymptote reflects the ability of the organism to prevent shocks altogether. Other procedures have incorporated an asymptote greater

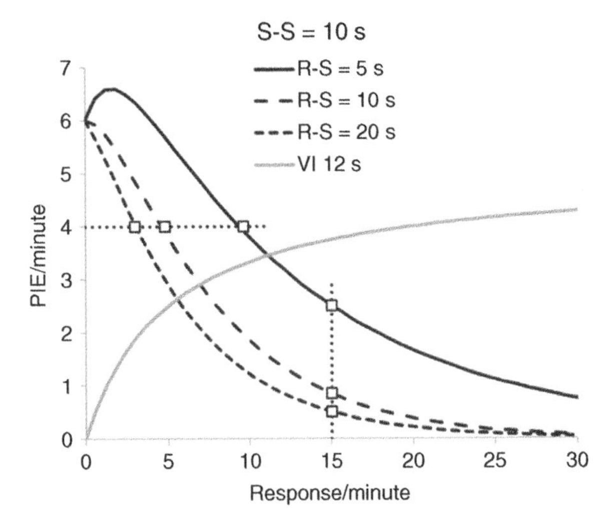

Figure 10.1 Feedback functions for avoidance
Note: The 3 decreasing curves show feedback functions for free-operant avoidance with a S-S interval of 10 s and R-S intervals of 5 s, 10 s, and 20 s. The gray curve illustrates an increasing feedback function as for a variable-interval 12 s. The dotted horizontal line indicates how the same shock rate would be obtained by different press rates on the 3 different schedules. The dotted vertical line indicates how the same press rate would generate different shock rates on the 3 different schedules.

than zero, preventing the pressing from eliminating shock entirely (e.g., Herrnstein & Hineline, 1966). The curves illustrate that the longer the R-S, the faster the shock rate decreases with increasing press rate. The second feature of the procedure appears in the solid curve, for R-S 5 s. At low response rates, if the R-S is less than the S-S, pressing can actually increase the shock rate. Then response rate must exceed a threshold before avoidance is effective.

Figure 10.2 shows Sidman's (1953) data. Rate of lever pressing appears as a function of R-S interval for various S-S intervals. The relations are roughly linear in these logarithmic coordinates, except that, for each rat, several data points fall away from the bulk of the points, which are connected by lines. The unconnected points represent procedures in which the R-S was shorter than the S-S and which failed to maintain adequate levels of avoidance; likely avoidance occurred sporadically. As shown in Figure 10.1, these are conditions in which low response rates actually increase shock rate above the baseline rate set by the S-S interval. The unconnected squares come from procedures in which the R-S was shorter than the S-S of 5 s. The unconnected triangles come from procedures in which the R-S was less than the S-S of 10 s. The unconnected Xs come from procedures in which the R-S was less than the S-S of 20 s. The unconnected circles and crosses come from procedures in which the R-S was less than the S-S of either 30 s or 50 s. In all of these conditions, the rate of shocks received approached or exceeded that programmed by the S-S. For purposes of understanding the maintenance of avoidance, I will focus on the procedures that maintained adequate levels of pressing.

Figure 10.2 shows that when a procedure maintained avoidance, the S-S interval made no difference. All the lines in the graphs fall on top of one another. The R-S interval was the crucial determiner of avoidance.

The theory that shock-rate reduction reinforces avoidance activity implies that avoidance lever pressing should depend directly on shock-rate reduction. Some researchers

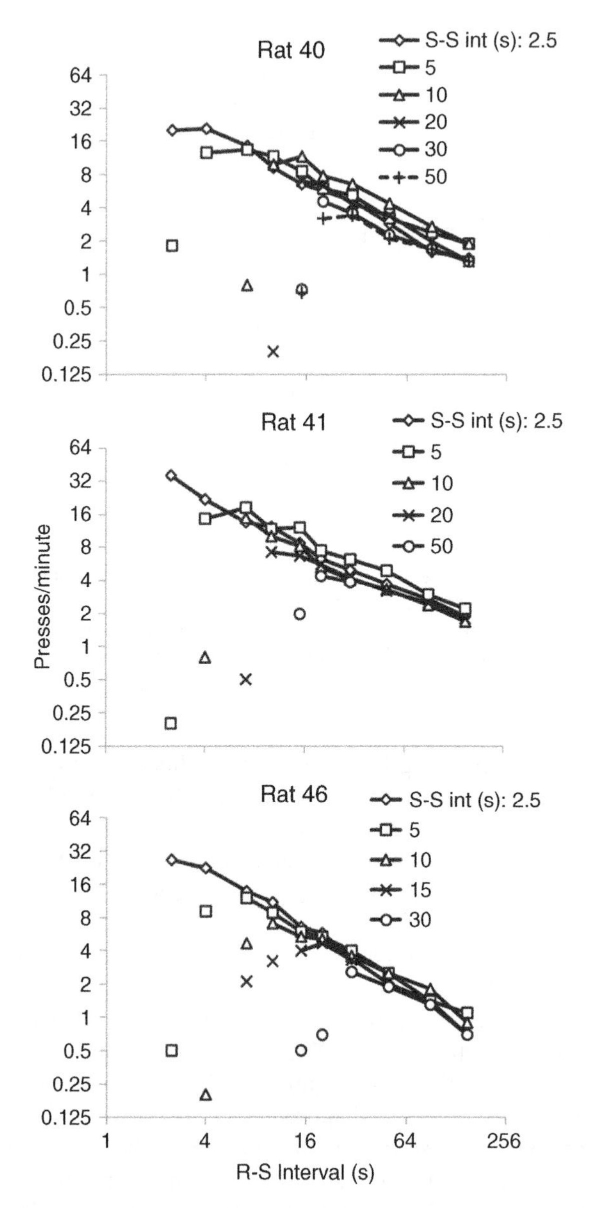

Figure 10.2 Sidman's (1953) raw data
Note: For each of the 3 rats, lever press rate is shown as a function of R-S interval for several different S-S intervals. The unconnected points that lie away from the rest represent conditions in which adequate avoidance was not maintained. Note logarithmic axes.

reported such relations (e.g., Courtney & Perone, 1992; de Villiers, 1974), but with limited data sets. As far as I know, Sidman's (1953) data set is the most comprehensive in existence, because he varied S-S and R-S independently and over wide ranges. Yet, as far as I know, no one has tested this reinforcement theory with Sidman's data.

Figure 10.3 shows lever pressing in Sidman's (1953) experiment as a function of rate of shock avoidance—that is, shock-rate reduction, calculated by subtracting received shock rate from baseline shock rate set by the S-S. All the procedures represented in Figure 10.2

that maintained adequate avoidance are represented in Figure 10.3. For Rat 40, 5 points were omitted, for Rat 41, 4 points were omitted, and for Rat 46, 8 points were omitted, as indicated by the unconnected points in Figure 10.2. The rate of avoiding shocks increased as the S-S interval decreased; the higher the baseline shock rate, the more shocks the pressing avoided. The feature of Figure 10.3 that stands out is the almost complete vertical arrangement of the points for the different R-S intervals paired with each S-S interval. These vertical arrays indicate that avoided shock rate—that is, shock-rate reduction—depended almost entirely on S-S, even though press rate depended almost entirely on R-S (Figure 10.2). For each R-S, press rate stayed at a level that avoided most shocks. Whatever the press rate, the shock-rate reduction was about the same. Shock-rate reduction did

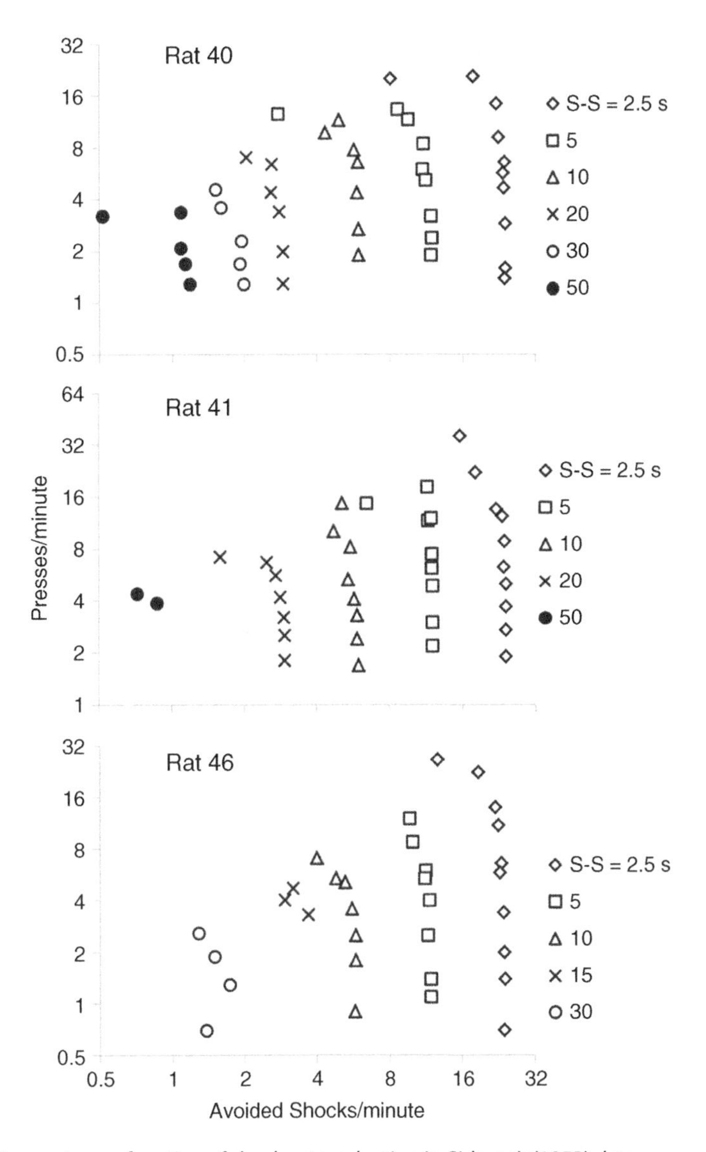

Figure 10.3 Press rate as a function of shock-rate reduction in Sidman's (1953) data
Note: Conditions that failed to maintain adequate avoidance (unconnected points in Figure 10.2) are omitted. Note logarithmic axes.

not unequivocally determine press rate. No direct relation exists between press rate and shock-rate reduction. Thus, although shock-rate reduction must be part of the account, in the form of feedback functions like those in Figure 10.1, shock-rate reduction by itself cannot explain the press rates that Sidman obtained, and the reinforcement theory is incorrect or at least incomplete.

Figure 10.4 shows press rate as a function of received shock rate for all of Sidman's conditions that maintained adequate levels of avoidance—i.e., all the conditions shown in Figure 10.3. The different symbols represent the different S-S intervals. Despite some unsystematic variability across S-S intervals, all the data are pretty well fitted by the broken lines shown. Variance accounted for (r^2) ranged from .73 to .79. Analysis of residuals

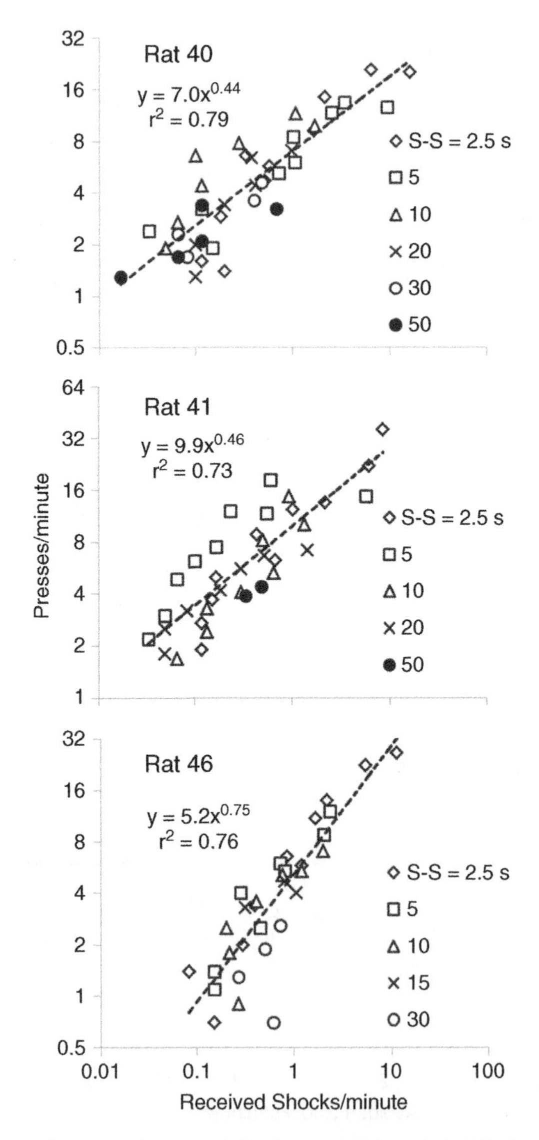

Figure 10.4 Press rate as a function of received shock rate in Sidman's (1953) data
Note: Different symbols represent different S-S intervals. Conditions that failed to maintain adequate avoidance (unconnected points in Figure 10.2) are omitted. The lines were fitted by least-squares regression. The equation of each line appears near it. Note logarithmic axes.

revealed no systematic deviation from the lines. In these logarithmic coordinates, the lines represent power functions, and the equation of each function appears in the graph. The exponent of the power function (sensitivity in Equation 10.1) varied across rats (0.44 to 0.75), but all were less than 1.0. Thus, despite some unsystematic variation, the rates of lever pressing tracked the shocks received. The shocks induced the lever pressing. Avoidance was maintained by its failures.

Figure 10.4 reveals a noteworthy feature of Sidman avoidance: lever pressing never reduced shock rate all the way to zero. As noted earlier, this suggests that the received shocks were crucial to maintaining pressing while at the same time pressing was reducing shock rate. The combination indicates that the avoidance activity was regulatory, decreasing the shock to a certain level and maintaining it there. In other words, avoidance may be understood as maintaining an equilibrium between shock-rate reduction and induction. Figure 10.5 illustrates this concept. It shows the same three avoidance feedback functions as in Figure 10.1 (black lines) and induction by a power function as in Figure 10.4 (gray curve). The squares indicate the equilibrium points. The arrows illustrate that any deviation from equilibrium is corrected and performance drawn back to equilibrium. If press rate falls, shock rate increases, and the increased shock rate induces an increase in press rate back to equilibrium. If press rate rises, shock rate falls, and less pressing is induced, again bringing press rate back to equilibrium.

Another way to portray shock-rate regulation in avoidance describes it as a behavior-environment feedback system (Baum, 1973, 1981b, 1989, 2018a, 2018b). Figure 10.6 shows a diagram of the equilibrium shown in Figure 10.5 as a feedback system. In the top box (process), shock rate r induces avoidance activity B according to a functional relation represented as $f(r)$, which here is a power function (Figure 10.4; Equation 10.1). The set point B^* represents the equilibrium press rate. Press rate B is compared with B^*, and ΔB, the amount of mismatch or "error," results. The lower box (process), represents the feedback function imposed by the experimental programming, which is the organism's environment. In avoidance, g is a decreasing function of press rate ($B + \Delta B$). If ΔB is positive, press rate B falls below B^*, and shock rate r increases, tending to induce more pressing and to decrease ΔB toward zero. If ΔB is negative, press rate exceeds B^*,

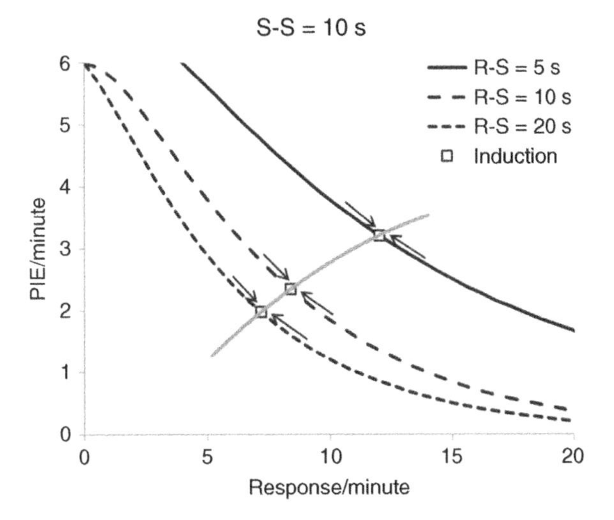

Figure 10.5 The same 3 feedback functions as in Figure 10.1 with induction added
Note: The gray curve indicates induction of avoidance activity by shock rate. Each square represents an equilibrium point, indicated by the arrows.

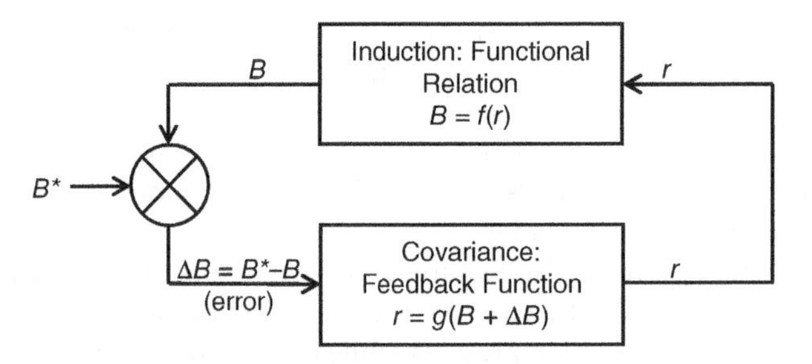

Figure 10.6 Representing avoidance equilibrium as a feedback system
Note: Shock rate r induces press rate B, which compares with the equilibrium rate B^*, generating error ΔB. The feedback function g determines shock rate r. The system acts to reduce ΔB to zero.

and shock rate r decreases, tending to induce less pressing and to bring ΔB back to zero. Thus, press rate B is always corrected back to the equilibrium rate B^* with its associated equilibrium shock rate, which may be denoted r^*. The feedback system maintains the performance at a point (B^*, r^*), as shown by the squares in Figure 10.5.

Hineline (1978)

Keeping R-S equal to S-S, Hineline (1978) exposed rats to a range of Sidman avoidance procedures, varying these intervals from 5 s to 60 s. The top graph in Figure 10.7 shows Hineline's results, with press rate as a function of R-S interval. Except for Rat 3F with R-S

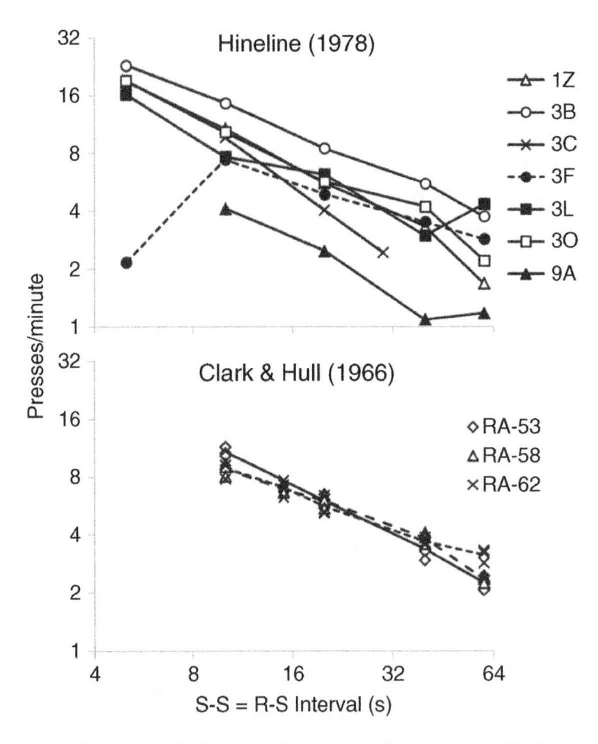

Figure 10.7 Press rate as a function of R-S interval in two studies in which R-S interval was equal to S-S interval as both were varied
Note: Different symbols show data from different rats. Top: press rates from the 7 rats in Hineline's (1978) study. Bottom: press rates from the 3 rats in the study by Clark and Hull (1966). Note logarithmic axes.

equal to 5 s, the pattern of relations resembles that seen in Sidman's (1953) data in Figure 10.2. The one outlying data point does not appear to represent a complete failure to maintain avoidance, because the low press rate (2.17 presses/minute) still kept received shock rate (4.68 shocks/minute) far below the baseline rate of 12 shocks/minute, although Rat 3F's received shock rate was the highest among the 5 rats exposed to R-S of 5 s. One may speculate that this rat developed some temporal discrimination in that condition, making its measured press rate lower.

The top graph in Figure 10.8 shows Hineline's (1978) results with press rate as a function of received shock rate, as in Figure 10.4. Each rat's press rates were fitted with a power function, shown by the line through its points. The equation of each line appears next to it. Although the rats' performances varied, they are all well described by the power functions; variance accounted for (r^2) ranged from .85 to .95. Exponents (sensitivity to shock in Equation 10.1) varied considerably: two exceeded 1.0, but the rest were comparable to those in Figure 10.4, ranging from 0.47 to 0.83. Thus, Hineline's results support power-function induction by received shocks.

Clark and Hull (1966)

The lower graph in Figure 10.7 shows results Clark and Hull (1966) obtained in Sidman avoidance with 3 rats. R-S was held equal to S-S and varied from 10 s to 60 s. The press rates varied with R-S interval similarly to Sidman's (1953) data in Figure 10.4.

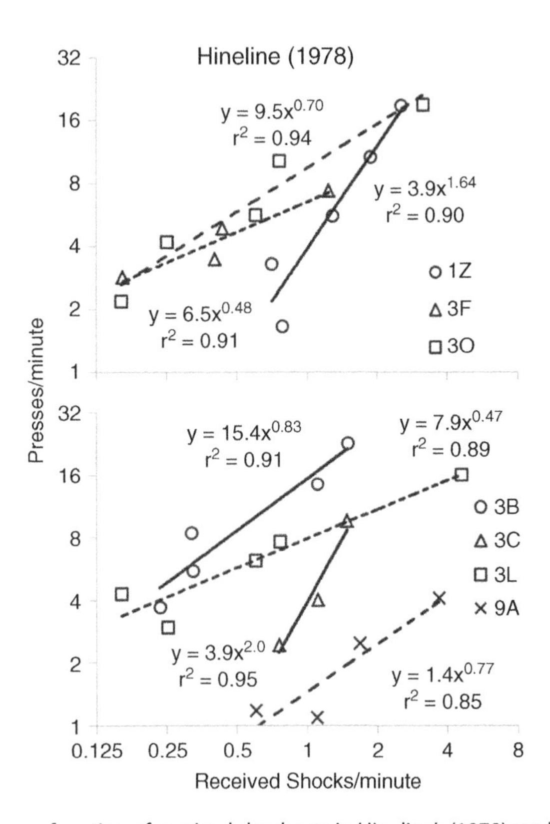

Figure 10.8 Press rate as a function of received shock rate in Hineline's (1978) study
Note: Different symbols indicate press rates for different rats. Lines were fitted by least-squares regression. The equation of each line appears near it. Note logarithmic axes.

Figure 10.9 shows the press rates obtained by Clark and Hull (1966) as a function of received shock rate. The lines represent power functions fitted to the press rates, and the equation of each power function appears alongside it. All the fits are excellent: variance accounted for (r^2) ranged from .96 to .99. These results also support power-function induction by received shocks.

de Villiers (1974)

de Villiers (1972) invented a schedule of free-operant avoidance differing from Sidman's (1953). The procedure is based on a VI schedule of avoidance. If no responses occur, shocks are delivered at variable intervals, setting a baseline shock rate. Whenever a

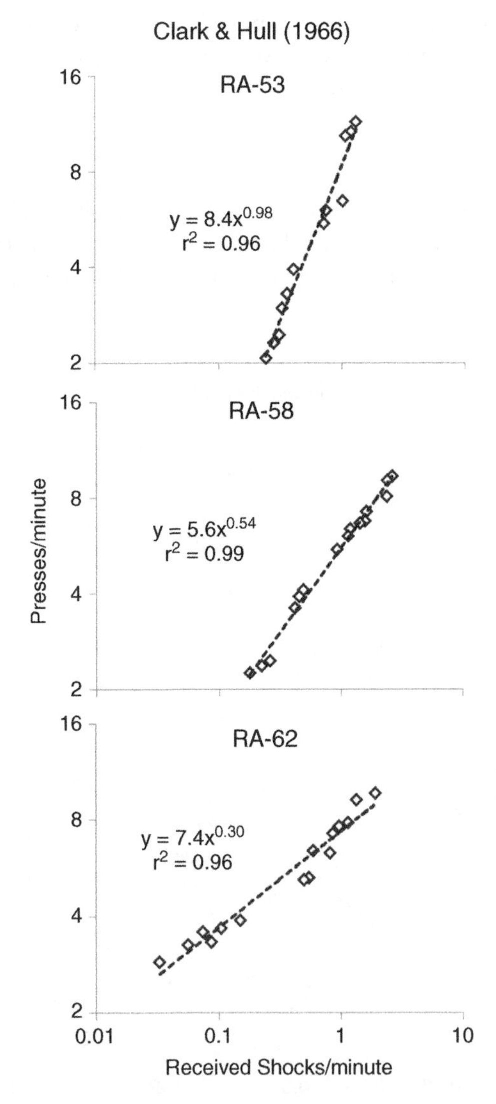

Figure 10.9 Press rate as a function of received shock rate in the study by Clark and Hull (1966)
Note: Lines were fitted by least-squares regression. The equation of each line appears near it. Note logarithmic axes.

response occurs, it cancels the next shock scheduled. As in Sidman avoidance, if the organism responds often enough, the received shock rate can be reduced to zero. de Villiers (1974) studied 4 rats' lever pressing with this procedure and using several base VI schedules varying from 15 s to 60 s. Figure 10.10 shows de Villiers's results with press rate as a function of received shock rate (diamonds and lines). The equation of each power function appears near it. The sensitivity to received shocks (s in Equation 10.1) varied across rats from 0.21 to 0.88. Analysis of residuals revealed no systematic deviations from the lines.

[Herrnstein and Hineline (1966) studied a procedure analogous to de Villiers's with unequal S-S and R-S VI schedules, but their raw data are no longer available (Hineline, personal communication).]

In an earlier paper (Baum, 2018a), I overlooked the possibility that de Villiers's (1974) results might be accounted for simply as a power function of received shock rate. Instead, I derived an equation from matching theory:

$$\frac{B_1}{B_1 + B_0 + B_N} = \frac{c_1 r_1^{s_1}}{c_1 r_1^{s_1} + c_0 r_1^{s_0} + r_N} \quad (10.2),$$

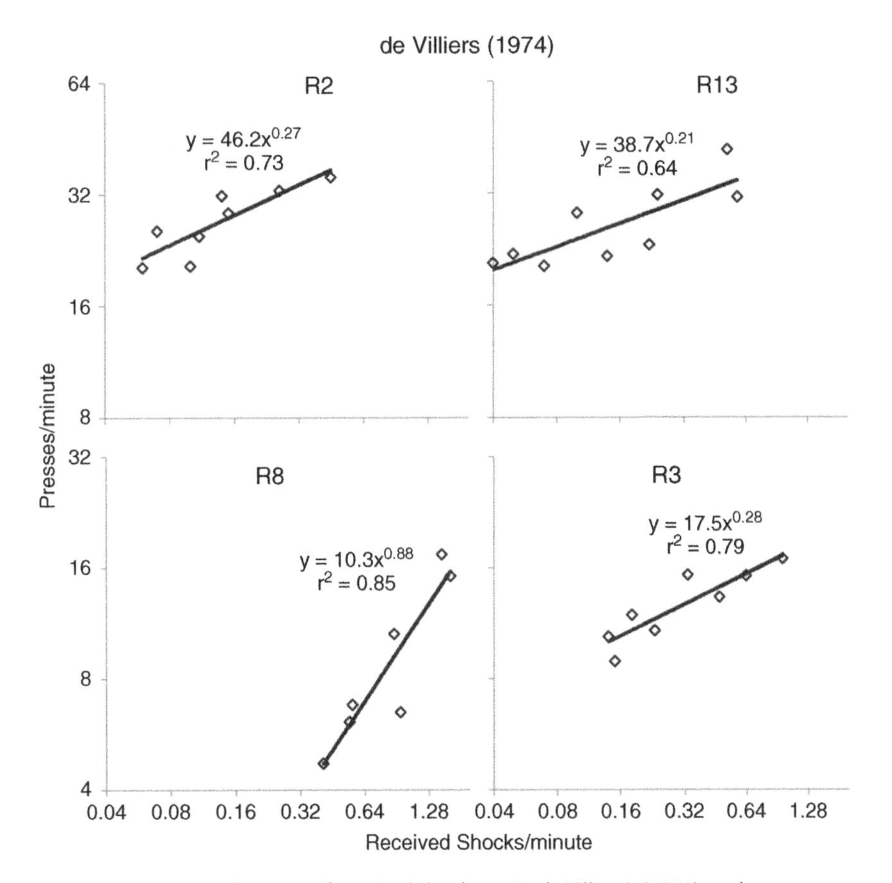

Figure 10.10 Press rate as a function of received shock rate in de Villiers's (1974) study
Note: Diamonds show press rate as a function of received shock rate. Lines were fitted by least-squares regression. The equation of each line lies near it. Note logarithmic axes.

where B_1 is rate of operant behavior—avoidance here—induced by the PIE (shock) rate r_1 according to power-function induction, B_0 is rate of non-operant activity induced also by the PIE (shock) rate r_1, B_N is activity unrelated to the PIE, and r_N represents the environmental outcomes associated with B_N. According to matching theory, the denominator on the left side of Equation 10.2 remains constant and is usually set equal to K, because B_1, B_0, and B_N take up all the time available. As r_1 increases, B_1 and B_0 both may increase at the expense of B_N, which decreases. I overlooked the possibility that the denominator on the right side of Equation 10.2 also might remain constant. As r_1 increases, r_N would decrease along with B_N, because r_N depends on B_N. If the denominator on the right remains constant, the ratio of the two denominators equals C, and Equation 10.1 applies. Thus, the results in Figure 10.10 both support power-function induction and are compatible with matching theory.

de Villiers (1972) presented VI avoidance schedules also in multiple schedules. In the presence of two different discriminative component stimuli, two different schedules maintained different press rates. One may compare this result with multiple VI schedules of food delivery. Different food rates in two signaled components maintain different response rates. If the component signals are omitted, different food rates nevertheless induce different rates of operant activity. Baum and Grace (2020) studied pigeons' key pecking in sessions that consisted of 7 different VI schedules randomly presented with no component signals. Response rates varied with food rate and were accounted for by matching theory combined with power-function induction. In their experiment, the food induced pecking, but in usual multiple schedules the component signals also induce pecking in accord with the food rate. The relation between a discriminative stimulus and the activity it is said to "control" is the relation of induction, and both the PIE and the discriminative stimulus work in concert.

Davison and Baum (2006, 2010) tested the comparison between a signal and a PIE by providing pigeons with both unsignaled concurrent food rates (VI schedules) and response-produced brief stimuli that signaled which alternative was the richer of the two. These brief stimuli were never paired with food and were produced at different rates by pecking at the two response keys, which also produced the food. When a high rate of signals indicated the key pecked was the richer key, the signals induced more pecking on that key. When a high rate of signals indicated that the key pecked was the leaner key, the signals induced more pecking on the other key. Davison and Baum interpreted their results to show that food and food-related signals function similarly—as inducers, according to Baum (2012a). One may see de Villiers's (1972) multiple schedules as similarly sharing induction by component signals and PIEs (shocks).

Logue and de Villiers (1978)

Logue and de Villiers (1978) studied several pairs of concurrent VI avoidance schedules. Two rats were presented with conditions in which a rat could press two levers, each of which was associated with a VI avoidance schedule: VI 60 s, VI 40 s, VI 120 s, VI 35 s, or VI 210 s. The schedules were paired so that they programmed an overall shock rate of 2 shocks per minute. Figure 10.11 shows the results as reported by Logue and de Villiers. The left graph shows the ratio of presses as a function of the ratio of avoided shocks. The different symbols represent data from the different concurrent schedule pairs. The filled symbols are for Rat 7, and the unfilled symbols are for Rat 9. The lines fitted to the ratios (solid for Rat 7, broken for Rat 9) are good fits—variance accounted for (r^2) equal to .89 and .92. Although I used the numbers Logue and de Villiers published, the fits here differ

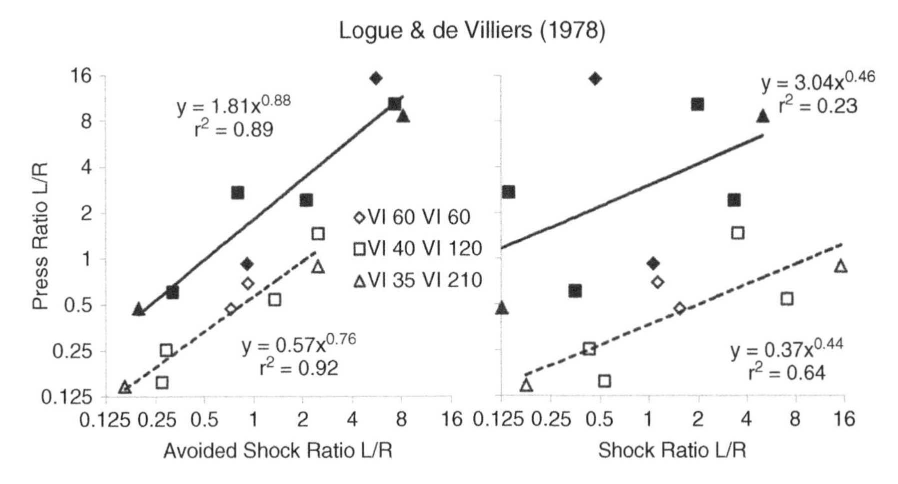

Figure 10.11 Press ratios in concurrent pairs of VI avoidance schedules as presented by Logue and de Villiers (1978) with symbols added to distinguish the different concurrent pairs
Note: Left: press ratio as a function of shock-rate-reduction ratio. Right: press ratio as a function of received shock ratio. Filled symbols show press ratios from Rat 7, and unfilled symbols show press ratios from Rat 9. Lines were fitted by least-squares regression. The equation of each line appears near it. Note logarithmic axes.

a bit from the fits they reported, but only trivially. The graph on the right shows the same press ratios as a function of the received shock ratio. The fits are much poorer than in the left graph. Logue and de Villiers concluded that they found matching to the ratio of shock-rate reduction but not to the ratio of shocks received.

Looking at Figure 10.11, one might think that the results support Sidman's (1962, 1966) theory that avoidance is maintained by shock-rate reduction. One would be mistaken, however, because the results are actually an artifact of the authors' computations.

As the first hint that the results shown in Figure 10.11 are suspect, we may note that the procedure presents an accounting problem (also known as "assignment of credit"). Although the experimenters could calculate shocks received from each of the schedules and shocks avoided by presses on each of the levers, when a shock occurred, neither the rat nor the experimenters could know to which alternative the shock ought to be assigned. A shock occurs—did it come from pressing too slowly on the current lever, or did it come from the schedule on the currently neglected lever? Experiencing only the shock, one would be at a loss as to which alternative produced it. This problem makes the experimenters' calculation of ratios of avoided shocks and received shocks inaccurate in unknown ways.

To help explain why the results in Figure 10.11 are artifactual, Figure 10.12 shows feedback functions for the various VI avoidance schedules. The top graph shows received shock rate as a function of press rate for the 5 VI schedules. Data from both rats are included, assuming the feedback depends only on the press rate. Although the exact shape of the feedback function for these schedules is unknown, the shock rates are well fitted by power functions. The lines are all approximately parallel (the line for VI 60 s would be parallel to the others except for an outlier). The exponents are all close to -0.5. Thus, shock rate was approximately inversely proportional to the square root of press rate. The bottom graph in Figure 10.12 shows feedback as avoided shock rate. Here, too, the lines are all close to parallel, and the lines are all closer to one another also. The exponents

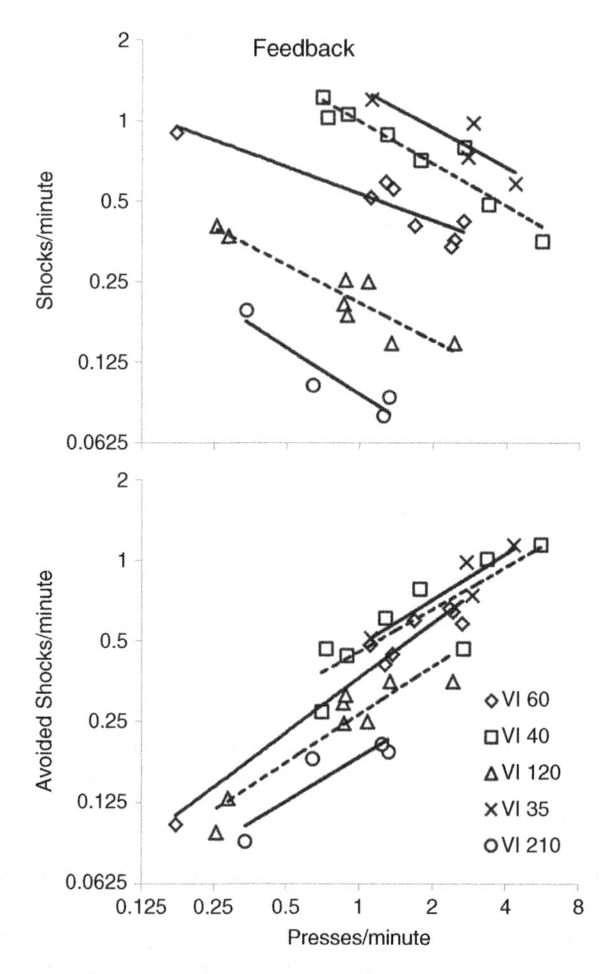

Figure 10.12 Feedback functions in the study by Logue and de Villiers (1978)
Note: Top: shock rate as a function of press rate. Bottom: shock-rate reduction as a function of press rate. Different symbols indicate functions for different VI avoidance schedules. Lines were fitted by least-squares regression. Note logarithmic axes.

are all close to a mean of 0.58. The coefficients range from about 0.2 to 0.5. Thus, both received shocks and avoided shocks were closely determined by the individual schedules.

To see how the feedback relations in Figure 10.12 would affect the ratios of shocks received and avoided, we may consider what happens when we take the ratio of two power functions like Equation 10.1. If received shock rate is related to press rate with an exponent of -0.5, then press rate is related to received shock rate by a power function with an exponent of -2.0. The ratio of press rates on any pair of the VI schedules will be related to the ratio of received shock rates by a power function with an exponent of -2.0. Similarly, if avoided shock rate is related to press rate with an exponent of about 0.58, then press rate is related to avoided shock rate with an exponent of 1.72, and the ratio of press rates will be related to the ratio of avoided shock rates by a power function with an exponent of about 1.72. These predictions are tested in Figure 10.13.

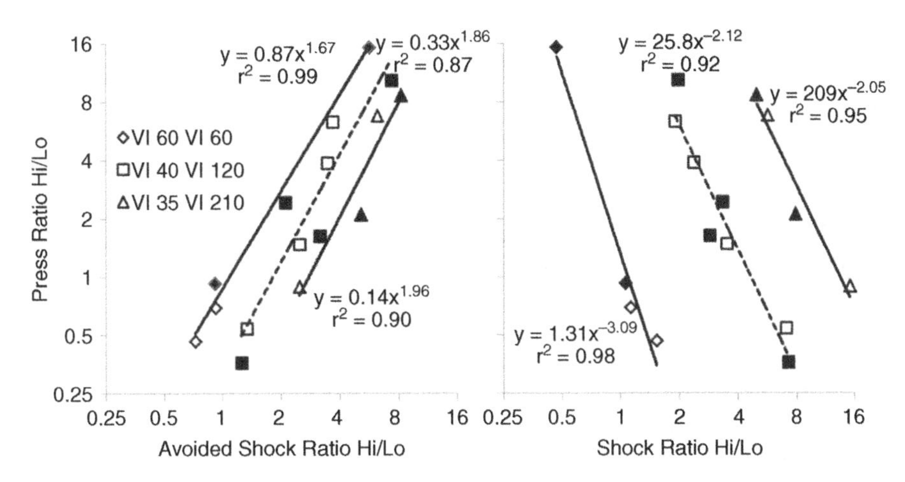

Figure 10.13 Press ratios derived from feedback functions in Figure 10.12 for the study by Logue and de Villiers (1978)

Note: Left: press ratio as a function of shock-rate-reduction ratio expressed as high shock rate divided by low shock rate (left/right for VI 60 VI 60). Right: press ratio as a function of received shock ratio. Filled symbols show press ratios for Rat 7, and unfilled symbols show press ratios for Rat 9. Lines were fitted by least-squares regression. The equation of each line appears near it. Note logarithmic axes.

Figure 10.13 shows the relations between ratios that derive from the feedback functions in Figure 10.12. For the condition with VI 60 s on each lever, the ratios are plotted as left/right. For the other two schedule pairs, the ratios are plotted as higher shock rate over lower shock rate. The filled symbols represent Rat 7 and the unfilled symbols Rat 9. A power function is fitted for each schedule pair, and the equations appear near the lines. In the graph on the left, press ratio is related to avoided shock ratio. The exponents range from 1.67 to 1.96, approximating the predicted exponent of 1.72. In the graph on the right, press ratio is related to received shock ratio. The exponents range from about -2.1 for the pairs of unequal VI schedules, and is -3.1 for concurrent VI 60 s VI 60 s, a bit steeper than -2.0 because of the outlier.

If we compare Figure 10.11 with Figure 10.13, we see how the feedback relations determined the outcome of the experimenters' computations. In the left graph of Figure 10.11, the two lower triangles are connected, and the two upper triangles are connected. The 4 lower squares are connected, and the 4 upper squares are connected. The 4 diamonds also are connected. All these represent the increasing relations shown in the left graph of Figure 10.13. Similarly, in the graph on the right in Figure 10.11, the 4 diamonds are connected in a decreasing relation, the lower triangles and the upper triangles are connected also in decreasing relations, and the lower squares and the upper squares also are connected in decreasing relations, as shown in Figure 10.13. Thus, the ratio relations in Figure 10.11 simply represent the feedback relations, and the apparently superior fit to the avoided shock ratios is just an outcome of these feedback relations.

The only reliable conclusion we may draw from the results obtained by Logue and de Villiers (1978) is that generally more pressing occurred on the lever with the higher baseline shock rate. The graph on the right of Figure 10.13 illustrates this summary: apart from the diamonds (equal VI 60 s schedules), only three data points lie below a press ratio of 1.0—the two lower squares and the lowest triangle.

Discussion

The analyses indicate that avoidance activity depends, not on shock-rate reduction alone, but on an equilibrium between shock-rate reduction and received shock rate. When Sidman (1962, 1966) pointed to shock-rate reduction, he had half the explanation: the negative covariance between shock rate and avoidance activity rate captured in feedback functions like those in Figure 10.1. Figure 10.3 shows that more is required, because avoided shocks failed to determine unique press rates. The other half of the explanation is the induction of avoidance activity by the shocks received (Figures 10.4, 10.8, 10.9, and 10.10). This induction was well-fitted by power functions. While the feedback function imposed by the programming equipment determines what shock rate will obtain at the equilibrium press rate, the shock rate determines that equilibrium press rate, as illustrated in Figures 10.5 and 10.6. These shocks are essential to maintaining avoidance. In other words, avoidance is maintained by its failures, because the received shocks—the shocks the avoidance activity fails to prevent—induce the avoidance. Just as food-maintained operant activity is induced by food, shock-maintained avoidance is induced by shock (Baum, 2012a, 2018a, 2018b).

If "reinforcement" means strengthening, and a "reinforcer" is supposed to strengthen any response it follows, then avoidance is not reinforced at all. When Sidman (1962, 1966) asked, "What is the reinforcement for avoidance?" the question itself was wrong. It presupposed that operant behavior is maintained by strengthening, by reinforcement. Operant behavior, including avoidance, increases or decreases depending on covariances (i.e., feedback relations) in the environment, both human-created and naturally occurring (Figure 10.1). PIEs induce operant behavior, along with non-operant activities (sometimes called "adjunctive" activities) induced as a result of phylogeny (Baum, 2012a, 2018a, 2018b). A threatening PIE like electric shock induces activities that mitigate or avoid it. Shock is not a "reinforcer," but it does induce avoidance. Nothing like "reinforcement" enters in.

Reinforcement has sometimes been considered much the same as reward. Skinner (1969d) rightly rejected "reward" as a technical term. The analyses presented in this paper suggest that neither food nor shock should be thought of as reward or reinforcement, but should be thought of more generally as PIEs, which induce activities, rather than reward or reinforce them. Food and shock are not reinforcers but inducers. Induction is a more general explanation of behavior than reinforcement, because induction explains both food-induced activities—operant and non-operant—and shock-induced activities, whereas reinforcement only explains food-induced operant activities.

If avoidance is induced by its failures—if pressing is induced by received shocks—then a number of predictions follow. Several phenomena associated with avoidance behavior become more understandable. I will take up some of these now.

Acquisition and Maintenance of Avoidance

In training a rat to press a lever with food, one customarily begins with the richest possible schedule, fixed ratio (FR) 1, which provides a steep feedback function that operates even in very brief time frames. Baum and Grace (2020), for example, found that even an extremely rich VI-1s schedule is discriminable from FR 1. Once responding is established with the FR 1, the schedule may be gradually extended, so that even relatively shallow feedback functions continue to maintain the operant activity. This is standard procedure, and even the lowest food rates still maintain responding well above the near-zero rate in extinction (Baum, 2012c; Baum & Grace, 2020).

When experimenters train rats to press a lever to avoid electric shock, they similarly begin with extremely steep feedback functions, discriminable from unavoidable shock in relatively brief time frames. This means pairing a short S-S with a substantially longer R-S. For example, Hineline (1978) began avoidance training with S-S of 5 s and R-S of 20 s, and after pressing was reliable, switched to S-S of 20 s. Sidman (1953) began with S-S equal to 15 s and R-S equal to 30 s.

Whether food-based or shock-based, the training establishes covariance between operant activity and a PIE: positive covariance between food and the activity; negative covariance between shock and the activity (Figure 10.1). After the covariance comes to maintain the operant activity, other feedback functions may be introduced, because now the food or shock induces the activity. Just as food-based activity is maintained by food received, so shock-based activity is maintained by shocks received.

Short-Term versus Long-Term Covariance

In everyday life short-term covariance between an activity and PIE may conflict with long-term covariance between that activity and more important PIEs. The short-term effects of an alcoholic's drinking conflict with keeping job and family; clothes shopping may conflict with managing credit-card debt; having an affair may conflict with preserving one's marriage; eating junk food may conflict with maintaining health. When the activity induced by short-term covariance is considered maladaptive, the activity is called "impulsive"; when the activity induced by long-term covariance is favored, it is called "self-control" (Logue, 1995; Rachlin, 2000).

In the laboratory, researchers can arrange choice between short-term covariance and long-term covariance. Typically, a subject is offered a choice between a small PIE (e.g., food or money) to be had after a short delay ("smaller-sooner") versus a larger PIE (food or money) after some longer delay ("larger-later"). If the subject prefers the smaller-sooner option, the failure of the larger-later option to be chosen is said to reflect "delay discounting," which may reflect a failure of the larger-later covariance to induce its choice. When a series of choices is offered in which the amount of the sooner option is varied, a point of indifference between the two options appears, and if this procedure is repeated with various longer-delay comparisons, one obtains a discount function relating the smaller amount at indifference to the delay of the larger amount (e.g., Odum, 2011). The curve marks the border line between induction by the short-term covariance and induction by the long-term covariance. The steeper the curve, the more the short-term covariance tends to dominate choice. If such choice is considered "impulsive," researchers look for ways to tip choice more toward the longer-later alternatives—which is to say they look for ways to improve induction by the long-term covariance.

Parallel conflicts between short-term and long-term covariance occur in avoidance of bad PIEs. When someone avoids the dentist, for example, only to have toothaches and root canals later, the short-term avoidance of the dentist is maladaptive when going to the dentist might have avoided worse PIEs in the long-term. Avoiding the dentist might be called "impulsive."

Maladaptive avoidance has been studied less than maladaptive choice of beneficial PIEs, but some experiments indicate that rats exposed to conflicts between short-term and long-term covariance between lever pressing and electric shock sometimes behave maladaptively. Hineline (1970) exposed rats to a procedure in which, in the absence of lever pressing, a 20-s cycle repeated: 10 s of lever inserted, 10 s of lever retracted, and a shock delivered at the 8th second. A press on the lever before 8 s retracted the lever and produced a shock after 18 s, creating a short-term negative covariance between pressing and shock. In one experiment, the shock-rate remained at 3 per minute regardless of whether a press occurred, because the 20-s cycle began when it would have had no press occurred, and presses occurred on about 80 percent of the cycles. Thus, in the absence of any long-term covariance, the short-term covariance sufficed to maintain pressing. In another experiment, pressing increased the shock rate because a press started a new 20-s cycle immediately, bringing the shock closer in time, and lever pressing ceased. When the long-term covariance between pressing and shock was positive, it acted like punishment, inducing activities other than pressing.

Gardner and Lewis (1976) followed up on Hineline's study by exposing rats to a baseline of variable-time (VT) 30 s shock (2 per minute) and a lever that when pressed produced a signaled 3-minute alternate condition in which 6 shocks were delivered at 1-s intervals after a delay. The longer the delay, the more presses occurred and the more time was spent in the alternate condition. Thus, Hineline's (1970) result was replicated, because the pressing never reduced the shock rate below 2 per minute. The short-term negative covariance sufficed to maintain lever pressing in the absence of any long-term covariance. Gardner and Lewis then increased the number of shocks delivered during the 3-minute alternate condition to 9, 12, and 18 shocks, delayed for 161, 158, and 152 s, increasing the shock rate during the alternate condition to 3, 4, and 6 per minute. Rats pressed the lever even when the shock rate increased from 2 to 3 or 4 per minute. Thus, the short-term covariance maintained pressing even in the face of the long-term positive covariance between pressing and shock. This result offers a model of "impulsive" avoidance, because the short-term covariance overrode the long-term covariance (see Baum, 2017c for a theoretical account). When 18 shocks occurred in the alternate condition, 4 out of 6 rats ceased pressing the lever; the long-term contingency proved severe enough to engage the behavior of these rats.

Lewis, Gardner, and Hutton (1976) studied a procedure like that of Gardner and Lewis (1976) with unchanged shock rate in a 3-minute or 5-minute signaled alternate condition in which the first one or two shocks were delivered on the baseline schedule (VT 30-s or fixed time 30-s) and the remaining shocks were delivered after a delay. Lever pressing was maintained, thus replicating Hineline's (1970) result that the short-term covariance sufficed to maintain pressing in the absence of any long-term covariance. Lewis et al. explained the result by suggesting that the transition from the baseline to the alternate condition may have functioned to maintain pressing because shock rate consists of integrated delays to shock, concluding, "Delays to all shocks are integrated, and the integrated delays to shocks in the absence of a response *versus* integrated delays to shocks following a response determines the value of the transition" (p. 386). This interpretation concurs with present view if one interprets "value" as inducing power.

Some research indicates that avoidance may be affected by longer-term covariance beyond the covariance existing in Sidman avoidance (Figure 10.1). Mellitz, Hineline,Whitehouse, and Laurence (1983) exposed rats to a situation with two levers, each programmed with the same avoidance schedule. One lever had the additional effect that every press shortened the session time by one minute. Pressing shifted in favor of the lever that shortened the session. When the session-shortening was switched to the other lever, pressing shifted too. Thus, the long-term covariance that decreased shocks received over the course of a day served to maintain increased pressing on the effective lever. Experiments with response-produced timeout from avoidance also suggest efficacy of longer-term covariance (e.g., Courtney & Perone, 1992). An experiment by Galizio (1999) showed induction by both short-term and long-term covariances. Galizio exposed rats to a situation with two levers. One lever postponed shocks on a Sidman avoidance schedule (S-S = 5 s; R-S = 30 s). The other lever produced 2-minute timeouts from avoidance according to a variable-ratio 15 schedule. When the rats had been well trained in these conditions, Galizio instituted extinction in two ways: (a) responses on the timeout lever became ineffective; or (b) electric shocks were discontinued. When presses on the timeout lever were no longer effective, pressing on that lever rapidly declined. When electric shock was discontinued, pressing on the avoidance lever declined rapidly, but pressing on the timeout lever declined far more slowly. Possibly the longer-term covariance between pressing on the timeout lever and electric shock was less susceptible to extinction because the removal of the longer-term covariance was less discriminable. The phenomenon might be compared to the longer time to extinguish of responding following low food rates (Baum, 2012c).

Extinction of Avoidance

Much more research has occurred on extinction following food-maintained behavior than following avoidance, but some parallels exist. First, just as when food is discontinued, when shock is discontinued, responding gradually decreases to low levels. For example, Sidman, Herrnstein, and Conrad (1957) trained 2 rhesus monkeys to lever press on a Sidman avoidance schedule with S-S and R-S intervals equal to 20 s and then discontinued shock. One monkey received daily 2-h sessions, and the other 6-h sessions. After about 15 sessions of extinction, responding dropped to levels 5-10-fold less than in avoidance.

Alternating training and extinction on a daily basis results in formation of a discrimination between the two conditions when responding is maintained with food (Baum, 2012c, Bullock & Smith, 1950). A discrimination theory of extinction predicts that the higher the food rate, the faster should be extinction, because higher food rates differ more from zero. In an experiment with pigeons key pecking on VI food schedules and daily alternation between VI and extinction, when the rate of food varied across conditions, this prediction was confirmed (Baum, 2012c).

In one experiment on repeated training and extinction of avoidance, Boren and Sidman (1957) found that extinction of avoidance, like extinction following food, occurred faster with repeated exposure. Powell and Peck (1969) trained rats in a variant of Sidman avoidance in which shocks occurred every 5 s and each lever press reduced the intensity of the shocks for 20 s. Following training, they exposed the rats to 1-h sessions in which the avoidance schedule operated for the first 10 minutes, following which shocks were discontinued for 50 minutes. The rats quickly came to discriminate between avoidance and extinction, responding at a high rate during the first 10 minutes and at a low rate during the 50 minutes of extinction. These results with repeated extinction offer evidence that extinction results from discrimination between a shock rate greater than zero during training and a zero shock rate during extinction. Shocks induce pressing when they are present, but do not induce pressing when they are absent.

Unavoidable Shock

If avoidance responding is maintained by induction, then electric shocks delivered independently of behavior ("free" shocks) may, at least temporarily, induce the responding. Sidman, Herrnstein, and Conrad (1957) trained 3 rhesus monkeys to lever press to avoid shocks (S-S and R-S equal to 20 s) and then discontinued the avoidance schedule while occasionally delivering free shocks. Lever press rate was extinguishing across sessions, but the free shocks increased lever pressing whenever they were present. Although pressing decreased to low levels, the level in the presence of free shocks always remained higher than when the free shocks were absent. When the free shocks were signaled with an auditory stimulus, the signal caused an increase in press rate, indicating that the signal, as a conditional inducer, acquired the ability to induce avoidance in addition to the induction by the free shocks.

Powell and Peck (1969) also studied the effect of free shocks following avoidance training and found that free shocks continued to induce lever pressing for a time. They varied the rate of the free shocks and found that the rate of pressing varied directly with the shock rate. These results also support the theory that avoidance rate is induced by received shock rate. One note seems particularly telling. Powell and Peck found that intensities of shock that failed to maintain avoidance nevertheless induced pressing when non-contingent. They remarked, "It would appear that responses depended on shock intensities that the animal could detect, rather than those that were aversive." (P. 1055.)

The experiment by Herrnstein and Hineline (1966) included unavoidable shock mixed with avoidable shock. In the absence of lever pressing, shocks occurred at a baseline rate at variable intervals. A press switched the program to a lower shock rate, but shock rate never equaled zero; the post-press shocks were unavoidable. When pre-press and post-press rates were equal, presses had no effect, and pressing ceased. As long as the post-press shock rate was lower, pressing was maintained, and the larger the differential between the two shock rates, the higher was the press rate. Since pressing stabilized at rates that reduced the shock rate but never reduced it all the way down to the post-press shock rate, the performance might represent equilibrium between shock-rate reduction and induction, as portrayed in Figures 10.5 and 10.6, but in the absence of data, the analysis cannot be done.

A possible way to view the Herrnstein-Hineline (1966) procedure would be to compare the mix of avoidable and unavoidable shock with procedures that mix response-contingent with response-independent food (Baum & Aparicio, 2020); Kuroda et al. (2013); Rachlin & Baum, 1972). Response-independent and response-contingent food in those experiments function as different sources of food, and they induce both operant and non-operant activities. Baum and Aparicio recorded some non-operant (adjunctive) activities in rats along with operant lever pressing and found that the operant and non-operant activities tended to compete with one another for time, in accord with matching theory. For example, when press rate was low because most food was response-independent, non-operant activities occurred at a high rate, but when press rate was high because most food was response-contingent, non-operant activities diminished. Operant activity was induced according to power functions of response-contingent food rate. If avoidable and unavoidable shocks function as independent sources of shock this way, then the Herrnstein-Hineline procedure might be analyzed according to the proportion of avoidable shock—that is, relative avoidable shock.

Punishment of Avoidance

The very same shock that induces avoidance responding also can serve to punish the responding. Powell and Peck (1969) conducted 60-minute sessions in which shocks every 5 s were non-contingent during the first 30 minutes and then were produced by the lever presses during the last 30 minutes. No signal indicated the switch. When the shocks switched from non-contingent to contingent, the rats quickly ceased pressing (Powell & Peck, 1969; Figure 7).

What happens if punishment is superimposed on avoidance? Presumably a conflict would arise between responding induced by the shocks not avoided and responding deterred by the contingent shocks. Baron, Kaufman, and Fazzini (1969) superimposed punishment on rats' lever pressing maintained by a Sidman avoidance procedure with S-S and R-S both equal to 60 s. They varied density of shock punishment from 2 per minute to 0.2 per minute and varied delay of contingent shock from zero to 60 s. Both density and delay of punishment affected avoidance lever pressing; the shorter the delay and the higher the density, the more lever pressing was suppressed. When delay was long and density low, avoidance responding was unaffected. As press rate decreased, shock rate due to the avoidance schedule increased. The results indicate that when a positive covariance between pressing and shock rate was detectable, it suppressed pressing to some extent, as might be expected if shock is a PIE. When the covariance was strong (i.e., when delay was zero), it suppressed pressing substantially, as happens with lever pressing induced by contingent food (Azrin & Holz, 1966). Most likely, the means of suppression is induction of other activities incompatible with pressing.

Punishment of avoidance may be compared to the procedure with pigeons known as negative automaintenance (Williams & Williams, 1969). A light on a response key is paired with food, as in autoshaping (Brown & Jenkins, 1968), with the effect that the light induces pecking at the key, but in negative automaintenance, a peck at the key eliminates food presentation. The positive covariance between the light and food conflicts with the negative covariance between pecking and food. The more the pigeon pecks the key, the less the light is paired with the food, and the less the pigeon pecks the key, the more the light is paired with food. The negative covariance, when operative, decreases peck rate substantially (Sanabria, Sitomer, & Killeen, 2006). Thus, the food comes to induce other activities than pecking the key (often pecking next to the key).

Paradoxical Shock-Maintained Behavior

With the right preliminary training, a seemingly paradoxical performance occurs, in which lever pressing produces electric shock and yet is maintained by the shocks produced. The phenomenon has been demonstrated in squirrel monkeys (e.g., Kelleher & Morse, 1968; Malagodi, Gardner, Ward, and Magyar, 1981) and cats (Byrd, 1969), and usually with shocks occurring on a fixed-interval (FI) schedule. The preliminary training consisted either of avoidance or food-maintained pressing with shock added and then food omitted.

This shock-maintained pressing may be compared to food-maintained pressing. When food covaries with pressing, the food induces the pressing, and the pressing produces the food, completing a feedback loop. In shock-maintained pressing, shock covaries with pressing, the shock induces the pressing, and the pressing produces the shock, similarly completing a feedback loop. The preliminary training is crucial, because it establishes the initial covariance between pressing and shock, causing shocks to induce pressing. Avoidance does this as we have seen in Figure 10.4, for example. Alternatively, correlating shock with food that is produced by pressing causes the shock to become a conditional inducer (or discriminative stimulus) that induces pressing (Kelleher & Morse, 1968). Either way, once shock reliably induces pressing, switching to a situation in which the pressing produces the shocks closes the loop: shock induces pressing, pressing produces shock, shock induces pressing, and so on. If the shock schedule is a FI, then a temporal discrimination is superimposed, and as time elapses in the FI, later time induces pressing, creating an increasing rate comparable to the FI scallop observed with food (Malagodi et al., 1981).

Although the paradoxical shock-producing performance may be explained by induction, it challenges any theory of reinforcement based on strengthening. Shock is supposed to be a punisher, an aversive stimulus that weakens responses on which it is contingent. How could it strengthen behavior that it follows? In contrast, induction explains the performance and renders it no longer a paradox. The prior establishment of covariance between pressing and shock apparently causes the shock to induce pressing, and then the pressing producing the shock completes the feedback loop to maintain pressing.

Warmup

Researchers on avoidance usually record performance only in the later part of a session—the last two-thirds or the last half—because responding at the beginning of the session is lower than later on in the session. The tendency for avoidance responding to increase after the beginning of the session is known as the "warmup effect." All the data sets we have considered in the present paper excluded the warmup period at the beginning of sessions and reported only performance late in sessions. How should we explain the lower response rate at the beginning of the session?

Hineline (1978) studied the warmup as well as the later performance. He found that the warmup effect was greater for shorter S-S intervals and decreased as S-S interval increased. The greater the shock rate, the greater was the change from beginning to end of the session and the longer warmup lasted. Hineline considered some possible explanations, such as a "motivational" effect and habituation, but concluded that these predicted the opposite effect: that the warmup should be less for the higher shock rates.

Power-function induction may provide an explanation for the variations in warmup that Hineline (1978) reported. If shock induces, not only avoidance responding (pressing) but also other non-operant activities like freezing and jumping about, then the induced activities would compete with pressing for time and reduce press rate. The greater the shock rate, the more these non-operant activities might be induced at the beginning of the session, and as time passes the shocks induce avoidance, and pressing begins competing more effectively, thereby increasing press rate and decreasing the non-operant activities.

The question remains as to why warmup occurs in the first place. Why doesn't avoidance responding begin immediately? The answer probably lies in the unique stimulus conditions of the beginning of the session. Typically, the animal is taken from its home cage and transported to the experimental apparatus. The beginning of the session is uniquely paired with shocks and may induce the non-operant activities that compete with avoidance. Its effects would wane as time in the session elapsed. In keeping with this explanation, Hoffman, Fleshler, and Chorny (1961) reported that exposing rats to unavoidable shocks in the absence of the response lever during the first half of sessions eliminated the warmup that occurred during the first half when avoidance training began at the beginning of sessions. The unavoidable shocks may have induced non-operant activities that ceased as soon as the lever was introduced.

This explanation of warmup resembles an explanation that was offered for spontaneous recovery (Kimble, 1961). Even if, following a change in contingencies, performance adjusts within a session, the beginning of the next day's session, following a stay in the home cage, sees a partial recurrence of the previous day's initial performance. This occurs with extinction, for example, when the beginning of the second session of extinction shows a recurrence of responding, even though responding had extinguished at the end of the first session. Mazur (1997) reported a similar recurrence in choice when concurrent schedules changed. The unique stimulus conditions at the beginning of the session induce the same performance at the beginning of the second session as they did at the beginning of the first session.

Avoidance and Matching Theory

The power functions shown in Figures 10.4, 10.8, 10.9, and 10.10 present a puzzle for matching theory. Equation 10.2 shows the law of allocation (also the "generalized matching law") for just one operant activity in relation to PIE rate (Baum, 1974a, 2018a, 2018b). The law assumes that behavior is continuous, meaning that the activities of an organism take up all the time available, and that the time available is limited. These assumptions together imply that the various activities of an organism must compete with one another for time. Since time budget is limited, if one activity increases, others must decrease, and if one activity decreases, others must increase. Thus activities take up time relative to one another, compete with one another for time, and depending on environmental factors such as rates of PIEs (e.g., food and injury), the mix of activities will tend toward stability in a stable environment (Baum, 2012a, 2018a, 2018b). The law of allocation specifies time allocation among activities:

$$\frac{T_j}{\sum_{i=1}^{n} T_i} = \frac{V_j}{\sum_{i=1}^{n} V_i} \quad (10.3),$$

where T_i or T_j is time taken up by Activity i or Activity j, one of n activities, and V depends on PIE rate, amount (e.g., Davison & Baum, 2003; Cording, McLean, & Grace, 2011), and immediacy (or delay; e.g., Chung & Herrnstein, 1967; Davison & Baum, 2007). Going beyond strict matching, incorporating Baum's (1974a, 1979; Sutton, Grace, McLean, & Baum, 2008) modifications that introduced bias and sensitivity, we may substitute power functions of r, PIE rate, or a product of a number of power functions of other PIE-related variables, such as amount and immediacy:

$$V_i = \prod_{j=1}^{m} c_j x_{ij}^{s_j} \quad (10.4),$$

where x_{ij} is a variable such as rate, amount, or immediacy (up to m such). V_i in Equations 10.3 and 10.4 represents the competitive weight of Activity i. Equation 10.3 states that the relative time taken up by any activity equals (matches) the relative competitive weight of the activity. Baum and Rachlin (1969) used Equation 10.4 to propose a version for two alternatives that when generalized leads to Equation 10.3.

Equation 10.3 may be rewritten in various ways (Baum, 2012b). Equation 10.3 works well for explication of matching theory, but logarithmic forms work better for fitting to data (Baum, 1974a, 1979; Baum & Rachlin, 1969). Accordingly, analyses based on the law of allocation are often represented in logarithmic coordinates, as in Figures 10.4, 10.8, 10.9, and 10.10.

We may simplify Equation 10.2 by assuming that r_N is negligible except when r is extremely low. Then, dividing top and bottom on the right by $c_1 r_1^{s_1}$ we obtain:

$$B_1 = \frac{K}{1 + c r_1^{-s}} \quad (10.5),$$

where c equals c_0/c_1 and s equals $s_1 - s_0$. As long as s_1, the sensitivity of the operant activity, exceeds s_0, the sensitivity of non-operant (adjunctive) activity, Equation 10.5 describes an increasing concave-downward curve.

In an earlier article (Baum, 2018a), I fitted Equation 10.5 to de Villiers's (1974) data (Figure 10.10). Figure 10.10 shows that simple power functions actually fit de Villiers's data better than Equation 10.5; analysis of residuals indicated no systematic deviation from the power functions. When I attempted to fit Equation 10.5 to Sidman's (1953) data (Figure 10.4), no satisfactory fits were possible. The same holds for Hineline's (1978) data (Figure 10.8) and the data of Clark and Hull (1966; Figure 10.9). Thus, if Equation 10.2 were to apply to avoidance, one must conclude that the denominator on the right is constant.

The law of allocation in Equation 10.3, however, implies that the total competitive weight on the right should vary with the overall richness of the environment. It indicates that the denominator on the right in Equation 10.2 ought to vary with the PIE rate r_1. How could it remain constant?

One way that the denominator on the right of Equation 10.2 could remain constant and a simple power function could apply would be that an implicit assumption underlying the law of allocation (Equations 10.3 and 10.4) is violated: that the activities and competitive weights are independent of one another—that is, they are induced independently. On this assumption, a single operant activity and its concomitant non-operant (adjunctive) activity would compete with one another. Equation 10.5 implies that the ratio of competitive weights, not the sum, determines allocation between operant and adjunctive activities. This assumption appears to hold for food-maintained activities, because Equation 10.5 fits single-operant response rates well (Baum & Grace, 2020). Baum and Aparicio (2020) studied a situation with mixed response-contingent and non-contingent food, and found constant total competitive weight, but in that experiment the overall food rate was constant.

Possibly in these experiments, with limited space and no possibility of escape, the activities induced by electric shock are induced as a suite and are not independent of one another. The non-operant activities such as lever biting, jumping, and freezing (Pear et al., 1972; Pear et al., 1978) might be physically linked to the operant avoidance activity. One might suppose this is true with avoidance in a shuttle box; it might be less obvious when the activity is lever pressing. Further research on avoidance might shed light on this and other possible explanations.

Conclusions

We place a hungry rat in a chamber that contains a lever and a food hopper. Lever pressing occasionally produces a food pellet in the hopper. Lever pressing increases along with other activities like licking and sniffing in the hopper (Baum & Aparicio, 2020). If food rate increases with press rate, press rate increases. A steady rate of food induces steady rates of both pressing and hopper activities. We call the pressing "operant" activity, and we call the hopper activities "adjunctive" activities.

We place a healthy rat in a chamber that contains a lever and a stainless-steel grid floor. We deliver brief, inescapable electric shocks to the grid floor every few seconds. Lever pressing reduces the rate of the shocks. Lever pressing increases along with adjunctive activities like freezing, jumping about, and attacking the lever (Pear et al., 1972; Pear et al., 1978). If shock rate decreases, press rate decreases. If press rate decreases, shock rate increases. A steady rate of shock induces steady rates of both pressing and adjunctive activities—equilibrium, as in Figures 10.5 and 10.6.

Neither of the accounts above makes any appeal to strengthening or reinforcement. Induction takes the place of reinforcement and more besides, because it applies not only to food-induced operant activities, but also to adjunctive activities and avoidance. If one tried to equate reinforcement with induction, however, one would run into paradoxes. Would we want to say that electric shock reinforces avoidance responding? Adjunctive behavior by definition is not reinforced, so how could one say it is reinforced?

We face the question, "What is reinforcement?" A nagging question that never went away was, "How can a consequence, following a response, act backward in time to strengthen the response?" This question remains unanswered so long as we consider reinforcers to be consequences. Induction allows us to see that a PIE produced by an operant activity is not only a consequence but is also an antecedent. The distinction between consequences and antecedents collapses when we think of behavior and environment as temporally extended and activities as induced by PIEs (Baum, 2012a).

Examples of the collapse of the distinction between consequences and antecedents abound. When a rat's lever presses produce food pellets, the food is a consequence, but for the rat's continued pressing, the food is an antecedent—an inducer. If the food is discontinued, the pressing stops. As long as the pressing and food covary in a temporally extended correlation, the pressing continues. If a person is paid weekly for work, the pay is a consequence of last week's work, but an antecedent (inducer) for next week's work. Work and pay covary. If the pay were discontinued, the work would stop. If Ted and Jane are in a loving relationship, Jane's loving actions induce loving actions in Ted as a consequence, and Ted's loving actions, as an antecedent, induce further loving actions in Jane. If loving actions of one person stop, the other person's loving actions also stop. This continuous cycle of maintenance is perhaps best represented by a feedback system, rather than open-ended causality (Figure 10.6 here; Baum, 1973, 1981b; 2012a).

An earlier paper (Baum, 2018a) proposed three laws of behavior that together account for most of what we know about behavior. The laws are: (a) the law of allocation (or the generalized matching law); (b) the law of induction; and (c) the law of covariance. The concept of reinforcement served well in its time; its weaknesses could be overlooked as long as a better alternative was unavailable. Now a better alternative is available.

Appendix

The Shape of the Avoidance Feedback Function

Deriving a feedback function for Sidman avoidance requires examining the interaction of interresponse times (IRTs) with the schedule parameters. Lever pressing in Sidman avoidance is often far from random. Pressing occurs in bursts separated by longer IRTs that might be called "pauses." For example, de Villiers (1972) reported, "Bursts of very short interresponse times (<1.0 s) were found immediately after a shock was presented, but not at any other time" (p. 500). Such bursts may reflect the rat's attacking the lever—that is, induced aggression (Pear et al., 1972; Pear et al., 1978). Frequency distributions of avoidance IRTs show the shortest IRTs are the most frequent (Sidman, 1966), and this mode represents post-shock bursts. Food-maintained operant activity in rats also occurs in bursts and pauses (Brackney & Sanabria, 2015), and the pauses begin and end approximately randomly (Shull, Grimes, & Bennett, 2004). In deriving the avoidance feedback function, we focus on the pauses, because the pauses most affect shock rate.

Figure 10.A1 illustrates one possible approach to the avoidance feedback function that emphasizes the interaction of the pauses with the R-S and S-S intervals (hereafter *R* and *S*). It shows a frequency distribution of times to a response assuming a response has just occurred—that is, IRT. For all IRTs less than *R*, shock rate is zero. Their probability is the area under the curve from zero to *R* (area *a* in Figure 10.A1). For all IRTs greater than *R*, the expected shock rate is about *1/R*, and its probability is the area under the curve to the right of *R* (10 s in Figure 10.A1). For all IRTs greater than *R+S*, the expected shock rate is about *2/(R+S)*, and its probability is the area under the curve to the right of *R+S* (20 s in Figure 10.A1). And so on. The contribution of each expected shock rate to the overall average shock rate is the product of the probability times the shock rate. In general:

$$r_n = P(x > R+(n-1)S)\frac{n}{R+(n-1)S} \qquad (10.A1),$$

where x is the IRT, and n is the number of shocks. Equation 10.A1 expresses constituent shock rate r_n as the expected rate times its probability.

Equation 10.A1 may approximate constituent shock rate r_n, but two considerations must apply. First, the equation specifies an infinite series that could not be computed exactly. The series converges rapidly, however, as the probability (i.e., area under the curve) decreases. We can approximate the mean shock rate out to infinity with just the first few terms, because the remainder is small enough to be negligible. Second, if the longer IRTs ("pauses") terminate randomly, according to a Poisson process (i.e., their termination probability is the same across time), then their durations are distributed according to an exponential frequency distribution. That means the probabilities in Equation 10.A1 may be approximated as:

$$P(x > R+(n-1)S) = e^{-a(R+(n-1)S)} \qquad (10.A2),$$

where the expression on the left is the area under the curve from *R+(n-1)S* to infinity. The rate constant *a* determines how quickly the probability falls and is directly proportional to the overall press rate *B*. We may set *a* equal to λB and expect λ to vary from rat to rat and possibly from schedule to schedule if patterns of interaction with the lever change.

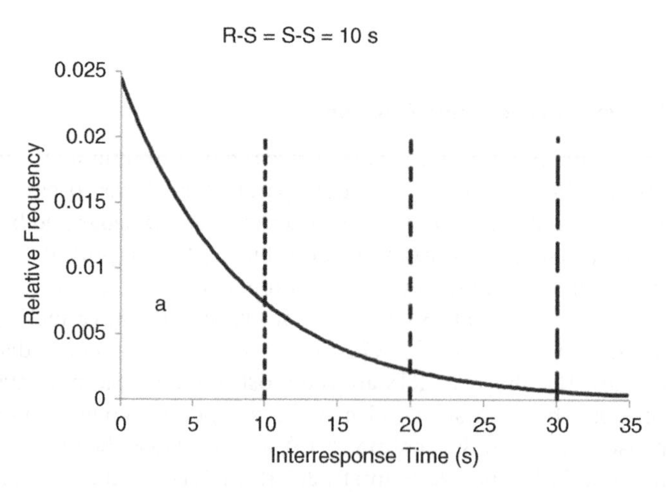

Figure 10.A1 A frequency distribution of interresponse times (IRTs) in Sidman avoidance
Note: The particular schedule has R-S and S-S intervals (*R* and *S*) both equal to 10 s. Vertical lines demarcate *R*, *R+S*, and *R+2S*. See text for more explanation.

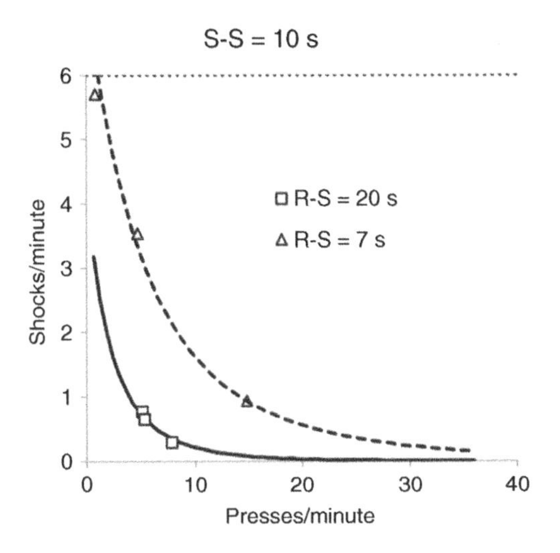

Figure 10.A2 Feedback functions for Sidman avoidance
Note: The two functions shown are for S-S interval equal to 10 s, with R-S interval equal to 20 s (solid line) and with R-S interval equal to 7 s (broken line). The data points show Sidman's (1953) results for his 3 rats.

To approximate the overall shock rate, I averaged the first 4 terms given by Equation 10.A1, using Equation 10.A2 to give the probabilities. Figure 10.A2 shows feedback functions for two schedules with S-S equal to 10 s. The squares show Sidman's (1953) results with R-S equal to 20 s—greater than S-S. The triangles show Sidman's results with R-S equal to 7 s—less than S-S. The parameter λ equaled 1.0 for R-S 20 s and 0.65 for R-S 7 s. All points lie close to the feedback functions as expected. When R-S was greater than S-S, all three rats' shock rates lay substantially below the baseline rate of 6 per minute. When R-S was less than S-S, shock rates were higher, and one rat's shock rate, paired with a low press rate, came close to the baseline of 6 per minute.

Since Sidman's (1953) was the only data set to include schedules with unequal S-S and R-S, Figure 10.A2 shows only Sidman's results. Figure 10.A3 shows results from Sidman's,

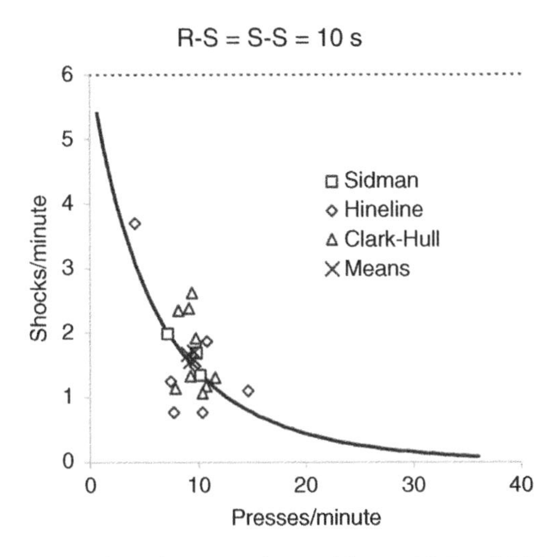

Figure 10.A3 Feedback function for Sidman avoidance with equal R-S and S-S intervals
Note: Data points show individual rats' performances in the experiments of Sidman (1953), Hineline (1978), and Clark and Hull (1966).

Hineline's (1978), the Clark-Hull (1966) experiments with R-S and S-S equal to 10 s. Although the individual rats' performances varied considerably around the feedback function, the deviations do not appear to be systematic, and the mean performances across rats (Xs) fall close to the curve. The parameter λ equaled 0.6 for the curve shown in Figure 10.A3, but presumably varied from rat to rat.

Figures 10.A2 and 10.A3 show a sample of feedback functions. Other schedules produced comparable results.

11

Multiscale Behavior Analysis and Molar Behaviorism: An Overview

Foreword

This paper aimed to summarize and integrate various threads from research and theory up to 2018. It made the link to evolutionary theory through the concept of induction and elaborated on the way the induction replaces concepts like "reinforcement" and "strength." In the molar view, behavior consists of processes (activities) serving functions that ultimately serve reproduction. Empirical studies that came after 2018 added to the plausibility of viewing behavior as process and underlined the necessity of understanding behavior as extending through time. Induction allows one to quantify avoidance (Baum, 2020), competition between activities (Baum, 2021; Baum & Aparicio, 2020), and changes within activities (Baum & Grace, 2020).

Abstract

In the context of evolutionary theory, behavior is the interaction between the organism and its environment. Two implications follow: (a) behavior takes time; and (b) behavior is defined by its function. That behavior takes time implies that behavioral units are temporally extended patterns or *activities*. An activity functions as an integrated whole composed of parts that are themselves smaller-scale activities. That behavior is defined by its function implies that behavior functions to change the environment in ways that promote reproductive success. Phylogenetically important events (PIEs) are enhanced or mitigated by activities they induce as a result of natural selection. Induction explains all the phenomena that have traditionally been explained by reinforcement. This multiscale view replaces discrete responses and contiguity with multiscale activities and *covariance*. A PIE induces operant activity as a result of covariance in the form of a feedback relation between the activity and the PIE. A signal (conditional inducer) induces PIE-induced activities as a result of covariance between the PIE and the signal. In an ontological perspective, behavior is a process, and an activity is a process individual. For example, ontological considerations clarify the status of delay and probability discounting. A true natural science of behavior is possible.

Keywords: molar behaviorism, multiscale behavior analysis, time allocation, phylogenetically important event, induction, contingency, class, individual, process

Science and Philosophy of Behavior: Selected Papers, First Edition. William M. Baum.

Source: This paper contains portions of the English version of a book chapter published in Portuguese in D. Zillo and K. Carrara (eds.), *Behaviorismos*. Vol. 2. Sao Paulo, Brazil: Paradigma. The author thanks Howard Rachlin and Tim Shahan for thoughtful comments on earlier drafts. Originally published in *Journal of the Experimental Analysis of Behavior*, 110 (2018), pp. 302–322. Reproduced with permission of John Wiley & Sons.

This paper aims to bring together a few disparate lines of thought into a single cohesive framework for behaviorism and behavior analysis. I originally called the view I was developing "molar" behaviorism, but came to discover that the label "molar" was misleading, because people seemed to assume it only applied to phenomena at long time scales and could not apply to phenomena at short time scales. Following Phil Hineline's suggestion, I began calling it the "molar multiscale view," with the intention that I would eventually just call it the "multiscale view." By 2013, in a paper, "What counts as behavior: The molar multiscale view," I was able to put together the time-based view with scale, choice, and evolution. Much of what I have to say has appeared before in print in various places, but I will try in a brief space to weave together the concepts and observations of those earlier writings. The following section summarizes the main concepts of multiscale behavior analysis and the topics I will enlarge upon in the paper.

Prolegomenon

The importance to behavior analysis of making contact with evolutionary theory can hardly be overstated. Behavior analysis is properly part of biology. It is not a part of psychology, but an alternative to psychology. For psychology, behavior is a superficial phenomenon that must be understood by inferences to a "deeper" level: the mind or the brain. As long as behavior is not considered a subject matter in its own right and behavioral phenomena are considered secondary, a true natural science of behavior is impossible. Biologists often are naïve about the mind and consciousness, but they have no trouble thinking about behavior as real and primary. When asked, biologists who I have met agree that behavior is an organism's interaction with the environment.

The organism is not the agent of its behavior, but the medium of its behavior. Organisms and behavior go hand in hand, because they both enhance the fitness of the genes that promote them. Organisms and behavior would not exist if the genes making for organisms were not selected by having greater reproductive success as a result of being located in organisms.

The connection to evolution and natural selection allows a rethinking of the concept of reinforcement. Once we recognize that ethologists' "fixed action patterns" and the notion of operant behavior are equally relevant to understanding behavior, we can bring the two together, as Segal (1972) showed, with the concept of *induction* (Baum, 2012a; Tinbergen, 1963). Events impacting fitness, *phylogenetically important events* (PIEs), induce activities that enhance good (fitness-increasing) PIEs and mitigate bad (fitness-reducing) PIEs and also induce operant activities correlated with these PIEs. The operant activities that produce or avoid the PIEs are induced along with the unconditionally induced activities, including fixed action patterns. An activity may be called an "operant" activity to the extent that its ontogeny and maintenance depend on feedback between that activity and the inducing PIE. Events correlated positively with PIEs become proxies for them and induce many of the same activities as the PIEs themselves induce.

Multiscale Behavior Analysis

At the beginning of the twentieth century, scientists studying behavior relied on only two concepts: reflexes and associative bonds. Both concepts entailed discrete events and contiguity between the events. Pavlov's (1960/1927) conditional reflexes (called "conditioned" due to a translating error) depended on contiguity between a conditional stimulus and an unconditional stimulus (which he also called a "reinforcer"). Pairing the two stimuli was supposed to result in a bond between the conditional stimulus and a conditional response. Before Pavlov, nineteenth-century philosophers and psychologists considered ideas to be connected by associative bonds. The associative bond, when combined with the reflex, became a bond between stimulus and response, or an S-R bond. Ethologists invented a similar concept, in which a sign stimulus was said to "release" a fixed action pattern (Tinbergen, 1963). Thus was born the vocabulary of stimulus, response, and reinforcer.

The early behaviorists Watson (1930) and Thorndike (2012/1911) theorized about S-R bonds. Although Watson considered S-R bonds sufficient, Thorndike added to the associative laws, such as the law of contiguity, another law, which he called the "law of effect." According to the law of effect, an S-R bond is strengthened when a satisfying event closely follows the S-R sequence.

Skinner (1938) introduced a new concept with his invention of operant behavior. In 1938, he tied it to the reflex, but he soon recognized that operant behavior cannot be characterized by S-R bonds, because no identifiable stimulus precedes each occurrence of the response. He followed with two innovations: (a) measuring behavior as response rate; and (b) stimulus control. With these two new concepts, Skinner left S-R bonds behind. Instead, he thought of response rate as the primary measure of behavior, and a discriminative stimulus as exerting "control" by modulating response rate. Thus, stimulus control replaced the eliciting of the response by the stimulus that characterized the reflex. Skinner's innovations pointed in a direction away from discrete responses and contiguity, but he never made a further move in that direction because he never went beyond the "operant" as a class of discrete responses or the theory that an immediately following reinforcer "strengthens" an operant response.

Critique of the Molecular View of Behavior

The view that behavior consists of discrete responses that are strengthened by closely following (contiguous) reinforcers may be identified as the *molecular* view of behavior (e.g., Skinner, 1948). It seems to explain the observation that response rate increases when responses produce reinforcers (e.g., food). That is about all it explains, however. It does not explain even the most basic phenomena in behavior analysis. For example, the molecular view cannot explain why ratio schedules maintain extremely high response rates, whereas interval schedules maintain response rates that are moderate—that is, lower but not extremely low (e.g., Baum, 1993a). In attempting to explain the rate difference, molecular theorists cite differential reinforcement of relatively long interresponse times (IRTs) on interval schedules. Morse (1966), for example, showed that on an interval schedule IRTs followed by a reinforcer generally exceed IRTs not followed by a reinforcer. The reason is that the longer the IRT, the more likely an interval will have timed out during the IRT, setting up reinforcer delivery for the next response. Differential reinforcement of long IRTs explains why rate on an interval schedule should be lower than rate on a ratio schedule, because IRT is the reciprocal of response rate.

One trouble with this IRT theory is that it predicts something incorrect. If the key to lower rate on interval schedules is that the probability of reinforcer delivery increases as IRT increases, then IRTs should increase until the probability equals 1.0. For every response to produce a reinforcer, response rate on an interval schedule would have to be extremely low, but response rates on interval schedules, though lower than rates on ratio schedules, are still moderately high. When I have pointed out this theoretical failure, some molecular theorists answer by suggesting that such long IRTs would tend to increase the inter-reinforcer interval. That is so, but it is not part of the theory. In particular, because IRT is the reciprocal of response rate, and inter-reinforcer interval is the reciprocal of reinforcer rate, the suggested addition actually introduces an extended relation between response rate and reinforcer rate.

The moderately high rates on interval schedules cannot be explained without reference to reinforcer rate. When response rate is low on an interval schedule, increases in response rate produce large increases in reinforcer rate. As response rate rises to moderate levels, reinforcer rate ceases to increase. This relation is captured in the interval schedule's feedback function, which is negatively accelerated and approaches an asymptote (Baum, 1992).

Not only does the IRT theory fail to explain why interval response rates are as high as they are, it also fails even more obviously to explain the extremely high rate on ratio schedules, because in a ratio schedule, no relation exists between IRT and reinforcer probability. When one considers that the feedback function for a ratio schedule is simply an increasing straight line, an explanation in more extended terms appears. Increases in response rate always increase reinforcer rate; the only limit is the organism's ability to respond quickly. Someone trying to defend IRT theory might point out that, typically, responding occurs in bursts alternating with pauses, and on a ratio schedule a high-rate burst may be more likely to produce a reinforcer than responses spread out in time (i.e., low rate). Such an explanation departs from IRT theory, however, by pointing to differential reinforcement of response rate. It also implies that high response rates shorten the interval between reinforcers—which is to say that high response rates increase reinforcer rate, because the inter-reinforcer interval is the reciprocal of reinforcer rate. Not differential reinforcement of IRTs, but differential reinforcement of response rate by increasing reinforcer rate explains the extreme response rates that ratio schedules maintain.

Another phenomenon that molecular theory cannot explain is negative reinforcement, particularly avoidance. Suppose Tom, a divorced man with a grown son, Sam, receives a phone call from Sam inviting Tom to his wedding. Tom declines the invitation because Sam's mother, Tom's ex-wife, will be at the wedding, and Tom doesn't want to see her. Thus, Tom avoids his ex-wife, but why? Declining the invitation produces no immediate reinforcer; it only insures that something will not happen. The molecular view has no way to explain this, because it cannot appeal to any immediate reinforcer, although so-called "two-factor theory" would postulate an implausible and invisible "fear" of the ex-wife that is reduced by the declining. Instead, we can view Tom's declining as part of an extended pattern of avoiding his ex-wife: he not only turns down invitations to events at which she will be present, but he in general avoids places where she might be. He might not always be successful, but his avoidance activities reduce the likelihood that he will have to see her.

This explanation of Tom's behavior jibes with the explanation of free-operant avoidance in the laboratory. Sidman (1966) suggested that rats press a lever that postpones electric shock because pressing the lever reduces the rate of shocks received. Herrnstein (1969) elaborated on this appeal to extended relations and pointed out the inadequacy of the molecular view as adopted by Skinner and some other behavior analysts. Another view, also based on extended relations, explains avoidance as due to induction of the operant activity by failures—shocks received or encounters with the ex-wife—rather than shock-rate reduction (Baum, 2018a).

Some behavior analysts, notably Herrnstein and some of his students (e.g., Hineline, 2001, and Rachlin, 1994, 2014), moved ahead in the direction that Skinner had pointed out—toward temporally extended phenomena and theories. A major step was the discovery of the matching relation (Herrnstein, 1961). Generalizing this discovery leads to a law of behavior: the Law of Allocation.

The Law of Allocation

As Herrnstein (1961) originally presented it, the matching relation stated that the proportion of behavior allocated to an alternative tended to match the proportion of reinforcers obtained by that alternative:

$$\frac{B_1}{B_1 + B_2} = \frac{r_1}{r_1 + r_2}, \tag{11.1}$$

where B_1 and B_2 are rates of behavior allocated to Alternatives 1 and 2, such as pecking at two response keys, and r_1 and r_2 are the rates at which reinforcers, such as bits of food, are obtained. Equation 11.1 represented a major step, because it introduced reinforcer rate as a valid independent variable for understanding response rate. Just as Skinner had recognized an extended measure, response rate, as a dependent variable, the matching relation introduced an extended measure, reinforcer rate, as an independent variable, and together they indicated that behavior and its controlling relations could be seen as extended in time.

Herrnstein (1970) generalized Equation 11.1 to any number, N, of alternatives:

$$\frac{B_j}{\sum_{i=1}^{N} B_i} = \frac{r_j}{\sum_{i=1}^{N} r_i}. \tag{11.2}$$

Equation 11.2 goes beyond the original observation (Equation 11.1) to state a law. It builds on the insight of the necessity of temporal extension to offer a more general framework.

From the recognition that the matching law implies temporally extended variables and relations, only a short step was required to write matching more generally in terms of time (Baum & Rachlin, 1969):

$$\frac{T_j}{\sum_{i=1}^{N} T_i} = \frac{V_j}{\sum_{i=1}^{N} V_i}, \tag{11.3}$$

which states that the proportion of time taken up by one activity j matches V_j relative to the total of V_i across all alternatives, and each V_i is a composite measure of reinforcer variables, such as rate, amount, and immediacy, that determine the relative time. This is truly a law, and is a tautology, as all laws are (Rachlin, 1971). Equation 11.3, perhaps, might be called the "generalized matching law," but that term nowadays denotes a more specific version, even though that equation is not really a law in the usual sense, but an empirical generalization (Baum, 1974a, 1979; Poling, Edwards, Weeden, & Foster, 2011). Though I originated the equation, I did not name it as such, but the usage has become common (Poling et al., 2011).

Equation 11.3 may be rewritten in a variety of ways (Baum, 2012b), but it is general enough for present purposes to be called the Law of Allocation. It has been used to explain impulsive choice (Aparicio, Elcoro, & Alonso-Alvarez, 2015) and resurgence—the reappearance of extinguished responding when an alternative activity is extinguished (Shahan & Craig, 2017). These applications draw on the implication that no activity exists apart from the context of other, competing, activities. If one activity increases, others must decrease, and if one activity decreases, others must increase. Extinction, for example, consists of a shift from an allocation high in operant activity to one that includes little of the operant activity (Baum, 2012c).

Like any scientific law, the law of allocation embodies and depends upon a number of assumptions or axioms. They might be taken as guidelines for experimenting and theorizing about behavior. These were discussed less formally in an earlier paper (Baum, 2013; see also Baum, 2018a).

Axiom 1: Only whole organisms behave.

Axiom 1 applies to all organisms: multicellular—humans, dogs, pigeons, fish, cockroaches, or hydras—and unicellular—Paramecia or Amoebae—and archaic—bacteria and viruses. As we will see below, these are all individuals that interact with their surrounding environment.

For behavior analysis, Axiom 1 excludes inanimate things, because behavior analysis deals with living organisms. An exception might be robots with artificial intelligence, if their interactions with the environment become too complex to predict from their programmed algorithms.

In associating behavior only with whole organisms, Axiom 1 rules out behavior by parts of an organism. My heart's beating may be part of my physiology, but it is not part of my behavior. In particular, Axiom 1 denies that the brain behaves (Bennett & Hacker, 2003). Bennett and Hacker (2003) explain the logical reason that only whole organisms behave. For example:

> Psychological predicates are predicable only of a whole animal, not of its parts. No conventions have been laid down to determine what is to be meant by the ascription of such predicates to a part of an animal, in particular to its brain. So the application of such predicates to the brain . . . transgresses the bounds of sense. The resultant assertions are not false, for to say that something is false, we must have some idea of what it would be for it to be true—in this case, we should have to know what it would be for the brain to think, reason, see and hear, etc., and to have found out that as a matter of fact the brain does not do so. But we have no such idea, as these assertions are not false. Rather, the sentences in question lack sense. (p. 78)

According to multiscale behavior analysis, what Bennett and Hacker (2003) say in this quote about "psychological predicates" applies to behavior in general, not just thinking, reasoning, seeing, and hearing. To speak of the behavior of parts of living things—anything other than whole living organisms—"transgresses the bounds of sense." The brain does not perceive, choose, or sense, any more than the brain can walk or talk; these are activities of whole organisms. In this view, the brain is an organ of the body and participates in behavior, but does not behave itself.

A more important reason for Axiom 1 derives from evolutionary theory. From the perspective of evolutionary theory, behavior only exists because organisms exist. Organisms exist because the genes that make for organisms reproduce more successfully than competing genes that would undo organisms—that is, the genes that produce and reside in organisms have higher fitness than any competitors. The competition continues now, just as long ago. Multicellular organisms continually face challenges by less organized life forms, particularly bacteria and viruses. These threats are countered by evolved mechanisms, such as the immune system, symbiosis with micro-organisms in the gut and on the skin, and practices such as treating water before drinking it. The success of the organism-making genes relies on the organism's interaction with the environment around it, because the organism's actions change the environment in ways that are, on average, advantageous to survival and reproduction. Often the environmental changes feed back to affect the organism's further actions. The organism's actions are the organism's behavior. (See Baum, 2013, for further discussion.)

Axiom 2: To be alive is to behave.

Axiom 2 says that so long as an organism is alive, it behaves continually. It immediately implies that behavior takes up all the time available. If one observes an organism for an hour, a day, or a year, one observes an hour's worth, a day's worth, or a year's worth of behavior. If behavior is allocated among various activities, those activities each take up some of the time, and together take up all of the time. Moreover, time is limited. Every living thing lives within a time budget, because lifetimes are finite, a day contains just 24 hours, and opportunities for behavior are ephemeral. As a result, activities compete with one another for time, in accord with Equation 11.3. The key task of behavior analysis is explaining the allocation of time among all the organism's activities.

The connection to evolution further supports the central principle that behavior takes up time, because interaction with the environment can only take place over time. The phrase "momentary interaction" is an oxymoron, because interaction can only be extended. That behavior cannot occur at a moment tells us that the historical concept "momentary response" was logically and theoretically flawed.

Indeed, no activity can be identified at a moment. A snapshot of a person holding an open book tells almost nothing about what activity is occurring; the person might be reading, looking for something in the book, pretending to read, and so on. Only by observing for some time, before and after the moment, can the activity be identified as reading or pretending or something else. Similarly, a snapshot of a rat with its paws on a lever tells almost nothing of what activity is occurring; one has to see what went before and what came after to decide if the rat is pressing the lever at a high rate, at a low rate, pressing at all, exploring the chamber, or something else. (See Baum, 1997, 2013, for further discussion.)

The impossibility of identifying behavior at a moment implies that attempting to explain behavior at a moment is not only unrealistic or quixotic, but also impossible.

Behavior is continuous—a flow—not a series of momentary discrete "responses." The only discrete events might be switches from one activity to another, and perhaps those might be predictable. For the rest, explaining and predicting behavior consists of explaining and predicting the amount of time spent in various activities, on whatever time scale works.

Viewing behavior as consisting of discrete responses leads inevitably to hypothetical constructs that can only mislead. For example, Killeen and Jacobs (2017) argued for the importance of considering an organism's physiological state in predicting its behavior, but they moved from physiological states to "dispositions" as causes of behavior. A disposition is nothing, however, if it is not behavior. Physiological states may be measured, but a disposition that does not result in measurable action is a phantom. The rate at which an activity occurs *is* its disposition to occur. Attributing the activity to its disposition is circular and is a category mistake (Ryle, 1949).

One might assert the converse of Axiom 2 also: to behave is to be alive. Not only bacteria, which have a cell or plasma membrane, are considered alive because they reproduce and interact with the environment around them—secreting chemicals, attacking cells, and exchanging genetic material—but also viruses, naked molecules lacking any membrane, are considered alive because they reproduce in and interact with the bacteria and cells they encounter. Prions, smaller protein molecules that only replicate, are not considered to be alive. Thus, behavior is inextricably tied up with life and characterizes what are considered "live organisms."

Axiom 3: Every activity is composed of parts that are themselves activities.

Axiom 3 introduces scale into Equation 11.3. It says that the time taken up by any one activity may be subdivided into the less-extended, smaller-scale activities of which it is composed, and that the time taken up by those parts adds up to the time taken up by the more-extended, longer-scale activity of which they are parts (Baum, 1995a; Baum & Davison, 2004). If I play tennis for an hour, during that hour I am serving shots, returning shots, keeping score, exchanging remarks with my opponent, and so on. Together these activities constitute playing tennis, and together they take up the whole hour of my playing tennis. If a pigeon pecks at keys in concurrent schedules, its performance has parts: pecking at the right key, pecking at the left key, and background activities other than pecking. Its pecking might be organized into long visits to the preferred key ("fixing" on the rich key) alternating with brief visits to the non-preferred key ("sampling") plus background activities (Baum, 2002; Baum, Schwendiman, & Bell, 1999). Thus, Equation 11.3 may apply at any time scale, to the parts of playing tennis or to the activities of a day, one of which is playing tennis, and to the allocation of pecking between keys or to the pattern of pecking and switching between keys. It may apply even at time scales of fractional seconds, to the parts of a pigeon's key peck or a rat's lever press (e.g., Smith, 1974). Axiom 3 underpins what I call multiscale behavior analysis. (See Baum, 2018a, for further discussion of laws of behavior.)

The Behavior-Environment Feedback System

Some earlier papers suggested that the interaction of behavior with the environment may be compared to a feedback system (Baum, 1973, 1981b, 1989, 2016a). To say that behavior interacts with the environment implies that behavior and environment constitute a feedback system. Figure 11.1 shows a diagram of the feedback system for one activity. It depicts a simplification, because no one activity can be isolated from others. (A more

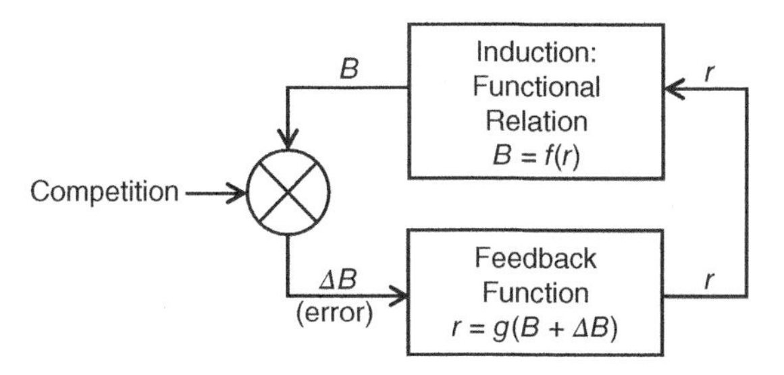

Figure 11.1 Behavior and environment as a feedback system
Note: The Law of Allocation ("Competition;" Equation 11.3) determines the set-point. A feedback relation in the environment translates error (ΔB) into a PIE rate (r). An induction relation through the organism translates r into rate of the operant activity (B), as, for example, in Equation 11.4.

detailed attempt may be found in Baum, 1981b, Figure 1.) "Competition" appears in the role of a set-point, but cannot be thought of as a set-point in the same sense as a thermostat's set-point. Instead competition from other activities constrains the effect of fed back events in r on activity B. Competition accords with the Law of Allocation; one may think of it as Equation 11.3. It incorporates the current rate of the activity, B, and the deviation or "error" equals ΔB, which is input to an environmental relation. The function g represents a feedback function—a property of the environment. The output of the feedback function, r, is a rate of consequences, PIE rate (e.g., food rate). The rate r is input to an organism-based functional relation and represents induction, discussed below. Some evidence suggests that this relation may be a power function, at least for relating food rate and pigeons' rate of key pecking (e.g., Baum, 2015; Baum & Davison, 2014b):

$$T_j = b_j r_j^{s_j} \tag{11.4},$$

where T_j is time spent pecking, r_j is rate of food, s_j is sensitivity of T_j to r_j, and b_j is a coefficient. Equation 11.4 states that r_j induces time spent pecking according to the power s_j and in proportion to b_j. In principle, r_j need not be only rate of food; for example r_j could represent amount of food or immediacy of food (Baum & Rachlin, 1969). Equation 11.4 combined with Equation 11.3 results in an extremely general version of the Law of Allocation (Baum, 2012b).

The function f in Figure 11.1 may be thought of as Equation 11.4, with B equal to T_j. The system stabilizes when ΔB equals zero. That equilibrium is often called "stable performance." Although local variation never ceases, allocation may be considered stable when it ceases to exhibit a trend across time (Baum, 2017c).

One might notice that the model in Figure 11.1 omits anything that might be called "reinforcement." It approaches a stable, but variable, state, but nothing like strength appears in the model, any more than a thermostat controlling the temperature in a room entails strengthening of room temperature. Induction replaces reinforcement. For this reason, some researchers have suggested that matching, as embodied in Equation 11.3, is "unconditioned" or "innate" (Gallistel et al., 2001, 2007; Heyman, 1982). Behavior is selected, but by competitive induction, not by strengthening.

As a model of behavior, Figure 11.1 is extremely general. To apply it to any particular situation, one must decide how to measure feedback r and activity B. A number of researchers have suggested various approaches to measuring r when it refers to PIE ("reinforcer") rate. Gallistel and associates suggested an algorithm that examines cumulative records to estimate reinforcer rate, change in reinforcer rate, and abrupt changes in responding (Gallistel, Mark, King, & Latham, 2001; Gallistel et al., 2007). Davison and I, using pigeons, were able to examine short-term effects of individual food deliveries (Davison & Baum, 2000). Examining transitions from one long-term concurrent schedule pair to a new pair, I found that, after sufficient feedback—food deliveries—from the new schedule pair, transitions in behavior were abrupt (Baum, 2010; Baum et al., 1999).

The need to invent ways to estimate the controlling variable *r* leads to a question: If we discover a method of estimating *r* that successfully predicts some behavior, does that mean the organism also uses that method? Although no warrant exists to make such an inference, some researchers seem to make it nonetheless. Gallistel, for example, writes that the organism under study makes the computations, rather than his algorithm (Gallistel et al., 2001, 2007). Shahan (2017) wrote, ". . . having something somewhere accumulated or stored up—such preservation is a logical and physical necessity for solving the problem of bringing the past into the present" (pp. 111–112). No physical necessity exists, because we have no idea as yet how the past is brought into the present or even if that is the correct way to think about the past; the method that the physiology of the organism uses may have nothing to do with accumulating or storing. The necessity might be "logical" in the sense that bringing the past into the present *means* accumulating and storing, but that makes Shahan's reasoning circular. The experimenter estimates the variables and, with luck, finds some order.

Multiscale Behavior Analysis and Evolutionary Theory

Axiom 3 above introduces the fundamental property of scale. If driving to work is part of working, then driving to work occurs on a smaller time scale than working. One may say that working takes longer than driving to work. Driving to work is an activity composed of yet shorter activities such as driving on the highway and driving on town roads (see Wallace, 1965 for a detailed discussion of driving to work). Variation in an activity consists of variation of its parts. Some days I may drive to work by the highway and other days by back roads; sometimes I may drive at high speed other times at low speed.

At the longest time scale for an individual organism, only one activity occurs. We may call it "living." Recalling the logic of evolutionary thinking, according to which multicellular organisms only exist because of the success of the genes they carry, we may conclude that living serves one function: reproducing. All other activities, whatever their scale, are ultimately parts of reproducing. In particular, surviving is often a necessary part of reproducing. Exceptions exist—for example, male mantids and spiders that are eaten by the female after copulation, providing a good meal for the female that benefits the male's offspring. Parents of many species sometimes risk their own lives for the sake of their offspring. Individuals of our own species, which is uniquely cooperative, sometimes even risk their lives for complete strangers.

Surviving is a necessary part of reproducing the same way that getting out a mixing bowl is necessary to making a cake; the longer-scale activity cannot be completed without it. Though necessary, however, surviving is not sufficient for reproducing. Other parts, like mating and caring for offspring, more directly related to reproducing, must also occur. Surviving only has to provide opportunities for these other parts of reproducing on average and in the long run. Evolutionary arguments always contain this proviso, either explicitly or implicitly. A beneficial gene may be selected in a population even though some members of the population possessing the gene die without reproducing, because the gene confers advantage to offspring on average and in the long run. Similarly, an operant activity may be selected even though its consequences are sometimes bad if the consequences are better than competing variants on average and in the long run. Camping outdoors may usually be an exhilarating experience, but sometimes is ruined by a rainstorm; people still go camping. Indeed, an activity may be maintained that only serves a function in the long term, but serves no function each time it occurs (Rachlin, 1992). For the sake of my dental health, I brush my teeth every morning and night, even though, on a daily basis, the activity is just an effort that takes up time.

The perspective offered by evolutionary theory, that organisms exist to reproduce, may be summarized as, "Organisms are the means by which DNA makes more DNA." It helps to answer many questions about life in general and human life in particular. For example, why do organisms age and die? Lifespan is tied to generation time; once a generation of parents has produced offspring, the parents may no longer have a function and, rather than live on and compete with their own offspring, they die—genes are selected that result in this built-in obsolescence. Human beings present a challenging puzzle: the phenomenon of menopause. In most species, both males and females continue to be fertile as long as they live, but in our species only the males remain fertile. A possible reason lies in the uniquely long period of dependence of our offspring. This dependence creates a "grandmother effect," in which menopause may be selected by its benefit to women's genes in their grandchildren. If a woman ceases reproducing and helps care for her grandchildren, her genes may be more likely to be passed on through the grandchildren. Genes making for this pattern would be selected if the beneficial effect on the grandchildren outweighs any advantage of producing more offspring.

Evolutionary theory helps to understand why many human activities exist that otherwise would have no explanation. Even though activities like art, music, and religion might seem to have little connection to reproducing, they can be fitted into the larger context of evolution. A highly social species like ours lived all its evolutionary history in groups, and many shared practices (i.e., operant activities), collectively known as "culture," belong to the group (Baum, 2017a). Some practices serve the individual person's reproductive success, and some practices serve the group as a whole. Avoiding poisonous plants serves the individual, but ingroup-outgroup discrimination serves only the group as a whole. Art, music, and religion may provide ways to enhance one's status within a group and thereby open opportunities for mating and gaining resources. Practices with less-obvious function often serve the group as a means of maintaining group cohesion—for example, wearing certain tattoos or clothing, speaking a certain language dialect, and attending a certain church. Since group membership is fundamental to human life and survival, most human activities tie less directly to reproducing than to surviving.

Surviving, like any other activity, has parts. The parts are not always easy to identify as such. In the past, I suggested three long-scale human activities: maintaining health, gaining resources, and maintaining relationships (Baum, 1995b, 2017a). All three promote survival, and this division is useful for discussion, but these parts sometimes overlap. One usually needs to be healthy to gain resources, and sometimes resources make for good health. Earning a living by holding a job requires getting enough sleep, but having income allows one to have the shelter needed to get enough sleep. Relationships may help with gaining resources, but sometimes resources allow formation of new relationships. A friend may lend you money, but having money also may open doors that might otherwise be shut. Despite the overlap, Axiom 2 above tells us that behavior takes up all the time available and cannot take up more time than is available. The overlap, along with Axiom 2, leads to what may be called the "accounting" problem—that is, the problem of deciding when one activity begins and another leaves off in order to measure the time spent in each activity.

The Accounting Problem: Defining and Measuring Activities

In the laboratory, we can arrange conditions so as to prevent overlap between activities. We define activities so that they are readily measured. For example, research with concurrent schedules has produced support for viewing choice as allocation of time among activities. An experiment by Bell and Baum (2017) studied concurrent variable-interval (VI) variable-ratio (VR) schedules of key pecking in pigeons. Although the two types of schedules maintained qualitatively different patterns of pecking, Bell and Baum were able to measure the time spent at each alternative, and the time allocation between them provided the best description of the choice relations as relative reinforcers obtained varied across the alternatives. The accounting problem appears to be solved because no pecks are possible at one key while pecking is occurring at the other key.

Yet, even in the laboratory ambiguity arises. As Herrnstein (1970) noted, a pigeon in an experimental chamber is not limited only to pecking keys. Every organism brings with it unmeasured activities like grooming, scratching, and exploring. That is why he added a term r_0 to the version of Equation 11.1 that described responding at a single programmed alternative. Subsequent research indicates that such "background" activities separate into those that are induced by the reinforcer (PIE; e.g., food) and those that occur independently of the reinforcer. Analysis by Davison (2004) suggests that several different background activities occur alternately.

The accounting problem is less challenging in the laboratory than in more naturalistic settings, with humans or other animals, inside or outside the laboratory. When doing research, one must define activities so that they are mutually exclusive. Once the definitions are clear, one may tackle measurement. The best approach is to record behavior and have two or more observers code the videographic recordings (e.g., Simon & Baum, 2017). That approach, however, is labor-intensive. Another approach with humans is self-report; one simply asks a person how much time they spend in various activities, but this method relies on people to be accurate in their estimates.

Defining activities

Skinner (1938) introduced the definition of operant activity by its function. Evolutionary theory explains why definition of behavior by function is indispensable. Because the function of organisms is to reproduce, behavior exists ultimately as interaction with the environment in the service of that function. Behavior consists of activities that serve functions that ultimately serve reproducing. Thus, when a rat's lever pressing produces food, that activity may serve the function of feeding (along with other parts, like consuming the food); pressing the lever is then part of feeding, and feeding is essential to surviving and reproducing.

In more naturalistic situations, defining activities depends on deciding which functions they serve. Depending on one's research interest, having lunch with a friend may be construed as an activity that maintains a relationship—a variant of socializing—or as an activity that maintains health—a variant of eating. A third possibility, if one wanted to separate socializing from eating, would view having lunch with a friend as multitasking (Baum, 2013). Research on multitasking indicates that it entails rapid switching back and forth between two activities (resulting in poorer performance on both than either by itself; e.g., Caird, Willness, Steel, & Scialfa, 2008).

Proper definition of activities allows one to study practical problems. For example, suppose one wished to study work-life balance in someone's life. Defining "work" and "life" plausibly would be crucial. If the two activities occur in two different locations, definition might be relatively simple. Even then, however, overlap might occur, as when a person gets a work-related phone call at home or a family-related phone call at work. Defining the activities so that they are mutually exclusive might be a bit inaccurate; the more natural the setting, the lower tends to be the accuracy with which it can be studied.

Measuring activities

For measuring activities, we gain clarity by distinguishing between episodes and constitutive parts. All activities are episodic. Suppose I drive to work every weekday. Each drive to work is an episode of the activity "driving to work." All the episodes of driving to work over the course of a month or a year together constitute an aggregate, which we may liken to a population. In evolutionary theory, the aggregate of members of a species makes a population. One may be interested in the population as a whole—its size and geographical distribution—or one may be interested in the variation across the members of the population—their physical characteristics or reproductive success. Similarly, one may be interested in the population of episodes of an activity—their number or total—or one may be interested in variation across episodes—their duration or their constituent parts. If I were just interested in the aggregate, I might want to know how much time I spend driving to work. If I were interested in the variation among episodes, I might note that some of my drives to work include driving through Smithtown (a part), whereas others might avoid Smithtown. At a smaller time scale, I might be interested in the population of my drives through Smithtown—for example, some might adhere to the speed limit whereas others might exceed the speed limit, attracting the attention of local police.

In laboratory research on behavior, populations and episodes are modeled by measuring bursts, bouts, or visits (e.g., Aparicio & Baum, 2006, 2009; Baum, 2017c; Baum & Davison, 2004; Bell & Baum, 2017; Shull, Gaynor, & Grimes, 2001). Operant activity, like all behavior, divides into bouts interspersed with pauses that represent time spent in other activities (Davison, 1993; Gilbert, 1958). Those bouts or visits may be thought of as episodes of the operant activity, and their aggregation constitutes a population. One might examine the variation in their duration for clues to initiation and termination of the bouts, or one might examine their function, as in the pattern called "fix and sample" (Baum et al., 1999), in which operant activity fixes on the richer of two choice alternatives and takes the form of brief samples at the leaner of the alternatives.

Laboratory research also occasionally raises variation in constituent parts of patterns of operant activity. An example may be seen in food-induced activities that compete with the operant activity (e.g., Breland & Breland, 1961). These activities figure into Equations 11.2 and 11.3 and other expressions of the Law of Allocation. For example, Baum and Davison (2014b) factored in induced activities to explain apparent deviations from the matching law. Conceiving of behavior as composed of multiple activities provides a plausible and elegant approach to measuring behavior.

Reinforcement and Induction

For over 100 years, a common interpretation of reinforcement was that reinforcers increase the "strength" of responses that produce them. Questions about this interpretation arose from time to time and increasingly in the 1970s. The conception that behavior consisted of discrete responses seemed to require the concept of strength, because explaining an increase in rate of a discrete response through time seemed to require some hypothetical temporally extended variable. When we move to conceiving of behavior as composed of temporally extended activities, however, the increase involved in "reinforcement" may be seen as an increase in the amount of time taken by an activity that produces the "reinforcers." If a rat is trained to press a lever, the activity of lever pressing takes up more time than before training. The need for "strength" disappears. The activity simply increases.

Explaining the increase in behavior without reference to strength invites relating such increases to evolutionary theory. If we think of an organism interacting with its environment, certain events hold special significance because they relate relatively directly to surviving and reproducing. These events may be called "phylogenetically important" because they impact reproductive success (fitness). A good phylogenetically important event (PIE) promotes fitness by its presence (e.g., mates, prey, and shelter). A bad PIE threatens fitness by its presence (e.g., predators, illness, and parasites). These PIEs need to be responded to well for a species to thrive. Individuals that respond in ways that enhance good PIEs like feeding or mating opportunities and avoid bad PIEs like predators or toxins are more likely to survive and reproduce than their less-adept conspecifics. Hence, we expect that selection will result in populations of organisms that respond to PIEs in certain reliable ways (Baum, 2012a).

Some authors have misinterpreted the concept of PIE, confusing "phylogenetically important" to mean locally efficacious. For example, Killeen and Jacobs (2017) argued for the need to introduce an organism's current physiological state into our analyses of behavior, because an organism exposed to the same contingencies may nevertheless behave differently at one time than another, depending on its physiological state. If you have just eaten a big meal, you are unlikely to act so as to eat more, but when you haven't eaten for several hours, you willingly take on more food. Discussing this, Killeen and Jacobs (2017) write that when in a certain state, a "consequential response, R_C . . . will be a *satisfier* (Thorndike)—an *important event* (Baum)" (p. 20; italics in the original). Removing the adverb "phylogenetically" creates ambiguity about the meaning of "important." Later, Killeen and Jacobs confuse the meaning further:

> What puts the I in the PIE? The state of the organism in that context, O, (a phylogenetically important motivation (PIM)?). It is *that* that powers induction (p. 26; italics in the original).

Cowie and Davison (2016), who generally include the adverb "phylogenetically," fall into the same ambiguity, writing, ". . . environmental events may, under current organismic conditions, be PIEs that are currently important to the organism (events such as food and water may not always be important, but pain is always important)." (p. 265.) No doubt the organism's physiological state is important to predicting when an organism will initiate feeding on a small time scale, but the meaning of "important" in "phylogenetically important" depends on a much longer time scale. Let us consider a couple of illustrative examples.

A pride of lions, having just gorged on a kill, now ceases hunting. After some days, the lions' physiological states have changed, and now they hunt again to feed again. Songbirds in winter must take in enough food during the day to last through the night. When starving, they shift from being risk-averse to risk-prone (Krebs & Kacelnik, 1991).

On one time scale, feeding for the lions and the birds is cyclic. For the lions the cycle occurs across several days; for the birds one day. Feeding depends on the passage of time, and as time passes hunting or foraging becomes more likely, which is to say that the PIE becomes locally efficacious; the passage of time or the physiological state—if we could measure it—induces the activity that produces the PIE. The relation between time and feeding, however, is local and within an organism's lifetime, whereas the relation between feeding and surviving occurs on the time scale of phylogeny. Lions must eat enough to survive, and selection favors those lions that hunt effectively and hunt when they are hungry; the others are selected against. As a result of natural selection, lions will both hunt effectively and hunt when they haven't eaten for some days. Similarly, a songbird that fails to decrease risk-sensitivity when starving is less likely to survive, and phylogeny results in birds that make the shift. Feeding is phylogenetically important in this sense: surviving (and reproducing) depend on feeding. A phylogenetically important event is an event that has been important across generations of lions, birds, and other species. As a result, a PIE, when efficacious, induces its PIE-related activities.

What sorts of events are phylogenetically important? An earlier paper suggested that they are the events traditionally thought of as reinforcers, punishers, and unconditional stimuli (Baum, 2012a). Premack (1965, 1971) argued that the events are behavioral events, that reinforcers and punishers are behavior of high and low probability. Logic and research indicate that Premack's generalization does not always hold. Under most circumstances, we assume that food must be eaten to function as a PIE, but experiments on "latent learning" in rats show that food uneaten may nevertheless change behavior later when the rats are hungry and eat the food (Kimble, 1961). This delayed change is not unusual. Many mammals and birds cache food for later consumption, sometimes months later (Krebs & Davies, 1993). Experiments on "sensory preconditioning" (Kimble, 1961) suggest that so-called "neutral" stimuli—tones and lights—may not be neutral, particularly if one considers that organisms may depend on signals in their environment to find food or avoid predators. Experiments with "sensory reinforcement" further support the non-neutrality of "neutral" stimuli, because they show that a variety of visual, auditory, and tactile stimuli may function as reinforcers (Kish, 1966).

Premack's generalization particularly falls short with aversive, bad PIEs, such as electric shock (injury), illness, and predators. These would be stimuli, not behavior. They happen to an organism, and they induce avoidance or defense, rather than consummatory behavior.

Events that are phylogenetically important may be behavioral events, like eating, drinking, and mating, but they may also be stimuli, like food (prey), water, and a mate. They may function as traditional reinforcers and punishers, but from the perspective of multiscale behavior analysis, they function as inducers, either inducing appetitive activities or avoidance. Premack's (1971) generalization falls short because it lacks any connection to evolutionary theory, which incorporates PIEs of all sorts.

As an explanation of behavior, the concept of reinforcement was always incomplete. The question remained, "Where did the behavior to be reinforced come from?" Segal (1972) proposed the concept of induction to explain the provenance of the to-be-selected behavior and, at the same time, integrate operant behavior with PIE-induced behavior (also known as "adjunctive," "interim," and "terminal"). A PIE like food or a mate induces a suite of activities like capture and consumption (food) and arousal displays and courtship (a mate). Thus, if a resting hungry pigeon starts to receive bits of food, the pigeon becomes highly active and, among other activities, it pecks. A hungry rat treated this way also becomes active and, among other activities, noses around in the food hopper.

Pigeons explore and exploit food sources in their natural environment by pecking (typically seeds, but sometimes bread and even chicken bones on city streets). Rats explore and exploit food sources in their natural environment by nosing and chewing. PIEs that are crucial to an organism's surviving and reproducing in its natural environment induce activities relevant to the PIEs. Thus, phylogeny is crucial for understanding species-typical behavior, but phylogeny is also crucial for understanding ontogeny of behavior in individual organisms.

In explaining provenance of operant behavior, induction indicates that induced activities provide the "raw material" for selection to act upon (Segal, 1972; Staddon & Simmelhag, 1971). One implication is that variation is neither "undifferentiated" nor "minimal," as Skinner (1984, p. 670) suggested in one article, but rather follows the "natural lines of fracture," as Skinner (1935/1961) suggested elsewhere. A PIE induces a variety of whole extended activities that are not random with respect to the PIE, but rather are the result of phylogeny. Timberlake (1993) gave a similar view with his concept of behavioral system, which refers to all the various activities induced by a PIE.

Relating induced activities to operant behavior requires only one further step: recognizing the role of contingency or, more generally, *covariance*. Skinner (1948) regarded close temporal proximity of a reinforcer following a response as the relation that increases the rate of a response. Subsequent research showed that contiguity is neither necessary nor sufficient. Studies of delay showed contiguity is not necessary for reinforcement (e.g., Baum, 1973; Lattal & Gleeson, 1990). Studies in classical conditioning and adjunctive behavior in which contiguity is decoupled from contingency indicate that contiguity is not sufficient to explain response increase (e.g., Rescorla, 1968; Staddon & Simmelhag, 1971). Instead, Rescorla (1967), discussing proper controls for classical conditioning, pointed out that a contingency required that two events both occur together and be absent together. In most experiments with classical conditioning, this requirement is met by comparing what happens in a trial with what happens in the inter-trial interval. In a trial, a conditional stimulus (CS) occurs along with an unconditional stimulus (US), whereas in the inter-trial interval both are absent. The CS is something like a tone, and the US is something like food, a PIE. In classical conditioning, the key requirement is that a CS should covary with a PIE. In positive covariance, they should be absent and present together or, more generally, their rates should vary together. In negative covariance, the inverse should hold. Operant conditioning relies on covariance in which the PIE depends on the activity, and the covariance may be positive or negative too (Baum, 2012a; 2018a). In a feedback relation, as the rate of the activity varies, the rate of the PIE increases or decreases with it.

What is true for a CS is also true for a discriminative stimulus (S^D): the S^D comes to increase the rate of an activity because it covaries positively with a PIE. A tone that covaries with food takes on some of the functions of food; in particular the tone induces many of the same activities that the food induces. Thus, positive covariance between a stimulus (S^D or CS) and a PIE changes the function of the stimulus to resemble that of the PIE (Davison & Baum, 2010).

A question arises about a stimulus (CS or S^D) that becomes such a proxy for a PIE: how similar is the behavior induced by the proxy to the behavior induced by the PIE itself? Is this concept a resurrection of Pavlov's (1960/1927) discredited theory of stimulus substitution? Pavlov's theory required that the conditional response and unconditional response be identical, a prediction that was often refuted (Kimble, 1961, pp. 204–205). Whether the activities induced by a CS or S^D are the "same" as those induced by the PIE itself is, at least to some extent, an empirical question. Not all activities induced by a proxy are identical to those induced by the PIE, but some undeniable resemblances occur. A few observations may serve to clarify. First, however, one must distinguish induction from elicitation. Although food may elicit salivation, the activities that food induces go beyond salivation. For example, Zener (1937) noted that unrestrained dogs responded in a variety of ways to a tone paired with food, particularly by approaching the food bowl. A rat or pigeon cannot eat in the absence of food, but food, presented either contingently or non-contingently, induces food-related behavior between presentations, and some of this behavior is directly included in eating. Breland and Breland (1961) observed this, and laboratory research (Staddon, 1977) shows that rats and pigeons fed periodically or irregularly in an experimental chamber engage in activities that are parts of feeding—e.g., drinking (rats) and pecking (pigeons). In a striking demonstration, Jenkins and Moore (1973) reported that when a key light is paired with food, pigeons peck at the key as they would when eating grain, but when the key light is paired with water, pigeons operate the key with motions like those of drinking. In the days when they were made of brass, response levers became green and encrusted with rats' dried saliva and often had to be replaced as they were gradually chewed away.

When we come to operant behavior, the activities induced by a discriminative stimulus are hardly distinguishable from the activities induced by food. That is why we say the S^D "modulates" response rate. If pecking at a green key sometimes produces food, and pecking at a red key does not, the green key induces pecking, just as the food does. Resemblance occurs also in operant avoidance. Regardless of what reflex responses may be elicited by an aversive PIE, the PIE will induce avoidance if avoidance is possible (Baum, 2018a; Sidman, 1966). If a signal covaries positively with an aversive PIE, the signal induces avoidance—e.g., in a shuttlebox—("negative reinforcement;" Baum, 2012a). If the signal induces avoidance, however, the activity it induces is the activity that avoids the PIE, not the signal (Herrnstein, 1969); Tom avoids his ex-wife, not his son.

What I call "proxies" of a PIE depend on positive covariance with the PIE, but negative covariance matters also. If a pigeon is trained to peck at a green key and not at a red key, the red key covaries with food negatively. The red key induces activities other than pecking (Baum, 2012a; Staddon, 1982; White, 1978). Food itself may covary negatively with food, as when food production alternates between two alternatives (Cowie & Davison, 2016). Similarly, if a signal covaries negatively with an aversive PIE, the signal may be called "safety" and induces activities other than those induced by the PIE, including avoidance.

Event-event covariance is one important sort of covariance affecting behavior; as indicated earlier, the other is PIE-activity covariance. When the rate of a PIE depends on the rate of an activity, as in a "reinforcement schedule," the activity (e.g., lever pressing) comes to be induced by the PIE (e.g., food or electric shock; Baum, 2018a). In other words, the activity becomes operant activity. We know that operant activity is induced by the PIE from observations such as reinstatement, in which non-contingent presentation of the PIE revives an extinguished activity (Reid, 1958). The lever pressing produces the food, the food induces the lever pressing, which produces the food, and a feedback loop occurs that selects lever pressing out of all other activities and maintains lever pressing at a high level. As Figure 11.1 indicates, depending on the shape of the feedback function, the maintained rate will be extreme on a ratio schedule, which imposes a linear feedback function, but moderate on an interval schedule, which imposes a feedback function with an upper limit to PIE rate (Baum, 1993a; 2015).

Figure 11.2 diagrams the relation between a PIE and its induced activities. On the left, a PIE (*S*; e.g., food) occurs in the environment (*E*) and impinges on the organism (*O*).

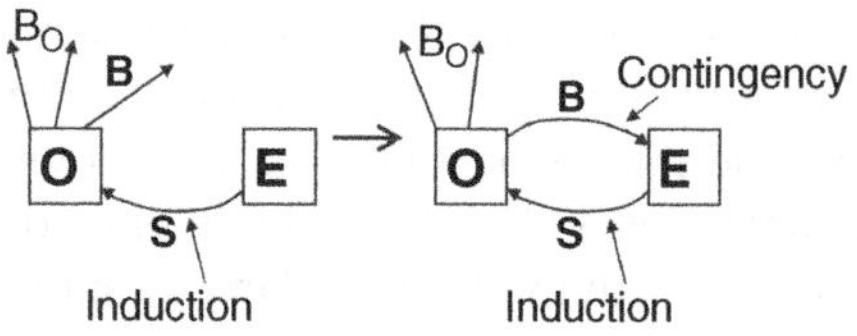

Figure 11.2 Contingency closes a loop
Note: Left: Without any contingency, the environment provides a phylogenetically important event (PIE; *S*), which impinges on the organism (*O*) and induces several activities *B* and B_O. Right: A contingency arranges that *B* produces the PIE and closes a loop, in which the PIE induces *B*, and *B* produces the PIE. The other activities denoted as B_O continue to be induced and compete with *B*.

Several activities are induced, including one labeled *B* (e.g., touching a lever) and others that compete for time with *B*, labeled B_O. On the right of Figure 11.2, a contingency has been introduced that makes *B* produce the PIE. Now the loop is closed, and the PIE induces *B*, *B* produces the PIE, and the PIE covaries with *B*. *B* was not initially operant activity, but has become operant and can now be shaped by changing the requirements for it to produce the PIE. The diagram shows, however, that the PIE still induces the activities denoted as B_O. They may still compete for time with the operant activity *B*, as indicated in Figure 11.1. Observation and theory support the assumption that the other activities compete with the operant activity (Baum, 2015; Baum & Davison, 2014b; Breland & Breland, 1961; Herrnstein, 1970).

Thus, covariance explains both conditional induction by a CS or S^D and conditionally induced (operant) behavior. A full theory of behavior will take into account phylogenetic sources of behavior such as PIEs and the activities they induce, and also event-PIE covariance (i.e., signaling) and activity-PIE covariance (i.e., feedback).

Instead of using "operant" as a noun, we may use it as an adjective and recognize that activities may be partially the result of development and partially operant. Activities given by development often are refined by feedback (e.g., Hailman, 1969; Tinbergen, 1963). In humans, verbal behavior offers a striking example, because infants respond to others' vocalizations from birth and later imitate the vocalizations they hear (e.g., Moerk, 1992). Induction is nowhere more relevant than in verbal behavior. When Ben and Jan are in conversation, Ben's utterances induce Jan's utterances and vice versa. Much remains to be understood about conversation (e.g., Simon & Baum, 2017).

Patterns of behavior—activities—repeat from time to time. When an activity is maintained, the PIEs that maintain it not only follow the activity, but precede it. The distinction between consequence and antecedent collapses, because the occurrences of the PIE keep the activity going in the way that a flywheel on an engine keeps the engine going. The inducing PIEs accompany the activity rather than precede or follow it. The PIE and the activity engage in a feedback loop, as in Figures 11.1 and 11.2.

New activities enter behavioral allocations with some frequency. In non-human animals, they enter by imitation and other accidents. In humans, they enter mainly through rule following, when a rule (a verbal discriminative stimulus) induces some novel action on the part of a rule-following person (Baum, 1995b, 2017a). When I start playing tennis, that activity becomes a regular part of my weekly activities. It repeats, and the Law of Allocation tells us that it must displace some other activities.

The complexity of human behavior in the everyday world presents a challenge for behavior analysis. The concepts developed within a science of behavior should apply readily to human behavior. The Law of Allocation makes a start, because it illuminates the way people spend time, problems like "work-life" balance (Baum, 2017a), and some forms of psychotherapy (Córdoba-Salgado, 2017). Induction allows us to think about behavior that we call "rational" and also behavior we call "irrational" or "emotional" (Segal, 1972). In contrast with the molecular view, concepts like induction and temporally extended activities begin to allow constructing explanations and treatments that are plausible and functional.

The Ontological Status of Activities

The central precept in multiscale behavior analysis is that the units of behavior are temporally extended. Other features, such as the matching law or the Law of Allocation stem from this concept. Apart from the inadequacy of the molecular view of behavior, two grounds for adopting the central precept appeared earlier: (a) that the interaction between the organism and the environment cannot be observed at a moment; and (b) that no behavior can be defined at a moment (Baum, 1997, 2013). The first is a theoretical argument, and the second is epistemological. I believe one may also derive the temporal extension of behavioral units more directly on ontological grounds. If so, behavioral units must be temporally extended.

Two ontological distinctions are helpful in thinking about activities: (a) between objects and processes; and (b) between classes and individuals (Baum, 2017b). They are not entirely independent of one another, but I will take up each in turn.

Object versus Process

In everyday parlance, an object is any distinctive feature of the world that is seemingly stable—a tree, a house, a river, a star. Their apparent relative stability translates into repeatability, as when I return to a room I left yesterday and find that all the objects are still in their places. Their repeatability arises because they may be labeled and classified. Atomic particles, for example, may be classified according to their energy levels. To the extent that discrete responses are treated as repeatable and classified according to fixed criteria, discrete responses are treated as objects. If a response is classified as any movement that depresses a lever a certain distance and with a certain force, the response is being treated as an object. Behavior, however, manifestly constitutes process.

In contrast with the stability of objects, processes are changes through time—movement, deterioration, transformation, metamorphosis, growth. Some objects undergo notable change and are spoken of that way, as when we speak of a child becoming an adult. When we recognize that behavior is interaction with the environment and that behavior takes time, we recognize that behavior is process. A rat's lever pressing, a child's crying, a person's reading—these are processes, although their full definition as activities requires incorporation of their functions. The rat's lever pressing might be part of feeding, the child's crying may serve to summon a caretaker, and the person's reading might serve to inform. Thus, activities are processes.

Like any process, an activity may occur for longer or shorter time intervals. I may read a book for one minute or one hour. No matter how brief a bout of lever pressing—even if it results in only one operation of the lever—lever pressing still entails movement and takes time; lever pressing is still a process and not a discrete response (Baum, 1976).

Class versus Individual

A class singles out objects or processes according to a set of defining attributes. Dog is a class of which specific concrete dogs are instances; my dog Fido is an instance. Deterioration is a class of which the wear and tear on my house and the progress of a disease are instances. Skinner's (1938) definition of an operant as a class meant that the movements that met the criteria of the class were instances of the class.

Classes cannot change and cannot do anything. They may have more or fewer instances, but they are fixed by their defining attributes. Change the force required for a lever press, and you change the class. A class cannot do anything, because it is an abstraction; only concrete particulars can do things. "Dog" cannot come when I call, but my dog Fido can come when I call. An operant, as a class, cannot do anything; only the concrete movements that are instances can get a lever pressed.

In contrast to classes, individuals can change while still retaining their identity. An individual could be either an object or a process. An individual is a concrete particular with parts that function together as an integrated whole and is situated in time and space. Classes are eternal, whereas individuals have beginnings and endings. Whereas classes have instances, individuals have parts. The relation between instance and class contrasts with the relation between part and whole; individuals are instances, but they have no instances. Individuals can be described, but they cannot be defined. Abraham Lincoln was an individual, and my dog Fido, but also the chair on which I am sitting and the Rocky Mountains; they are all concrete particulars, they all function, and they all have integral parts that function together. Although organisms are spoken of as individuals, they are not the only ontological individuals. A baseball team is an individual, insofar as the players function together and win or lose as a whole. As Ghiselin (1997) explains, species are individuals; their members are their parts, and their function is to evolve.

Processes occur in individuals. When an individual changes, the individual goes through a process. Abraham Lincoln grew from a baby into a boy and into a man. When we talk about behavior, however, our language for talking about processes may be misleading. When we say Abraham Lincoln grew, we mean only that the process of growth occurred in him. When we say that Rat 5 pressed the lever, we also should mean only that lever pressing occurred in Rat 5—yet an additional element creeps in: agency.

When we say that Abraham Lincoln delivered the Gettysburg Address, from the perspective of a science of behavior we mean that the speech delivering occurred in Abraham Lincoln. Confusion might exist if one thought that Lincoln's growing was a different sort of process from Lincoln's speech delivering because Lincoln did not *do* the former, whereas he *did* the latter—that is, the speech delivering involved agency. (See Baum, 1995c, for further discussion of agency.) Recalling that behavior is interaction with the environment, we may think of the organism as the medium of its behavior—thus, if we are trying to be precise and avoid confusion, we might say the behavior occurs *in* the organism. Of the many meanings of "in," the usage here is that in phrases like, "a statue cast in bronze" or "an agreement in writing" or "Speak in English." We don't usually speak about behavior this way, because the usual construction of saying, "The organism did such-and-such," is so familiar.

As behavior, activities are processes that occur in organisms. Not all processes that occur in organisms are activities, only the ones that affect the environment. The heart's beating, though essential to keeping the organism intact, is not behavior, because the heart is only a part of the organism (see Axiom 1 above) and does not affect the environment. Putting on a coat to stay warm counts as an activity, because staying warm is an interaction with the environment. Tom's avoiding his ex-wife is an activity, because it functions to keep him from seeing her.

An earlier paper (Baum, 2002) suggested that activities themselves may be seen as individuals (Baum, 2017b). Like any other individual, an activity is an integrated whole constituted of parts that work together to serve a function. As cells constitute an organ, and organs constitute an organism, an activity like playing tennis is constituted of activities: serving, returning, keeping score, and so on. The activity lever pressing is both a process that occurs in Rat 5 and an individual constituted of parts that are also activities—pawing the lever, biting the lever, licking the lever, and so on. The activity baking a cake is both a process that occurs in the baker and an individual constituted of parts that are also activities—getting out a bowl, adding ingredients, mixing, and so on. Thus, an activity is both a process and an individual—a process individual.

Why Ontology Matters

The insight that behavior is process and consists of activities that are process individuals offers the third of three arguments mentioned earlier against viewing behavior as composed of discrete responses and in favor of temporally extended units. This ontological argument concludes that process implies temporal extension and change, neither of which is captured by discrete responses that are instantaneous and unitary. Because behavior is process, its units must be temporally extended.

As an example, consider discounting, the idea that distant and uncertain rewards are less valued than immediate and certain rewards (e.g., Green, Myerson, Oliveira, & Chang, 2014; Odum, 2011). In a study of discounting, a person is given a series of binary choices, from which a number of indifference points between immediate small amounts and delayed large amounts are derived. These indifference points are plotted as a function of delay or odds-against, and either a hyperbolic function is fitted to the points or the area under the connected points is estimated. Either way, the method results in a measure of the relative steepness of the curve. Thus, discounting is a measure, not a process. This prompts the questions: what does discounting measure, and what process is involved?

The participant's behavior is the process: choosing or expressing preferences. The preferences may be indicative of the participant's preferences in everyday life. Discounting, as an index, measures choice—time allocation—between two extended patterns of behavior—i.e., two activities. One activity, often called "impulsivity," pays off in the short term, but is punishing in the long term. The other activity, often called "self-control," is punishing in the short term, but pays off in the long term (see Baum, 2016a, for further explanation). As the value-laden terms "impulsivity" and "self-control" suggest, allocation of time to the short-term-advantageous activity (e.g., drunkenness, smoking, overeating, selfishness) is assumed to be problematic. The long-term-advantageous activity (e.g., sobriety, healthy eating, cooperation) is assumed to be desirable. Thus, interventions aim to increase time allocation to "self-control," while (necessarily) decreasing time allocation to "impulsivity." This explication of discounting becomes clear only in the light of the insight that behavior consists of process individuals—that is, temporally extended activities with parts.

Molar Behaviorism

Behaviorism is the philosophy that underpins a science of behavior. The central premise in behaviorism is that a science of behavior is possible (Baum, 2017a). If a science of behavior were impossible, behaviorism would be unnecessary.

A science of behavior may be made impossible in a variety of ways. In psychology, the supposition that behavior is not a subject matter in its own right would make the science impossible. Particularly the assumption that behavior is done by an agent—an inner self, the mind, or the brain—makes the science of behavior impossible. Indeed, any notion that behavior is caused by internal, unobservable entities, such as a person's inner intentions, beliefs, desires, or thoughts, makes the science impossible or, at least, incoherent. Behaviorism may be called "radical" because it rejects inner causality and agency, and instead places causes in the environment.

Skinner (1945) made a mistake when he advanced private events to account for thoughts and feelings. He was responding to the criticism that behaviorism ignores the most important part of human life, our inner thoughts and feelings. He would have done better to challenge the traditional view that our behavior is caused by thoughts and feelings and to have stayed with the view that the origins of behavior (its "causes") always lie in the past and present environment. He and other behaviorists tried to save the inferences to private events by calling them "interpretation." Such "interpretation" bears no resemblance to explanation in other sciences, which always refer to empirical relations verified in observation. Skinner's "interpretations" resemble, not science, but poetry or literature.

Private events seem to be of two sorts: (a) sub-vocal speech (Watson, 1930, pp. 238–243); and (b) sensing or imagining. Almost anyone can talk to himself or herself without others hearing, and almost anyone can see images or hear sounds that others cannot. Artificial intelligence programs coupled with brain scans may render these private events, at least in part, to be no longer private, but public. Once our sub-vocal speech becomes public that way, it will become data—events to be explained. In ordinary circumstances, however, private events remain private and can only legitimately be offered as part of an explanation of behavior once they have been made public in those circumstances with instrumentation (Baum, 2011c). For example, we usually look to environmental conditions to decide if a person is lying, because usually only those are available, but if a computer and brain scan were able to give reliable indicators, we would have more evidence to bring to bear.

In contrast with private events, private stimuli do not exist. Positing private stimuli as causes of behavior denies the science of behavior, because science requires that causes be observable. For example, a person or animal said to be "in pain" exhibits pain behavior (Rachlin, 1985). The pain behavior is a response to a physical condition of the body, usually an injury. Pain fibers in the nervous system are stimulated when the skin is breached, or a muscle is strained, or a tooth is decayed, or the brain receives excessive blood flow, or a nerve is pinched. All these conditions are observable, if not with the unaided senses, then with the right instruments. If the pain is genuine, pain behavior is caused by the stimulation of the pain fibers. When a dog limps and whimpers, we look for a thorn in its foot. The injury is the cause of the limping and whimpering, not "pain," not a private stimulus. Similarly, when a person limps and says, "I have a pain in my foot," the cause of the limping and saying (if genuine) is the injury, not a private stimulus called "pain." The temptation to attribute behavior to private stimuli derives from everyday talk about behavior, but for a science of behavior private stimuli are unobservable causes. The everyday mistake that a science must avoid is inferring private stimuli from behavior and then, circularly, using them to try to explain behavior. When at a party Jan says to her husband Ben, "I'm tired; let's go home," she is not reporting on a private stimulus; her utterance comes from a long history with such utterances and their effects (perhaps escaping from uncomfortable situations). Verbal behavior depends primarily on the presence of a listener who is likely to respond; other aspects of the context may be important, too, but combine with the primary context. Jan may or may not be tired, just as someone may be in pain or fake being in pain; the listener's sympathetic response is what counts. Verbal behavior, like all other behavior, occurs because of past and present environment, not thoughts and feelings. (See Baum, 2011c, for further discussion of private events.)

A true natural science of behavior is possible, a science that makes no appeal to unobserved inner events or hypothetical inner objects or processes like response strength and inner computing. Such a science may be achieved by accepting that all behavior and the environmental covariances that affect it are extended in time. For example, if activity-PIE covariance turns an activity into an operant activity, we may not know the physiological mechanisms that make this possible, but nothing stops us from studying the dependence between the activity, which is measurable, and the covariance, which is also measurable. Measures based on information theory offer one possibility to measure covariance (Shahan, 2017). The physiology may follow and be far different from anything we could guess.

In a natural science of behavior, behavioral events are natural events. Natural events are explained by their relation to other natural events. For example, an increased frequency of hurricanes in the Caribbean is related to changes in water temperature, which are related to increased global temperature (i.e., climate change). Natural events, if thought of as "caused," are caused by other natural events.

In particular, natural events are not caused by agents; natural events just happen, they are not done by anyone (Baum, 1995c; Skinner, 1971, pp. 7–14). When a stone falls, it accelerates as it approaches the ground. No physicist today would say the stone accelerates because it (privately) wants to reach the ground. Saying it accelerates because of gravity would also be a mistake, because the acceleration is an example of gravity, and making gravity a cause would make it an unseen agent—committing what Ryle (1949) called a "category error." Similarly, no behavior analyst should say that a rat presses a lever because it has "memory" that pressing the lever produces food. The rat's pressing results from the past covariance between pressing and feeding, which was observable, in contrast to its inner "memory," which is not. As with gravity, one could say at best that the rat's lever pressing *is* its memory. The rat does not do anything, because it is only the medium of its behavior. No more than the rat are we the doers of our deeds.

Conclusion

Behavior takes time. This fundamental principle for understanding behavior is supported by logic, by theory, and by research. Its implications are profound. It puts aside the traditional molecular view based on discrete events and contiguity. It tells us that behavior must be understood as dynamic and extended in time, an insight that concurs with the view of behavior implied by evolutionary theory, that behavior is an organism's interaction with its environment.

Molar behaviorism and multiscale behavior analysis treat behavior as consisting of activities that are extended in time. They treat behavior at any time scale, whether milliseconds or years. An episode of an activity like a pigeon's peck, however brief, has temporal extent. Care should be taken to avoid confusing brief episodes of an activity with discrete responses; they are qualitatively (ontologically) different concepts.

Activities are processes and individuals. They function as integrated wholes and evolve through time as their parts (less extended activities) change through time and take up more or less time. Like species, activities have a beginning and may go extinct as other activities replace them—we change jobs, move to new neighborhoods, have children, and change spouses. This multiscale view applies plausibly both to behavior in the laboratory and to behavior in the everyday world.

12

Behavior, Process, and Scale

Foreword

Behavior analysts accustomed to thinking about behavior in a molecular paradigm sometimes incorrectly take the molar view as just the study of behavior in long time frames. The mistake is partly my own fault, because in 1973 ("The Correlation-Based Law of Effect"), before I understood that the molecular view and molar view are different paradigms, I suggested that the two might be complementary. Subsequently I realized my mistake and tried to correct it, particularly by pointing out the ontological differences between the two paradigms and by showing how time scale allows the multiscale molar view to apply to short-term processes and long-term processes. In a misguided and incoherent paper, Shimp (2020) attempted to support a molecular view of behavior by casting the difference between the molecular view and the multiscale molar view as only a difference of measures. According to the article, the molecular view focuses on moment-to-moment changes in behavior, whereas the molar view examines aggregates of behavior over long time scales. Understanding why this is incorrect requires understanding the paradigmatic nature of the difference, which Shimp's article overlooked. He portrayed the contrast between molar and molecular as simply a difference in time scale of measures of behavior, ignoring the ontological differences between the two paradigms.

The ontological differences lead to incommensurability. The molecular view sees behavior as composed of discrete responses and sees behavior–environment relations simply as temporal contiguity, whereas the molar view sees behavior as composed of processes (i.e. activities) necessarily extended in time and behavior–environment relations as covariance through time. They explain the same phenomena differently (e.g. avoidance) but also define different phenomena as worthy of study and useful for explanation (e.g. response strength versus induction and inter-activity competition). For example, the molecular view sees no role for feedback functions, whereas the molar view sees feedback functions as essential both for study and for explanation. To try to reduce the difference between the two paradigms to a difference of measures would be like trying to advance a geocentric view of the universe by saying that the planets' positions could be measured either in epicycles of the geocentric view or by a heliocentric model, as if the ontological differences did not matter. In the heliocentric view, however, the concept of "epicycle" is absent, whereas concepts like "solar system" and "satellite"—absent from

Science and Philosophy of Behavior: Selected Papers, First Edition. William M. Baum.

the geocentric view—are offered for study and explanation beyond what the geocentric paradigm could achieve.

The present paper responded to Shimp's (2020) article. It aimed to clarify the paradigmatic difference between the molecular view and the molar view, particularly the way that process accounts affect one's view, and to explain the importance of time scale in relation to the concept of an activity.

Abstract

If we study the behavior of organisms, we must understand the ontological status of both "organism" and "behavior." A living organism maintains itself alive by constantly interacting with the environment, taking in energy and discarding waste. Ontologically, an organism is a process. Its interactions with the environment, which constitute its behavior, are processes also, because the parts of any process are themselves processes. Processes serve functions, and the function of a process must be part of its identity. A process, by definition, extends in time. Time is the fundamental and universal measure of behavior. All processes have the property of scale. Activities of an organism have parts that are themselves activities on a smaller time scale. Scale varies continuously, and behavior may be studied on as large or as small a time scale as seems necessary. When researchers refer to the "structure" of behavior, they refer to smaller-scale activities. Attaching a switch to a lever or key is convenient, but one should never confuse operation of a switch with a unit of behavior. Shimp's (2020) "molecular" measures are small- scale measures. The molecular view based on discrete events has outlived its usefulness and should be replaced by a multiscale molar paradigm.

Keywords: process, scale, molecular view, multiscale molar view, activity

Source: Originally published as "Behavior, process, and scale: Comments on Shimp (2020), 'Molecular (moment-to-moment) and molar (aggregate) analyses of behavior,'" in *Journal of the Experimental Analysis of Behavior*, 115 (2021), pp. 578–583.

One cannot hope to achieve an adequate understanding of behavior without evolutionary theory. To understand what behavior is and why it even exists, one must understand what organisms are and why they exist, because when we say "behavior," we mean "behavior of organisms."

What is an organism? Living beings differ principally from nonliving things in that they are born, last for a while, and then die and disintegrate. While alive, an organism maintains itself far from the equilibrium state of death by constantly interacting with the world around. This interaction accomplishes gaining of resources required for life and excreting into the world the waste products of its metabolism. Without the metabolizing of resources like food, water, and oxygen, the organism soon ceases to be alive—obeys the second law of thermodynamics and reaches equilibrium in death.

Nicholson (2018), building on a seminal paper by Erwin Schrödinger (1956), "What is Life?" argues that, from an ontological viewpoint, the description above indicates that an organism should be seen not as a mechanism but rather as a *process*. One of the main reasons is the constancy of an organism's interactions with the environment. A machine has an on-off switch and may sit idle for indefinite intervals, but an organism has no such switch and must remain active all the time or cease to be. Nicholson remarks:

> Schrödinger explained that an organism stays alive in its highly organized condition by importing matter rich in free energy from outside of itself and degrading it in order to maintain a relatively low entropic state within its boundaries. The organism thus preserves its internal organization—thereby eluding (at least for a time) the inert, time-invariant state of thermodynamic equilibrium we call *death*—at the expense of increasing the entropy (in the form of heat and other waste products) of its external environment (pp. 143–144).

From Nicholson's (2018) exposition, we may draw conclusions about behavior. First, behavior is the interaction of an organism with the environment, the importing of energy into itself and the expelling of waste into the environment. Second, behavior is process. The parts of any process are themselves processes, and behavior is a part of an organism's process, along with physiological processes. The distinction between physiological processes and behavior may sometimes be unclear, because both are parts of an organism's process. For example, is digestion behavior? Drawing such a distinction becomes less important when we realize that digestion and foraging are both parts of living. Third, behavior is continuous. As long as an organism functions adequately—stays alive—it is behaving. To be alive is to behave. Fourth, behavior must be understood in light of its function. Processes always exist to serve functions and may be defined according to their functions. Broadly speaking, behavior exists to serve the functions of surviving and reproducing.

Structure versus Function

To talk about the "structure" of behavior injects a paradoxical incommensurability into our attempts to understand behavior. Pressing a lever and pushing a button differ, it is often said, in their "topography," but the distinction does more to obscure the difference than to elucidate it. What is "topography"? A topographical map shows the contours of the land. Presumably the topography of a response refers to the contours of the response, which seems just to mean the way it looks to an observer. The way an activity appears to an observer, however, is neither necessary nor sufficient to identify the activity. Not necessary for the reason that Skinner called the operant "generic;" many different topographies may be included. Not sufficient because appearances may be deceiving; a person sitting and gazing into a book might be reading or pretending to read. Topography alone cannot disambiguate the activity. Only ancillary observations on a longer time scale can decrease uncertainty—for example, does the person speak about the contents of the book later?

One cannot define an activity apart from its function. Reading a novel, for example, differs from reading a newspaper. Their functions differ. One is entertaining, and the other is informative. They will result in different subsequent behavior. The novel might induce one to praise the writing, whereas the newspaper might induce one to argue with one's neighbor. Shimp (2020) fails to grasp the importance of function when he presents a caricature of the multiscale molar view by conflating eating chips with eating beef Wellington.

Between button-pushing and lever-pressing, different muscles and motions are involved and different physiological processes too perhaps, but the difference may be thought of in functional terms with greater clarity. I use the word *activity* to denote a behavioral process. Lever-pressing and button-pushing are processes—that is, activities. In laboratory experiments these activities typically result in operating a switch attached to the lever or button. Functionally, they may be similar, even if the details of their quantitative properties differ (McSweeney et al., 1986). The parts of a rat's lever-pressing, for example, may be smaller-scale activities like moving a paw or shifting weight onto the hind legs.

Activities versus Discrete Responses

In the laboratory, we measure an organism's interactions with a lever, key, or button by attaching a switch and counting the number of operations of the switch. This method usually gives a good approximation to the time taken up by the interactions. It even allows analysis into bouts and pauses (Brackney & Sanabria, 2015; Shull et al., 2001; Shull & Grimes, 2003). In an experiment that was published in 1976, I attached running-time meters to the levers in a choice situation and found that the time a lever was depressed agreed exactly with the number of operations of the switch. The rats' interactions with a lever that produced food on a variable-interval (VI) schedule resulted in jiggling the lever and many short operations of the switch.

A number of findings indicate that time is the fundamental measure of behavior. A pigeon's interactions with a response key producing food according to a VI schedule differ from a pigeon's interactions with a response key producing food according to a variable-ratio (VR) schedule. The switch-operation rates on the VR are often twice the rates on the VI (Baum, 1993a). The difference is not simply that the same response is repeated more often on the VR than the VI. Rather, the interactions with the VR key are less "pecking" than "swiping" at the key (Palya, 1992). Yet counting the number of switch operations still gives a good estimate of the time taken interacting with the key. In fact, when the number of operations is compared with the time taken, choice between a VI and a VR schedule resolves in favor of the time (Bell & Baum, 2017).

Many activities inside and outside the laboratory provide no discrete responses for counting. In the laboratory, for example, running in a wheel can only be counted if some arbitrary unit is defined, such as a quarter-turn. Outside the laboratory, what is the discrete unit of watching television, playing tennis, or training for the Olympics? Discrete responses accord poorly with everyday life. Time is the universal and fundamental measure of behavior.

Behavior is composed of processes—activities—and processes by their very nature, by definition, extend in time (Baum, 2017b). The phrase "momentary process" is an oxymoron, a contradiction in terms. No activity can be identified at a moment because processes can only be identified through time. That is the reason that discrete responses do not exist. They only seem to exist because they are conflated with switch operations. A switch operation may detect activity, but it is not a unit of activity. Switch operations are a convenience, because they allow automation of laboratory experiments. When research goes on outside the laboratory, measurement usually requires coding of activities by observers in real time or in video recordings. Inter-observer agreement becomes important. However activities are recorded, the time taken by an activity is paramount (Simon & Baum, 2017).

Process and Scale

When a behavior analyst talks about the "structure" of an activity, we may infer that the topic is smaller-scale activities. Serving a tennis ball, for example, is an activity (process) that is part of the longer-scale activity (process) of playing tennis, along with returning serves and keeping score. The activity of serving a tennis ball similarly has parts, like tossing the ball, rising up, and swinging the racquet down on the ball. Those smaller-scale activities themselves have parts, and one may examine those parts in as much detail as seems useful. Perfecting one's serve, for example, requires attending to small-scale activities like placing one's feet and tossing the ball high enough.

All activities have the property of scale. At the extreme, surviving and reproducing are activities on the longest time scale. Their parts are the particulars of life. For a human, gaining resources functions to promote survival and has parts like holding a job and eating a healthy diet. Holding a job has parts like driving to work and the activities that compose "work." Wallace (1965) analyzed driving to work into the many activities that constitute its parts.

In his discussion Shimp (2020) uses the phrase "moment-to-moment" many times. Two usages stand out: moment-to-moment "shaping;" and moment-to-moment "sequential structures." Both usages appear to me to entail small time scale. The shaping seems to mean small-scale relations between activities and outcomes, like tossing the tennis ball (activity) such that the ball is driven hard over the net (outcome). The sequential structures seem to just be small-scale activities. The tennis serve seems to be an example: One adopts a stance, then tosses the ball, then raises the racquet, then brings the racquet down on the ball. The sequential nature of the activity may be more apparent than real and less intrinsic than imagined; the parts may overlap, but even more to the point, the tennis serve is an integrated whole.

Every activity is an integrated whole (Baum, 2002). Serving a tennis ball, playing a piece on the piano, eating a meal—these are all integrated wholes. I am unsure what Shimp (2020) intends by the phrase "moment-to-moment," but if it is meant to convey control at moments such that one discrete unit leads to the next and that complexity of behavior is built by accreting such discrete units, then it is clearly misguided. Lashley (1961/1951), in his classic paper, "The Problem of Serial Order in Behavior," pointed out the impossibility of characterizing behavior extended in time by such "moment-to-moment" analysis. His most compelling examples were verbal utterances. A spoken sentence cannot unfold word by word or sound by sound, because earlier elements in the sentence depend on later elements. In German, for example, the verb typically goes at the end of the sentence, requiring earlier words to anticipate it. Anyone familiar with a heavily inflected language like Russian can attest that earlier elements must be inflected to anticipate what is to come. The sentence takes time to be uttered, but it is an integrated whole.

Shimp's (2020) many different usages of "molecular" seem to me just to add confusion. One crucial distinction he fails to make is the difference between what he calls a "molecular measure" and what I call the "molecular view." The result is that he repeatedly misrepresents my criticism of the molecular *view* and conflates it with what he calls molecular *measures*. His molecular measures seem to me just to be measures that may be obtained in a short time frame. They are simply small-scale measures like interresponse time (IRT) or eye blink. When he contrasts "molecular" measures with "molar" measures, calling them different "levels," more confusion results, because "level" indicates a discrete difference. Scale, however, is continuous, and the distinction implied is a matter of degree; scale is relative, not absolute.

Molar and Molecular Paradigms

The difference between the molecular view and the multiscale molar view is a difference of paradigms (Baum, 2001). What I call the molecular view is the theoretical framework (paradigm) that assumes: (a) behavior consists of discrete units (an ontological claim); and (b) that immediately following (contiguous) "reinforcers" strengthen (increase the rate of) a discrete unit (a theoretical claim). The word "contiguous" means just closeness in time, and the assumption about contiguous reinforcers caused Skinner (1948) to claim incorrectly that noncontingent presentation of food must strengthen some behavior—whatever it was contiguous with. Subsequent research has shown that noncontingent food induces food-related activities (e.g., Baum & Aparicio, 2020; Baum & Grace, 2020; Segal, 1972; Staddon & Simmelhag, 1971). Moreover, subsequent research showed that response-reinforcer contiguity is neither necessary nor sufficient for favored activities to be selected (e.g., Baum, 1973; Kuroda & Lattal, 2018; Rescorla, 1967, 1968, 1988).

I have criticized the molecular view at more length elsewhere (Baum, 2018b, 2020; Baum & Aparicio, 2020). In brief, it fails to explain even the most basic laboratory phenomena, such as the high response rates engendered by ratio schedules. All IRTs are followed by food with equal probability in a VR schedule, yet response rate rises to approximate the highest possible—that is, to the shortest IRTs. A molar account explains it as selection of higher response rate by higher food rate according to a linear feedback function (Baum, 1981b, 1993a). Neither does the molecular view provide any plausible account of avoidance behavior. Explaining avoidance requires a molar view (Baum, 2020; Herrnstein, 1969).

On one hand, failing to distinguish between measures and theoretical stances creates confusion, but on the other hand, making a false distinction between "shaping" versus "strengthening" compounds the confusion; it is a false dichotomy. Every activity varies in its parts, and the different variants compete. If one variant increases, its competitors decrease. On a long time scale, working may compete with socializing; if time taken by working increases, time taken by socializing decreases and vice versa. On a short time scale, tossing a tennis ball weakly into the air competes with tossing it higher, and if tossing it higher results in a better outcome, subsequent attempts at serving will entail tossing it higher rather than weakly. Thus, shaping may occur on a long time scale or a short time scale, and it always entails strengthening some parts at the expense of others. That is selection by consequences (Baum, 2017a).

Finally, any theory of behavior ought to offer plausible accounts of everyday life, and a molecular view fails to do so. For example, if a person works week after week and is paid at the end of each week, how can contiguous reinforcers explain such behavior without resorting to imaginary unobservable events? Donahoe (2012) observed, ". . .molar relations are more easily communicated to an audience not intimately familiar with the experimental analysis of moment-to-moment, stimulus-response-reinforcer relations (that is, molecular processes)" (p. 197). I agree. The molecular view fails to provide plausible, comprehensible explanations of phenomena that, by their very nature extend in time—processes readily visible in everyday life.

Conclusion

Shimp's (2020) discussion of molar and molecular seems to me both confused and confusing. He clings to a molecular view at the expense of gaining clear explanations of behavior in everyday life and even behavior in the laboratory. The old molecular paradigm served to initiate the science of behavior. It served well enough for a time, until a better paradigm could be developed. Now we can drop the worn-out framework and adopt a more coherent and plausible paradigm: the multiscale molar paradigm (Baum, 2013, 2018b).

Part II

Molar Behaviorism

13

Radical Behaviorism and the Concept of Agency

Foreword

Two fundamental questions are: (1) "Who am I?" and (2) "What is all this?" They may really be one question, for the answer to one may entail an answer to the other. Possibly I and All This are one and the same, as Vedic teachings assert: "Thou art that" (*Tat tvam asi*). Possibly I is just a given, consciousness or existence that just is, and the question becomes, "Why is there something rather than nothing?" Whatever the answer, it will transcend everyday concepts and experiences.

The popular "solution" to existence assumes that each person has an inner self somehow associated with the body and that moves the body to action. It presupposes an inner realm called the "mind" where thoughts, feelings, and desires dwell and are expressed in outward behavior. This conception probably arises because our inner speech and imagery seem to be private—unavailable to others. Where this inner realm is located remains a mystery, because it is not *in* the body, as a person might be *in* a room. On examination, the popular view imagines two worlds: an inner, mental world, populated by immaterial things and an outer physical world populated by material things. It is built into languages like English, in phrases like, "I thought to myself" and "I saw it in my mind." This mental–physical dualism implies two kinds of existence—that is, two categories of existent things.

The great mystery attendant on this view is how an immaterial thing could affect a material thing. It is often referred to as the "mind–body problem." Many solutions have been proposed, but the problem remains insoluble as long as one assumes dualism. This is the reason that dualism is incompatible with science. Science assumes the universe to be comprehensible, and dualism makes it incomprehensible.

Dualism is particularly undermining for a science of behavior. Behaviorists writing about behavior with care avoid any explanation of behavior that might appeal to hidden mental causes and find ways to nullify the mind–body problem. Skinner suggested eliminating all mental terms when explaining behavior and appealing only to causes in the environment. When he wrote about thoughts and feelings, however, he appealed to private events, a move that undercut a natural science of behavior (Rachlin, 2018). Others, like myself, maintain the possibility of a true natural science of behavior, in which behavioral events are natural events and caused only by other natural events.

Science and Philosophy of Behavior: Selected Papers, First Edition. William M. Baum.

In *The Concept of Mind,* Gilbert Ryle (1949) cast the mind–body problem as a logical problem. He derided the popular view as the "Ghost in the Machine," likening it to the idea of an inner immaterial spirit enlivening a physical body. Instead he pointed out that mental terms like knowing, wanting, and believing actually refer to category labels of activities and that the activities themselves are concrete particulars that constitute instances of the category. The reason mental terms seem so elusive is that they are not concrete, but only labels. For example, to "know French" is to speak to others in French, read French books and newspapers, write letters in French, and so on. If a person does enough of these things, we say he knows French. Those activities, however, are not caused by "knowing French"; those activities are instances of "knowing French" as a category. To say that one's knowledge of French causes the activities is to make a category error—supposing that the label of a category is something distinct from the category itself. Tom does not know French *and* speak French; speaking French is an instance of knowing French.

Rachlin (1994) cast the mind–body problem as a confusion of causes. Instead of categories, Rachlin pointed to extended patterns of behavior and equated these patterns to mental terms. "Knowing French" is a pattern of behavior that occurs through time. The concrete activities—Ryle's instances—are parts that make up the whole extended pattern. Similarly, for example, believing that eating meat is wrong means asserting that, arguing with meat-eaters, joining protests at meat plants, giving money to vegetarian organizations, and so on. Believing something is an extended pattern of behavior of which these more specific activities are part.

In contrast to Ryle, Rachlin embraces extended patterns like "knowing French" and "believing that eating meat is wrong" as possible causes of the activity parts—just not efficient causes, but rather final causes. The extended pattern is the final cause of the activities that make up the pattern and fit into it as parts. Thus one can say that Tom speaks French *because* he knows French, as long as one understands that this only means that knowing French is the final cause of speaking French—that speaking French is part of knowing French, knowing French entails speaking French, and if Tom did not speak French he would not know French.

The present paper addresses one aspect of dualism: the concept of an inner self and its conflict with a natural science of behavior. A problem deeper than the mysteriousness of how an inner self could cause concrete behavior is the assumption that actions are done at all—the problem of agency. A natural science doesn't only rule out mysterious causes, it rules out any division of behavioral events into more than one category, particularly a division between natural events and actions. The subject–verb–object construction encourages such a division. "Jane kissed Bill" seems to be saying that Jane *did* the kissing, that the event was not like Bill getting kissed, which just happened to him. Although languages like English treat actions as a special category, a science of behavior requires that actions not be *done* but be natural events that just *happen* and are explained in relation to other natural events, which might be efficient causes or final causes. If a rat presses a lever, we can say that lever pressing occurs in the rat; the organism is the medium of the lever pressing, not the doer. No more than the rat are we the doers of our deeds.

Abstract

The central tenet that defines radical behaviorism is the idea that there is no such thing as agency, or stated positively, that behavior consists of ordinary natural events. Radical behaviorism denies the distinction, common to English speakers, that separates actions from natural events. Although an event's category cannot be determined from its label alone (e.g. "John fell"), its status is indicated by the presence or absence of surrounding mentalistic usage (e.g. "on purpose"). The linguist Benjamin Wharf called such categories *cryptotypes* and attributed the pervasiveness of mentalism in English to the fundamental cryptotype that distinguishes form from substance. Whorf argued that English cryptotypes are so poorly suited to science that scientists must talk "in what amounts to a new language." Radical behaviorists do this when they seek to include behavioral events in the cryptotype of natural events. In doing so, they omit the conventional notion of an individual self or doer and treat behavioral events as "just happening." Instead of explaining the events as "done," behaviorists see the events as occurring in a nexus of many other events both earlier and later. Although Skinner referred to this nexus as a "history," no special status attaches to the past, present, or future. The explanatory strategy resembles Rachlin's application of Aristotle's explanations by final cause. All of these root tenets of radical behaviorism—rejection of agency, the illusory nature of the individual self, behavioral events just happen, and happen in a broad nexus of other events—correspond to key tenets of Eastern mysticism. The unacceptability of behaviorism and mysticism to Westerners may arise from the same incompatibilities with standard English cryptotypes. Apart from differences in terms, when it comes to public stance, there is an affinity between behaviorists and New Age writers, who promote assimilation of Eastern mysticism into Western culture. In the public view, however, the New Age is unrespectable but nice, whereas behaviorism is respectable but not nice. The two might be blended into an account that is both respectable and nice.

Keywords: agency, behaviorism, Benjamin Whorf, cryptotype, mentalism, dualism, natural event

Source: Originally presented as the B.F. Skinner Memorial Address at the 1995 Convention of Behaviorology and published in *Behaviorology*, 3 (1995), pp. 93–106. Reproduced with permission of *Behaviorology* journal.

What's Radical about Radical Behaviorism?

If Skinner chose the name with an eye to etymology, radical behaviorism refers to a view that differs from others at its *root*—that is, in some basic tenet. Radical behaviorism is sometimes said to be distinctive because it includes private events, or verbal behavior, or operant behavior, but these are appurtenances, not roots. Sometimes it is said to be distinguished by its rejection of mentalism or its replacement of dualism with monism. This is closer to basic tenets, but still not quite there, because we still need to know why radical behaviorism rejects dualism.

There is a premise deeper, in the sense that eschewal of mentalism and dualism follows from it. Stated negatively, it is the idea that there is no such thing as agency. Stated positively, it is the idea that behavior consists of ordinary, natural events.

Science deals with natural events. All scientific questions can be phrased in the form, "Why does this happen?" For example, the question, "Why are there human beings?" translates into either, "How did this species evolve?" or "How does a fertilized egg develop into a whole organism?" This point seems uncontroversial, until it is applied to human behavior. Then it has remarkable results.

What is Agency?

Agency is the notion that actions are distinct from natural events. In Western culture, we find it easy to think that sunrise and sunset just happen, but if Liz walks to town, it seems some additional element enters in: There is an agent, a *doer*. Liz's walk to town doesn't just happen—it differs from sunrise and sunset—because Liz is there, *doing it*. This is the view that radical behaviorism contradicts.

The first column of Table 13.1 shows some simple labels of events. It is intended to be roughly graded in the likelihood that people will claim agency to be present. Almost no one would argue about sunrise. John's falling seems close enough to a rock's falling; his being animate makes no difference (unless, of course, he falls "on purpose"). John's sneezing might raise some questions, but most people, following Descartes, would classify this as automatic, involuntary, or reflex. With a label like "John rose," questions arise: Is he weightless? In a balloon? Did someone stimulate his legs with electrodes? With "John walked," we still ask whether it is conscious; after all, a cockroach may walk. By the time we get to labels like "John spoke," all uncertainty vanishes; unlike the sun's rising, John does this.

The ease of identifying ambiguous verbs (e.g., grow, sleep, hear) tells us that it is difficult to know where to draw the line between natural events, which just happen, and actions. There are no unambiguous markers to guide us, and different people will draw the line in different places. At other times, it was probably normal to draw the line elsewhere than today; before Descartes, a sneeze might have been taken to involve agency. Common though it is in our culture, there is no sure sign that distinguishes actions from natural events. If we believe Greek mythology or some Native American accounts of nature, it is possible to make no distinction and to take all events to involve agency. If the sun rises because Apollo draws it with his chariot, and thunder and lightning are acts of spirits or gods, then there are no natural events. Radical behaviorism also refuses to draw a line, but takes the opposite approach: There are only natural events.

Natural Events

To understand what is meant by a natural event, we can go back to the origins of science in the sixth century B.C.E. Farrington (1944/1980), in his book, *Greek Science,* discusses the philosopher Thales, widely considered one of the founders of scientific thought. The dominant cosmogonies of Thales's time were Egyptian and Babylonian. The Babylonian creation myth invoked a god named Marduk: "All the lands were sea. . . Marduk bound a rush mat upon the face of the waters, he made dirt and piled it beside the rush mat" (Farrington, p. 37). Farrington goes on to describe Thales's contribution:

> The general picture Thales had of things was that the earth is a flat disc floating on water, that there is water above our heads as well as all round us (where else could the rain come from?), that the sun and moon and stars are vapour in a state of incandescence, and that they sail over our heads on the watery firmament above and then sail round, on the sea on which the earth itself is afloat, to their appointed stations for rising in the East. It is an admirable beginning, the whole point of which is that it gathers together into a coherent picture a number of observed facts *without letting Marduk in.* (Farrington, p. 37; italics in the original).

From the beginning, science has always omitted hidden agents. In a sense, omission of hidden agents constitutes the very definition of scientific thinking.

"Natural events just happen" means natural events occur because of the way the universe is arranged. They occur because of the nature of things. In Farrington's words, the new idea in Thales and the other Ionian philosophers was that ". . .they did not think that life, or soul, came into the world from outside, but that what is called life, or soul, or the cause of motion in things, was inherent in matter, was *just the way it behaved*" (Farrington, p. 37; italics added). To omit agency from a science of behavior, then, is to say that behavior consists of natural events, events that just happen.

In talking to laypeople and students, I find that this idea is even more unacceptable than determinism. Students will often accept determinism, at least provisionally, because they recognize that the environment affects their behavior, but try to tell them there is no self, no person, somewhere inside them, that there is no doer, there are only behavioral events, and they become baffled.

Students, of course, are not the only ones. Professional psychologists often promote the idea of agency, even if they accept some form of determinism. Bandura (1989), for example, in a discussion of agency states, "In acting as agents over their environments, people draw on their knowledge and cognitive and behavioral skills to produce desired results" (p. 1181). He then puts forward a notion of determinism that excludes any conventional notion of free will, but allows that behavior influences the environment as well the environment influencing behavior: ". . . there is no incompatibility between human agency and determinism" (p. 1182), and "Selfgenerated influences operate deterministically on behavior the same way as external sources of influence do" (p. 1182). This resembles the opening statement of *Schedules of Reinforcement*, "When an organism acts upon the environment in which it lives, it changes that environment in ways which often affect the organism itself" (Ferster & Skinner, 1957). After accepting this kind of reciprocal control, however, Bandura closes with a strong endorsement of agency: "[People] may be taught the tools of self-regulation, but this in no way detracts from the fact that by the exercise of that capability they help to determine the nature of their situations and what they become. The self is thus partly fashioned through the continued exercise of self-influence." Behavior may be determined, but there is still a doer.

In teaching undergraduates, I skirt the ontological issues about self and agency by saying that we have a choice of how we talk about behavior, that the question comes down to a matter of talk. Scientists in all fields sometimes speak as if simple organisms and even inanimate things had agency. In a news article in a 1992 *Science* magazine, a bacteriologist named George Jacoby is quoted as saying, "Bugs are always figuring out ways to get around the antibiotics we throw at them." No doubt he would be surprised if someone criticized him for suggesting that bacteria are agents. Richard Dawkins (1989a) expressed surprise at criticism of his book, *The Selfish Gene*, on grounds that he imputed agency to pieces of genetic material. He commented, "This strategic way of talking about an animal or plant, or a gene . . .is a language of convenience which is harmless unless it happens to fall into the hands of those ill-equipped to understand it" (p. 278). This is so, of course, as long as one remembers that there is an alternative way of talking, and that, if asked, one could restate the matter in terms that omitted agency (Baum & Heath, 1992). When we come to human behavior, it is unclear how to achieve such restatement or even if it is possible in the English language.

Unwieldy English

The biggest obstacle to a science of behavior may be English grammar. Mecca Chiesa (1994), in her excellent book, *Radical Behaviorism: The Philosophy and the Science,* and Philip Hineline (1980) have discussed the unwieldiness of English usage for talking about events. They both draw on the work of linguist Benjamin Whorf (1956), who is known for the "Whorfian" hypothesis that language determines one's perception of the world rather than the other way around, or, in behavioral terms, that the way one talks about the world affects all one's responses and the categories of stimuli that control them. Whorf is known also for studying the Hopi language and comparing it with English. One of his conclusions was that English is suited to talking about things, whereas Hopi is suited to talking about events (or "eventing"; p. 147). English raises difficulty because every verb requires a subject, a noun, even though many events, particularly what I am calling natural events, properly have no subject. Whorf wrote:

> . . .we are compelled in many cases to read into nature fictitious acting-entities simply because our sentence patterns require our verbs, when not imperative, to have substantives before them. We are obliged to say 'it flashed' or 'a light flashed,' setting up an actor IT, or A LIGHT, to perform what we call an action, FLASH. But the flashing and the light are the same; there is no thing which does something, and no doing. Hopi says only *rehpi.* Hopi can have verbs without subjects, and this gives to that language power as a logical system for understanding certain aspects of the cosmos. Scientific language, being founded on western Indo-European and not on Hopi, does as we do, sees sometimes actions and forces where there may be only states (pp. 262–263).

Both Chiesa and Hineline include quotes like this and compare them to Skinner's concept of mentalism. The phrase "fictitious acting-entities" conjures up Skinner's rejection of explanatory fictions and homunculi. Yet the comparison misses the mark and fails to express Whorf's primary complaint.

There are at least two ways in which the comparison misses. First, English speakers do in fact distinguish between events that involve agency and events that do not—agentic events and natural events. However cumbersome English may be for talking about events, speakers still see no agency in "a light flashed." Nobody says the light did anything, the way someone might say John did something in "John flashed." Although the label for a natural event has the same form as the label for an agentic event, people discriminate between the two classes. Since they discriminate on some basis other than form, the problem must lie somewhere other than form. Second, the quote has more force with respect to physical sciences than a science that deals with living organisms. In a science of behavior, it is necessary somehow to tie a behavioral event to a particular organism. Although we may say that key-pecks occurred at a rate of 53 per minute, we still need to know *whose* key-pecks they are. Pigeon B22's key-pecks differ from Pigeon W96's.

Whorf's fundamental criticism that relates to mentalism has to do with what he calls the "habitual thought and behavior" of a culture. He points out that the class distinctions in a language may be both overt and covert. An overt class is marked by a difference in form. Gender in Latin, for example, is overt because masculine and feminine nouns differ in their endings. Plural nouns in English constitute an overt class, for the most part, because most of them end with *-s*. Ones like fish and deer that have no telltale ending, however, one can only distinguish by other usage around them (e.g., absence of articles and plural verbs). A covert class or *cryptotype*, is defined entirely by other usage. Transitive and intransitive verbs in English, for example, are cryptotypes; one cannot tell from the verbs themselves which type they are, only by seeing other usage around them (e.g., presence or absence of direct objects and passive voice). Whorf argued that the cryptotypes of a people's language are fundamental to their culture. The cryptotypes of English, for example, differ greatly from those of Hopi. Hopi includes no gender, no tenses, and distinguishes actual and potential instead of past, present, and future. Perhaps most fundamental of all is the difference Whorf describes between the way quantity is dealt with in English and Hopi. He argues that one of the most awkward aspects of English for scientific discourse is the way it divides the world for enumeration—that is, the way it speaks of objects.

Whorf blames the mentalism and dualism of English on the dichotomy (absent in Hopi) between form and substance. We speak of water, for example, as a substance apart from form, and when we wish to speak of a particular quantity of water, we resort to locutions like "a cup of water" or "the water in that cup," always combining a form with a substance. Hopi is different this way:

> It has a formally distinguished class of nouns. But this class contains no formal subclass of mass nouns . . . Nouns translating most nearly our mass nouns still refer to vague bodies or vaguely bounded extents. They imply indefiniteness, but not lack, of outline and size. In specific statements, 'water' means one certain mass or quantity of water, not what we call "the substance water." Generality of statement is conveyed through the verb or predicator, not the noun. . . . The language has neither need for nor analogies on which to build the concept of existence as a duality of formless item and form. (Whorf, 1956, pp. 141–142).

The result of this difference is that, whereas English distinguishes two worlds—an inner, mental world of forms without substance, apart from an outer world of forms with substance—Hopi conceives only one world. The result is an entirely different conception of thought and consciousness:

> It is no more unnatural to think that thought contacts everything and pervades the universe than to think, as we all do, that light kindled outdoors does this. And it is not unnatural to suppose that thought, like any other force, leaves everywhere traces of effect. Now, when WE think of a certain actual rosebush, we do not suppose that our thought goes to that actual bush, and engages with it, like a searchlight turned upon it. What then do we suppose our consciousness is dealing with when we are thinking of that rosebush? Probably we think it is dealing with a "mental image" which is not the rosebush but a mental surrogate of it. But why should it be NATURAL to think that our thought deals with a surrogate and not with the real rosebush? Quite possibly because we are dimly aware that we carry about with us a whole imaginary space, full of mental surrogates. To us, mental surrogates are old familiar fare. Along with the images of

> imaginary space, which we perhaps secretly know to be only imaginary, we tuck the thought-of actually existing rosebush, which may be quite another story, perhaps just because we have that very convenient "place" for it. (Whorf, 1956, pp. 149–150).

This is a point more relevant to a science of behavior than the necessity of a noun subject for each verb. It may be readily compared with Skinner's (1969c, 1974) criticisms of introspection and copy theory. What Whorf calls surrogates would nowadays be called representations. The forms without substance are spoken of as if they were things, as if they require a place, which, because it cannot be in the outer world of forms with substance, is spoken of as an inner imaginary space. This inner space so suffuses English that one cannot speak in English without using terms that seem inherently mentalistic and dualistic. Whorf points out how difficult it is for English speakers to talk about the world in nondualistic terms:

> From the form-plus-substance dichotomy the philosophical views most traditionally characteristic of the "Western world" have derived huge support. Here belong materialism, psychophysical parallelism, physics—at least in its traditional Newtonian form—and dualistic views of the universe in general. Indeed here belongs almost everything that is "hard, practical common sense." Monistic, holistic, and relativistic views of reality appeal to philosophers and some scientists, but they are badly handicapped in appealing to the "common sense" of the Western average man—not because nature herself refutes them (if she did, philosophers could have discovered this much), but because they must be talked about in what amounts to a new language. (Whorf, 1956, p. 152).

So it is with a science of behavior. We may appear to speak in English, but in many ways our talk about behavior pits us against some of the most basic cryptotypes of English. This puts radical behaviorists in the same boat as quantum physics, Hopi culture, and Eastern philosophy. As I shall explain below, New Age thinking and radical behaviorism are "badly handicapped in appealing to 'common sense'" because they both challenge traditional cryptotypes.

One such cryptotype is the class of agentic verbs. Nothing about the form these verbs distinguishes them overtly from non-agentic verbs (Table 13.1). They are distinguished by the other locutions that accompany them, particularly ones related to intention and purpose. No one asks whether a light flashed "on purpose," but everyone accepts the statement that John fell "on purpose." The cryptotype is distinguished by its association with mentalistic phrases.

Whorf's analysis implies that the use of an agentic verb sets the occasion for mentalism and dualism. The only ways out of this dilemma would be to avoid agentic verbs (rarely possible) or to "talk in a new language." Baer (1976) suggested that the organism might be viewed as the host of its responses (guests). This metaphor and some others, such as ownership (responses belong to the organism) and inclusion (this response is part of a larger pattern) help to guide writing and talking about behavior. They help to avoid locutions in which the organism is reinforced or punished, instead of its behavior (e.g., "the rat was reinforced for pressing the lever"). They encourage locutions more

Table 13.1 Phrases and verbs that induce various likelihoods of talking about agency.

Event
Labels
The sun rose.
The rock fell.
John fell.
John sneezed.
John rose.
John walked.
John spoke.
Verbs
Grow
Breathe
Run
Sleep
Awaken
Dream
Look – See
Listen – hear
Sniff – smell
Talk
Think

in keeping with other sciences in which the subject of study appears as the subject of a sentence (e.g., "a light flashed" or "the cell divided")—that is, locutions in which behavior, rather than organism, appears as the subject (e.g., "lever presses produced food" and "a press occurred"). With the use of possessives, one can often avoid making the organism the subject (e.g., "John's jumping increased"). Ultimately, however, the behaviorists' goal is to include behavioral events, however labeled, into the cryptotype of natural events.

Self-less Behaviorism

No agency means no self, at least in the traditional sense. Even if it might be honest to admit that radical behaviorism omits self, to do so might be impolitic. In an attempt to maintain surface contact with everyday English and avoid the unjust accusation that behaviorism has no account of self, behaviorists have sometimes offered behavioral views of self.

Self

Skinner (1953; 1974) proposed two different accounts of self: as repertoire and as locus. He begins each account by denying that self is an originating agent or a cause of behavior. In the first account, he proposes that "a self is simply a device for representing *a functionally unified system of responses*" (Skinner, 1953, p. 285; italics in original). Responses constitute a "functionally unified system" when they covary together as a function of environmental factors such as discriminative stimuli, reinforcement, and deprivation. He referred to such a system also as a repertoire: "A self or personality is at best a repertoire of behavior imparted by an organized set of contingencies" (Skinner, 1974, p. 149). There may be multiple selves (personalities), because different response systems come into play in different environmental settings. The selves may interact, as in self-knowledge—one response system for reporting on another response system.

Skinner's second account takes self as a locus: "A person is not an originating agent; he is a locus, a point at which many genetic and environmental conditions come together in a joint effect" (1974, p. 168). Every behavioral event must be located in time and space; that is a property of natural events. That location in time and space may be equated with the self.

Hayes (1984) takes the approach of trying to identify the sources of control over talk about "self" or "spirit." He puts aside the notion of self as an object, and, taking Skinner's point that the locus of one's own behavior remains always the same in some sense, he argues for self as a perspective:

> Consider the following question: "If you lost your arms and legs, would you still be you?" Given one sense of the question ("You as a physical structure") the question could reasonably be answered "no." If the sense of the question, however, refers to the verbally established you-as-perspective, then the obvious answer is yes (1984, p. 103).

In other words, self is the experience of a constancy, the continuity that allows one to answer "yes" to a question like, "Were you the same when you were 10 years old?"

Rachlin (1994), drawing on Plato and Aristotle, argues that the "soul" (which for present purposes I am treating as synonymous with self) consists of a person's pattern of overt behavior, described in terms of goals or functions. The sense in which the soul is in the body is the same as the sense in which health is in the body. To have a certain kind of soul is to function in a certain way. To understand the soul or self is to understand the complex patterns of a person's behavior. "To know the soul is . . .to understand how the body as a whole functions, as an expert driver would know a car" (Rachlin, 1994, p. 89). As we may say we know the functions of a car—"its cornering, its accelerating, its comfort, its stopping distance, and so forth"—so, if we know the pattern of overt choices in a person's life, we may say we know the person's self.

Taken together, these behavioral accounts of self have two aims: to account for verbal behavior involving terms like "self," "person," "spirit," or "soul" and to provide an alternative definition in terms of a functional analysis of behavior. Hayes and Skinner suggest that talk of self arises from the necessity that behavior have a locus, not because there is a self within. Rachlin and Skinner suggest that self might be equated with complex overt behavior, instead of with a self within.

The thrust of all these discussions is to retain contact with ordinary English while at the same time to deny the reality of the traditional notion of an inner self. However polite they may seem, none supports the inner self, except as an illusion. People may talk as if they had inner selves, but it is only talk. From an ontological perspective, behavior is real, but the individual self is illusory.

Behavior Happens

With no self, except as an illusion, behavior is never *done;* it just happens. Behavioral events are natural events. "The sun rose" and "John spoke" (Table 13.1) belong to the same sort of events. John's relation to his utterance is the same as the sun's relation to its rising. The "entity" involved in the event is part of the event's configuration—there cannot be a sunrise without a horizon, as there cannot be an utterance without a listener. As the sun's rising differs from the moon's rising, so John's saying "Hello" differs from Jane's saying "Hello." One could say the organism is only part of the definition of the event. To make this explicit, instead of "John said 'Hello,'" one would have to say something like, "'Hello' occurred, type John."

History as Nexus

A science of behavior departs most dramatically from a traditional view in its explanatory mode. Since behavior just happens, explanations rely, not on a hidden agent, but on other natural events. John's utterance "Hello" is understood in relation to a multitude of events in the past, beginning with John's birth into an English-speaking household, including reinforcement of utterances of greeting and farewell ("Hi" and "Bye-bye") in childhood, and, assuming a relationship exists, including events establishing a relationship with the particular person greeted. All of these events, taken with their environmental circumstances (discriminative stimuli) and their consequences, all together make up what Skinner called the behavior's "history of reinforcement."

Skinner argued that if we know enough about the history of reinforcement, we can predict the behavior. If we know enough about John's upbringing and interactions with Jane, we predict that he will say "Hello" when she appears. But what a lot there is to know! In this example and in general, there is no one crucial event; we need to know about many events over a long period of time. Moreover, we cannot predict the event is certain, only that it is likely, because John may be distracted or angry, or he may already have greeted Jane earlier in the day. To predict or to understand we look at John's behavior over an extended period of time, and we draw conclusions, not from individual events, but from the overall patterns.

In the analysis of behavioral patterns, the present has no particular status. We may be trying to understand behavior of a day or a year ago in the light of patterns extending still more remotely back. We have no hesitation to include subsequent behavior into our account. The full understanding of a behavioral event in the present may await events yet to come. To decide whether a child cries "on purpose," we examine past instances of crying in the light of circumstances and consequences, but our final conclusion may await observation of further instances before we are sure we see a pattern of circumstances (e.g., being put down for a nap), crying, and consequences (e.g., being picked up).

In this extended view of behavior the particular event (John's saying "Hello") is understood by its fitting into the larger patterns of John's life (and perhaps Jane's too), including his relationship with Jane, his behavior with friends, his social behavior in general, and even the overall pattern of his life (his "soul," according to Rachlin). All the events of the history are interrelated like the threads of a tapestry; none can be removed without affecting the whole. Taken in their entirety, they constitute a nexus, an interconnected whole, significant for its patterns, and therefore more than the sum of its parts.

Final Causes

Rachlin (1992; 1994) argues that explanation of behavioral events by their fitting into extended patterns corresponds to what Aristotle called explanation by final cause. Explanations in terms of efficient causes point to strictly preceding events, leading in an unbroken chain to the present, what is often called the "billiard-ball" view of causality. So committed did science become to efficient causes in the eighteenth and nineteenth century, that hardly any other type of explanation was even considered. Yet final cause is the model of explanation for selectionist accounts, both in evolution of species and in operant behavior. It is arguably the explanatory mode for complex, dynamic phenomena in any discipline, including physics (Chiesa, 1992; Waldrop, 1992). Accounts in terms of underlying mechanisms can supplement, but never replace, final-cause explanations. Understanding the mechanism of natural selection or reinforcement cannot tell you what species or what behavior will actually occur.

The search for patterns extended in time deemphasizes the distinction between past, present, and future, so essential to efficient-cause explanations. The conception of events required is reminiscent of the conception Whorf (1956) ascribes to Hopi. Instead of events occurring one after another in time, the present constantly slipping into the past, and the future constantly arriving into the present, the Hopi view is that all that ever happened or will happen *is*—exists in a great pattern of interconnected happenings. The Hopi experience is that some of this reality is manifest and some is coming to be manifest, but without linear time, this corresponds only roughly to the past and future of English. Whorf remarks, for example:

> In translating into English, the Hopi will say that these entities in process of causation 'will come' or that they—the Hopi—'will come to' them, but, in their own language, there are no verbs corresponding to our 'come' and 'go' that mean simple and abstract motion—they are 'eventuates to here' *(pew'i)* or 'eventuates from it' *(angqö)* or 'arrived' *(pitu,* pl. *öki)* which refers only to the terminal manifestation, the actual arrival at a given point, not to any motion preceding it (1956, p. 60).

When we explain behavior according to its patterns, to us also it seems that some patterns are manifest, whereas others are unfolding or becoming manifest.

Passivity

Behaviorism has often been accused of rendering the organism passive. This is true in the sense that behaviorism omits the active agent, that behavioral events are never done by the organism. Behaviorists deny the accusation, however, because they take it to mean that they view the organism as a sort of automaton, driven only by events of the present. To the extent that behavioral events are viewed as complex and dynamic—to the extent that final-cause explanations prevail—to that extent it might be said that behaviorism is an active, rather than a passive view (Chiesa, 1992).

Radical Behaviorism and Eastern Mysticism

The cluster of ideas we have identified as central to radical behaviorism—rejection of agency, that there is no doing, that the self is only a manner of speaking or an illusion, that behavioral events occur in a context of a whole nexus of events, a context that may be as broad as you please—all these ideas are to be found in Eastern mysticism. Alan Watts (1965)[1], known for his writings about Zen Buddhism, may have been the first to suggest an affinity between radical behaviorism and Eastern mysticism (see also Shapiro, 1978). Drawing primarily on *Science and Human Behavior*, he discusses the seeming contradiction between the idea that the environment controls behavior and the idea that people can control their environment. He attributes the appearance of contradiction to "a language system which incorporates . . .an outmoded conception of the individual—the individual as something bounded by skin, and which is pushed around by an environment which is not the individual" (Watts, 1965, p. 54), but he doubts that Skinner realizes the implications of his own ideas. Having set the context of control and counter-control, he focuses on the following passage:

> Actually, however, we are not justified in assigning *to anyone or anything* the role of prime mover. Although it is necessary that science confine itself to selected segments in a continuous series of events, it is *to the whole series* that any interpretation must eventually apply. (Skinner, 1953, p. 449; italics added by Watts, 1965, p. 54).

He calls this passage "the purest mysticism" and compares it to Mahayana Buddhism, which also denies that anyone or anything could have the role of prime mover. Watts comments:

> In Skinner's language, the popular conception of the inner self, the little man inside the head who is controlling everything, must be replaced by the whole system of *external* causes operating upon the individual, the whole network of causal relationships. . . .when we study the individual's behavior, we are studying a system of relationships, but we are looking at it too close up. All we see is the atomic events, and we don't see the integrated system which would make them make sense if we could see it. Our scientific methods of description suffer from a defective conception of the individual. The individual is not by any means what is contained inside a given envelope of skin. The individual organism is the particular and unique focal point of a network of relations which is ultimately a

[1] I am indebted to Phil Hineline for bringing this example to my attention.

> "whole series"—I suppose that means the whole cosmos. And the whole cosmos so focused is one's actual self. This is, whether you like it or not, pure mysticism (Watts, 1965, p. 55).

He is, of course, using the word "mysticism" in an approving sense, although he is aware that mysticism is usually viewed with suspicion by Westerners, particularly scientists. Referring obliquely to the unrespectability of mysticism and New Age thought, he comments ironically that Skinner states his conclusions "in veiled language, so that neither he nor his colleagues will see their disastrously unrespectable implications!"

Although it will be impossible to do full justice to the parallels between radical behaviorism and Eastern mysticism here, nevertheless I will sketch some parallels briefly, because they suggest the possibility of a sort of symbiosis between behaviorists and New Age advocates. If behaviorism is unpalatable but respectable and New Age thinking is palatable but unrespectable, there may be some combination that is both palatable and respectable.

Comparison

The central idea in the great majority of Eastern mysticism is that reality is one indivisible whole and that the appearance of many individual existences is an illusion. This one reality embraces everything, and therefore is limitless as well as indivisible. It is referred to in many ways: In the quote above, Watts calls it "one's actual self." Others call it the Tao, Self, God, paramatman, the Nameless. Another tenet, which we shall ignore here, is that this reality may be experienced in this life—that one may leave behind the ignorance of illusion for the knowledge of reality.

According to Meher Baba (1987), for example, "God alone is real; He is infinite, one without a second. The existence of the finite is only apparent; it is false; it is not real" (p. 384). According to D. T. Suzuki (1964), Zen, which prefers reality to be called the Nameless or nothing at all, holds: "When all these deep things are searched out there is after all no 'self'. Where you descend, there is no 'spirit,' no 'God' whose depths are to be fathomed. Why? Because Zen is a bottomless abyss" (p. 43). This idea in Eastern thought, that behind or beneath the superficial diversity and disconnectedness of appearances there is a reality that is perfectly connected and indivisible, explains why Watts was so impressed with Skinner's assertion that interpretation must eventually apply to the whole "continuous series of events." Skinner, of course, was no mystic, and Watts may be right that he was unaware of the implications of notions like the continuity of events, but he betrayed some awareness when he wrote, ". . .what one observes and talks about is always the 'real' or 'physical' world (or at least the 'one' world). . ." (Skinner, 1945/1961, p. 284).

If the doer is illusory, doing is illusory. This is made explicit, for example, in the *Bhagavad Gita,* a Hindu text, in which the term for the illusory limited self is the Atman:

Falsely he sees,
And with small discernment,
Who sees this Atman
The doer of action . . .
(Prabhavananda & Isherwood, 1944, pp. 161–162).

In Taoism and Zen this idea appears as the concept of "doing without doing" or simply "non-doing." Poem 47 in *Tao Te Ching*, by Lao Tzu, goes:

The world may be known
Without leaving the house;
The [Tao] may be seen
Apart from the windows.
The further you go,
The less you will know.
Accordingly, the Wise Man
Knows without going,
Sees without seeing,
Does without doing.

The concept of "doing without doing" corresponds to the idea that behavior "just happens," that behavioral events are natural events that occur in a "continuous series." Near the end of his autobiography, Skinner (1984) wrote, "If I am right about human behavior, I have written the autobiography of a nonperson" (p. 412). Apparently he thought of his entire life as nondoing.

Agency versus Communion

According to William James (1902), the loss of the sense of agency means the gain of the sense of communion. The less you think you are a separate active agent, the more you see the events of your life as connected with a larger whole. It is not necessary to belong to any traditional religion to feel "in a wider life" or "a sense of higher control" (as James puts it). This sense or feeling, which accords with the Hopi worldview, Eastern mysticism, and many other religious traditions, may be applied to radical behaviorism. The unreality of human agency represents, if you please, the ethical side to behaviorism. Not that there is *necessarily* an ethical side; just that such an ethical side is compatible with behaviorism.

The same moral lesson that John Donne drew when he wrote, "No man is an island, complete unto himself," may be drawn from behaviorism's assertion that all behavior occurs in a vast network of interconnected events. We belong to the world; the world does not belong to us. Therefore, we may conclude, we need not only to take care of one another but to take care of this whole world of which we are an integral part and on which we depend for all benefit and our survival.

Radical Behaviorism and the New Age

This sense of belonging to a larger whole—communion, as we are calling it—may be the point at which radical behaviorism comes together with the New Age, the movement that attempts to integrate Eastern mystical thinking into Western culture. The first assumption that Chopra (1993) calls upon his readers to reject is "There is an objective world independent of the observer, and our bodies are an aspect of this objective world" (p. 4). Instead, he suggests, "'The physical world, including our bodies, is a response of the observer. We create our bodies as we create the experience of our world" (p. 5). Instead of "As individuals, we are disconnected, self-contained entities" (p. 4), Chopra (1993) suggests, "Although each person seems separate and independent, all of us are connected to patterns of intelligence that govern the whole cosmos" (p. 6). Though couched in terms that few behaviorists would choose, Chopra's assumptions remain true to ideas like: perception is behavior, behavior cannot be separated from its environment, and the active person is a fiction. The compatibility between radical behaviorism and the New Age stems from the notion of interconnectedness, which, along with many of its implications, lies at the center of both schools of thought.

Interconnectedness

The primary concerns of New Age writers are love, fear, healing, and wellness. Insofar as you think of yourself as separate and fail to recognize your connection with the greater whole, you will experience fear and illness (Williamson, 1993). Insofar as you think of yourself as merged into the greater whole, you will experience love and wellness (Chopra, 1993). The transition from one to the other constitutes healing (Carlson & Shield, 1989).

For our present discussion, it is helpful to separate *recommendations* from *explanations*. The recommendations of the New Age usually translate readily into behavioral terms. Some come down to propositions as basic as your health depends on your behavior, a premise of behavioral medicine. New Age writers reject the concepts of credit and blame on both pragmatic and theoretical grounds comparable to those on which Skinner (1971) did (Williamson, 1993). The recommendation to cultivate love and shun fear translates into practicing positive reinforcement with others, arranging your environment so that your behavior is controlled by positive reinforcement rather than aversive contingencies, discovering the stimuli that make you afraid, and manipulating your environment so that there are alternatives to avoidance of those fearful situations. That these have to do primarily with our social environment changes nothing.

Sometimes recommendations are couched in language more difficult to translate. For example, "As the energy intensifies on the planet, and in our lives, the only way not to be pulled into the illusion is to maintain your center and cultivate your energy," requires extra interpretation. The meaning becomes clearer when we consider concrete examples: Skinner himself excelled at "maintaining center and cultivating his energy;" he arranged an environment in which he could be productive with minimal disruption. "Center" corresponds to equilibrium or stability of a certain pattern of activities, and "maintaining center" means acting (proactively) in ways that may maintain that pattern of activities. "Cultivating energy" might translate into behaving in ways that maximize reinforcement in the long run, instead of behaving in accord with short-term contingencies [what Rachlin (1994) would call self-control or, in his interpretation of Plato, choosing the good over the pleasant].

Although the affinities between radical behaviorism and the New Age suggest that the two are at least partially compatible, there appear to be points of incompatibility as well. If both are thoroughly deterministic, still the New Age view tends to be broader in scope because it spans many lifetimes, rather than just one lifetime. If both rule out the individualized inner self, still New Age writers often seem to attribute agency to the One Self. Yet, even these differences may be mitigated. If there is such a thing as Karma—a pattern to life events that spans many lifetimes—how might it be manifested in any one lifetime except in one's genes and environmental circumstances, the very matters that radical behaviorists point to as the determiners of behavior? New Age thinkers could accept control by genes and environment without compromise; Watts (1965) apparently does. Radical behaviorists could accept the possibility of reincarnation and Karma, even if they ignored them, also without compromise. Even if New Age writers seem to attribute agency to the Self, that may turn out to be just "strategic" talk, like Dawkins's

(1989a) attributing agency to genes. Williamson (1993), for example, writes "...the coincidences in our lives happen to guide us," sounding like an Intelligence sends events to us in order to guide us. From the radical-behaviorist perspective, and on examination from Williamson's perspective too, it is only important that the coincidences of our lives *do in fact* guide us (i.e., our behavior). It is a separate question, and one that behaviorists might consider, whether it is helpful to talk about the universe as intelligent or benevolent. Both Williamson (1993) and Chopra (1993) assign practical (i.e., healing) value to such talk, and behaviorists who study behavioral medicine might agree.

Acceptance and Respectability

New Age thinking is "unrespectable," as Watts implies in our earlier discussion, because the metaphors it relies on seem so foreign, not only to radical behaviorists, but probably to most Westerners. Talk about "maintaining center" and "cultivating energy" sounds at first like mumbo-jumbo. No doubt many people who encounter such strange talk are inclined to ridicule it. At the least, one might say with justification that it sounds like it comes from a foreign culture.

The problem with such talk, however, may lie more in its abstractness. The New Agers often seem to prefer to talk about life and the universe in broad terms and to have difficulty getting down to the concrete events of people's lives. When they are more concrete, as in many of their recommendations, behaviorists and others may resonate to what they say. The problem lies more in the explanations, because these are terribly abstract. Ideas like "energy," said to be "on the planet" and "in our lives," are hard to relate to concrete particulars.

The New Age may be unrespectable, but its public image is nice. This movement may seem wacky, but it is clear they are interested in brotherhood, peace, and freedom. Radical behaviorism, in contrast, may be respectable enough, appearing at least to fall within the purview of professional and academic psychology, but it is not nice. Among people who have heard of behaviorism, probably very few tie it to brotherhood, peace, and freedom.

If radical behaviorists were interested in improving their public image, it might be possible to blend the New Age with behaviorism and produce a presentation that would be both respectable and nice. Attempts have been made with modern physics (Capra, 1982, 1983). Although it is hard to imagine exactly what the blend would be like for radical behaviorism, we might , for example, adopt the view that every event—say, your reading this sentence—was ordained from the beginningless beginning. In practice this means that every event occurs in a web of interconnection with every other event, so that your reading this may be understood in relation to events in your life both before this and after this. Since time is deemphasized in this approach, instead of a "history of reinforcement," we might speak of the nexus of reinforcement, to better capture the dynamism and interconnectedness that inhere in the idea of behavior coming into contact with contingencies and being drawn by them. This is only a speculative suggestion; other people will have other ideas—that is, other clusters of reinforcement in the great nexus will include other suggestions.

Conclusion

Lay people have difficulty accepting radical behaviorism because it treats actions as natural events. The way behaviorists talk about behavior provokes bewilderment and resistance because it runs counter to an ancient covert category (cryptotype) common to

English and other Indo-European languages: agentic events. In this ancient way of talking about the universe, an imaginary inner space houses forms, images, and representations separate from the substantial objects of the outer world. It is home also to the inner self, agent, or doer of actions. Any way of talking that denies this separation faces an up-hill climb to acceptance. In this respect, radical behaviorism finds itself in the same boat with Eastern mysticism and its Western adaptation, the New Age. Behaviorism and the New Age both encounter resistance, but for different reasons: the one because it seems too harsh; the other because it seems too vague. In their search for a way to bridge the gap between their new way of talking and the ancient way, and particularly to avoid the appearance of harshness, radical behaviorists might find it useful to study the works of New Age writers.

14

Commentary on Foxall, "Intentional Behaviorism"

Foreword

Un-understanding critics have assailed behaviorism since its inception (Watson, 1913). They often engage in wishful thinking, saying that behaviorism is "dead." The criticism mainly takes two forms: (a) that behaviorism is wrong and morally repugnant; and (b) that behaviorism is incomplete because it only deals with surface behavior and fails to account for inner causes like intentions, desires, and beliefs. Both of these often arise from failure to understand the concept of verbal behavior and its implications.

The view that behaviorism is wrong arises from the need to exclude free will and agency from a science of behavior. Anyone is entitled to believe in free will and agency, but these concepts block explanations of behavior as natural events that are caused by other natural events; they make a science impossible. Someone committed to a belief in free will and agency rejects the idea that these concepts are nothing more than verbal usage and rejects the assumption of determinism. Evidence of environmental control might not convince such a person, because freedom might be a value—i.e. a preference (Hocutt, 2013)—about which no dispute is possible (*De gustibus non est disputandum*).

The criticism that behaviorism is incomplete arises, if not from a matter of taste, from failure to understand the incompatibility of dualism with science and what constitutes a valid scientific explanation. Inner–outer or mental–physical dualism arises from the belief that each person is a separate being, unconnected with others. From this arises the conception of the "ghost in the machine"—that each of us has a physical body driven by our own inner thoughts and feelings. This conception probably derives from the notion of privacy—that I alone am privy to my thoughts and feelings. As noted in the previous paper, one can hardly speak in any Indo-European language like English without affirming this conception, because it is built into the grammar and vocabulary. If one becomes familiar with Eastern philosophy, however, one may bring oneself to doubt this received view, with remarkable results, not least of which is that a true natural science of behavior is possible.

On the one hand, we have the view that a science of behavior can be had that focuses on behavior in relation to the environment—all observable matters. On the other hand, we have the view that our behavior originates from our inner thoughts and feelings. In *Behavior of Organisms*, Skinner (1938) espoused the former view. In 1945, writing for a symposium on operationism, Skinner took the occasion to respond to the criticism that

Science and Philosophy of Behavior: Selected Papers, First Edition. William M. Baum.

behaviorism only deals with outer behavior and fails to deal with inner thoughts and feelings. In doing so, he took the disastrous position of tying thoughts and feelings to inner stimuli and behavior, which he labeled "private." In other words, Skinner tried to have it both ways (Rachlin, 2018). The result has been endless confusion among behavior analysts, because private events, by definition, are not observable, and only observable events can be data for science (Baum, 2011a, 2011b, 2011c). If private events could be made public, then—no longer private—they would be data to be explained. This may happen, for example, when dentists or physicians use instruments to make physiological processes public, and hypothetically even inner speech might be so revealed. As long as an event is private, however, its status in a science must be at best hypothetical. If Skinner had stayed with his 1938 stance and explained that thoughts and feelings cannot originate behavior because all behavior originates in the environment, much confusion would have been avoided.

If thoughts and feelings are real, how can a science deal with them? An understanding of verbal behavior is crucial. If I say, "Tom's careless remark hurt Jane's feelings," what could possibly induce my utterance? My only basis could be hearing Tom and seeing Jane's expression and reply. One may resist the inference that I am guessing at something private, because all the events are public. Indeed from my perspective, Jane's facial expression and sharp reply *are* Jane's hurt feelings. Similarly, if I say, "Jane believes that democracy is good," my basis will be her (public) behavior: she states that democracy is good, says we must defend it, insists that everyone should vote, votes herself, and so on. If she didn't do these things, she wouldn't believe in democracy. Her engaging in these activities *is* her belief in democracy.

What about my own thoughts and feelings? The evidence is not exactly the same as for Jane's thoughts and feelings, but it is the same sort of evidence. I love my wife. How do I know? I kiss her often, I listen when she speaks, help her when she has a problem, wait for her to come to the table before eating, ask her to choose what we watch on television, and so on. If I did not do these things, I myself should doubt that I love my wife. When someone asks, "Do I love so-and-so?" the person is not looking within, but rather at their own public behavior. Our thoughts and feelings reside, not in some inner place, but in our own patterns of behavior—what we say and do.

Philosophers sometimes make a great deal over what they call "the incorrigibility of first-person statements." Statements like "I believe in democracy" only seem to be incorrigible because social conventions make disputing such a first-person statement rude. Even your best friends cannot doubt you without risking offense. If Jane says to Tom, "I feel tired and want to go home," Tom would be a fool to dispute her feeling tired; he had better take her home, because Jane may or may not be tired, but the point of her utterance is to get Tom to help her leave. Skinner's mistake was to treat an utterance like, "I feel tired" as a "verbal report," as if the speaker is reporting on a private event. Introspection has been shown over and over to be unreliable. Watson's (1913) "manifesto" was largely devoted to pointing out this unreliability. What people say about themselves may correspond but little to what another person may observe, and laboratory experiments have sometimes shown self-statements to be incorrect (Nisbett & Wilson, 1977).

The accusation of incompleteness may also arise from focus on momentary events. If Jane is in the supermarket and takes Cereal A from the shelf instead of Cereal B, Cereal A is discriminated from Cereal B by the labels on the boxes, but one might want to know why she chose Cereal A. Focus on this momentary act will make any explanation seem incomplete, because the causes of the act lie in the past. Rather than look at the past history with Cereals A and B, one might be tempted to invent an inner disposition or intention to buy Cereal A. Such an intention, however, would only be an unobservable surrogate for the observable history with the two cereals. The intention is inferred from the behavior it is supposed to explain and thus fails to explain it because it results in circularity.

Focus on momentary behavior tends to conjure causes present at the moment. If Jane reaches for Cereal A, the immediate cause might be the package label, but that doesn't explain why she was attracted to that label. If no cause is obvious, someone might be tempted to invent one, such as an inner preference for nutritious food. If Jane prefers nutritious food, however, she will exhibit many other choices in other contexts. Her preference is not hidden, it is manifest in her choices, and her choosing Cereal A is part of the extended pattern that we call "preference for nutritious food."

History, therefore, is not just in the past, it is continuous with the present. It consists of patterns of behavior that extend through time up to the present. Treating history as if it were an efficient cause in the present cannot make sense, because history is not in the present.

Aristotle distinguished between efficient causes and final causes, which are extended patterns into which fit more particular parts. Rachlin (1994) has argued effectively that a science of behavior must recognize that explanations of behavior often require final causes. For example, one could say that Jane's choosing Cereal A not only was part of her preferring nutritious food, but also that she chose Cereal A because she prefers nutritious food, taking preferring nutritious food as the final cause of her choosing Cereal A. This requires nothing unobservable.

Much confusion may be avoided if one understands verbal behavior as operant behavior. Verbal behavior may be vocal (speech) or gesture, but it typically induces behavior in another person (the "listener"). For this reason, verbal behavior is often said to be "reinforced" by the listener. The key point about verbal behavior is that it generates stimuli—auditory or visual—that induce behavior in the recipient or "listener." If someone says, "I have a headache," the utterance may induce sympathetic behavior in the recipient like fetching an aspirin, or may induce excusing the speaker from some onerous situation—an oblique way of asking to be left alone.

The present paper consists of comments on a paper by Gordon Foxall arguing typical un-understanding criticisms of behaviorism. Foxall's paper relied on philosophical approaches to language, including the notion that words "refer" to things and have "meaning." The concept of verbal behavior focuses instead on usage, actual utterances and the conditions that induce them. Reference and meaning simply do not come into the account. If one begins with a dualistic approach in which words seem to reach out to a real world outside, then one will conclude that a mental world must exist, a world of intentions. The problem lies in the beginning premise. Dualism ultimately renders a science of behavior impossible. Eschewing dualism allows a true natural science of behavior.

Abstract

Foxall's (2007) incorrect claims about behavior analysis arise from a failure to understand the stance of behavior analysis. Behavior analysis is the science of behavior; it is about behavior and not about organisms. It views behavioral events as natural events to be explained by other natural events. This view extends to verbal behavior. First-person statements and third-person statements, intentional or otherwise, are instances of behavior to be explained. Behavior analysis explains them by relating them to the history of context and consequences that might have led to their occurrence. Believing in Satan is an extended activity, of which statements about Satan constitute less extended parts; it is an error to suggest that the belief could stand as the efficient cause of its parts. That behavior repeats from time to time is no more mysterious than that other natural events repeat. Even if we do not know the physiological mechanism, filling in the temporal gaps with phony storage and representation is no help. Likewise, control by complex contexts is in no way illuminated by imagining phony processes within the organism.

Keywords: behavior analysis, behaviorism, intentionality, verbal behavior, representation, Dennett, Malcolm, Ryle, Skinner

Source: Originally published in *Behavior and Philosophy*, 35 (2007), pp. 57–60. Reproduced with permission of Cambridge Center for Behavioral Studies.

Foxall's paper "Intentional Behaviorism" (2007) makes several claims about behaviorism. Were they correct, they would be important criticisms. The most extraordinary claim, in my view, is that behavior analysis cannot "adequately" explain behavior without resorting to intentional terms. Foxall actually calls this a "fact" (p. 2). It is, however, false. I think that this incorrect assertion arises from a failure to appreciate what might be called the "stance" of behavior analysis.

In my book *Understanding Behaviorism* (Baum, 2005) I point out that, although behaviorists disagree among themselves about many issues, they all agree on one proposition: A science of behavior is possible. That is the core of behaviorism, and the science is usually called behavior analysis. It is not an area *within* psychology but an *alternative* to psychology, and although it has some relationship to philosophy of psychology (mainly critical), the philosophy of behavior analysis is behaviorism.

The assertion "A science of behavior is possible" implies, among other things, that behavioral events are natural events, to be understood and explained with the methods of natural science, which include confining explanations and interpretations of phenomena to talk about other natural phenomena. In some places, Foxall seems to understand this "stance," but he finally loses touch with it when he suggests that it is "optional" (p. 9) and that introduction of intentional terms is both desirable and necessary. The reasons he gives stem from confusion over the difference between ordinary language concepts and the concept of verbal behavior. He correctly asserts that verbal behavior is treated in the same way as other behavior, but then gets misled by accepting uncritically the views of philosophers like Daniel Dennett and Norman Malcolm about language and reference, which have little—if anything—to do with verbal behavior.

The distinction between first-person statements and third-person statements, of which philosophers make so much, is irrelevant for behavior analysis. If a person says, "I have a headache," "I am angry," "I felt like going home," or "I intended to portray someone getting married," that is behavior to be explained. If a person says, "She has a headache," "She is angry," "She felt like going home," or "She intended to portray someone getting married," that is behavior to be explained. To behavior analysis, none of these utterances *refers* to anything; they are natural events to be explained by other natural events, and none is uncaused, as Foxall suggests about the ones including the first-person "I." The behavior-analytic approach to explaining or interpreting these events is to talk about the history of context and consequences that might have led to them in the current context.

Thus the approach is in no way baffled by their inclusion of any particular words, particularly intentional terms, because those all have history and context to explain them. Misunderstanding this, Foxall makes the incorrect assertion that behavior analysis requires that the "information" (read context) occasioning the utterances including "I" must be identical with that occasioning the utterances including "she." The information must be of the same type (i.e., natural phenomena), but it is not the same for the one who says "I" and the one who says "she" any more than it would be necessarily the same for any two people who said "I" or any two people who said "she."

Foxall is misled here, not only by Dennett and Malcolm, but by Skinner himself. Skinner is sometimes the worst source for understanding radical behaviorism because he was often inconsistent and sometimes careless in his use of terms. Foxall includes a particularly unfortunate example (p. 12) in which Skinner uses the word "observe" in a way that suggests a person might observe events and act as a result. This leads Foxall to bring in Malcolm's argument about observing as a basis for intentional statements. Skinner, however, was using the word "observe" casually, and he probably understood well that observing is not an explanation of behavior but is itself behavior to be explained. Whether Skinner understood this or not, other behaviorists have built on Gilbert Ryle's (1949) explanations about the logical incoherence of terms like "believe," "know," and "desire" when they are taken to explain behavior. One does not "believe in Satan" and say "Satan roams the Earth like a devouring lion," because part of "believing in Satan" is to say "Satan roams the Earth like a devouring lion." Ryle calls the attribution of the behavior to the belief a category error because it confuses a category label (the belief) with one of its instances. I (Baum, 2002, 2004) might say, rather, that the utterance is part of a more extended activity (the belief) and that the mentalistic explanation mistakenly identifies a whole (more extended activity; the belief) as the efficient cause of one of its parts (less extended activity; the utterance), but the point is the same. Similar remarks apply to "observe." To observe my own headache or someone else's complaints about having a headache is to behave in the ways that we call by the label "having a headache." Parts of observing my headache would be events such as saying "I have a headache" and taking aspirin. Parts of observing another person's headache would be events such as saying "Do you have a headache?" and fetching aspirin for the person.

The point that Foxall gets wrong here is that verbal behavior has nothing to do with reference or content in any conventional sense. Intentional utterances are no more opaque than any others. Intentional utterances may be explained as are other utterances, they may be seen as occasioned by other behavior (one's own or someone else's, as in believing), but they are in no way necessary in the terminology of behavior analysis.

Failure to grasp the point that behavior analysis is about behavior leads Foxall to two other errors. In the section on the "continuity" of behavior he suggests that behavior analysis cannot be complete without an explanation of recurrence of behavior from one time and setting to another. This is an example of the question how an event at one time can affect behavior at a later time. We are ignorant of the mechanisms that allow this to happen, but even if physiology were to elucidate it, that would not necessarily affect the analysis of behavior. When Newton proposed the concept of gravity he had no idea how action at a distance might work, but that did not stop him from using the concept in understanding mechanics. Eventually an explanation was conceived by Einstein, but until that happened no gain in productivity could have resulted from filling in the gaps with phony mechanisms (such as invisible hooks). Similarly, nothing is gained by filling in the temporal gaps in behavior or trying to explain the temporal extendedness of behavior by inventing phony processes. Foxall brings up learning (one of the favorite phony processes) to try to talk about recurrence of behavior and asks, with Dennett, what is learned? The problem here is not the need for an answer; the problem lies in the question itself. It is based on the mistaken premise that if behavior reappears it must somehow be represented and stored. The recurrence of natural events in no way requires representation or storage. The sun rises and sets, an unsupported stone falls, combining sodium hydroxide with hydrochloric acid produces sodium chloride and water, a zygote of a sea urchin develops into a sea urchin, and none of these repeated phenomena involves representation or storage in any but the most metaphorical and tortured sense. When we understand the mechanisms by which behavior repeats we will no longer be tempted to speak of it being stored or represented.

By failing to understand that behavior consists of natural events, Foxall makes the other error in supposing that if behavior depends on multiple or complex contexts and consequences, the organism must have the capacities to process information and to select responses. The same remarks about filling in gaps in understanding with phony stopgaps apply, but another point is that behavior analysis is about behavior, not organisms. Behavior analysis is no more about the person or the pigeon than mechanics is about the Earth or the moon. The aim is to understand those phenomena that we call the behavior of organisms, and in that effort we do well to admit what we do not understand, live with our ignorance, and work toward real understanding instead of introducing unproductive stopgaps such as intentional terms.

15

Behaviorism, Private Events, and the Molar View of Behavior

Foreword

This paper makes the point that the science of behavior cannot include private events that occur somewhere internal to the organism. Private events are thought to include private stimuli and covert behavior, particularly sub-vocal speech. The main problem with private stimuli is that they often don't exist. If "private" means unobservable, then "private stimulus" is an oxymoron, because any event that affects behavior must be observable, at least with instruments. If a person viewing an optical illusion says that a straight line looks curved, should we look for a curved line somewhere inside? The whole array induces the judgment of curvilinearity, and the inner curved line does not exist. If "private" means "unobserved but observable," then an inferred private stimulus is hypothetical. If Jane behaves in the ways that we call "having a toothache," the injured tooth is hypothetical until her dentist discovers it. Then, however, the stimulus inducing her "having a toothache" is no longer private, but public. So, the only legitimate use of "private stimulus" would be "hypothetical private stimulus." If you insist that the toothache *itself* is still private, you have turned the toothache into an unobservable stimulus, which in a science of behavior cannot exist.

If her dentist fails to find any problem, one begins to ask if Jane is faking. How would we know? Only from her subsequent behavior. If she continues to engage in "having a toothache," either she is determined or the dentist missed the problem. If she soon stops "having a toothache," we may well suspect she was faking.

The trouble with covert behavior is that it is private, by definition. If we cannot observe it, even with instruments, then covert behavior cannot constitute data for the science. Were it made public, it would become data, behavior to be explained, but it would no longer be private. So, covert behavior is irrelevant because in a science behavior cannot be explained by unobservable causes. If, however, previously covert behavior were rendered public by instruments, then this public behavior might be worth studying.

We should be clear, however, that a person's public claims about covert behavior cannot be taken as firm evidence of covert behavior. The claims, being public, are available for study, but they cannot be taken to verify covert behavior, because they would be the product of introspection, which is always unreliable. When asked about inner processes, people just make up likely stories (Nisbett & Wilson, 1977).

Science and Philosophy of Behavior: Selected Papers, First Edition. William M. Baum.

Behavior analysts sometimes lose sight of the philosophical underpinnings of the science and backslide into dualism and mentalism. Some behavior analysts regard private events as existing somewhere within the body, a view that would apply to physiology, to muscular and neural events. Along with Skinner, unfortunately, they move on to posit private stimuli and covert behavior, to private events that cannot be inside the body. The result is that they fall back to dualism, to a dichotomy between inner events in some mysterious space and outer events outside of the body. This dualism renders the science at least incoherent and likely impossible.

The most common form of backsliding is to take introspection at face value. Behaviorists have long noted that introspection, as a method, is unreliable, because no way exists to corroborate its results. If a person says he feels sad, the statement is observable, but the inner state, if any, is not. We cannot simply conclude that the statement is correct. We might corroborate the statement with other behavior—loss of appetite, weeping, lassitude, and so on—but then we would say this pattern of behavior is what it is to "feel sad." If your friend tells you she meant to call you yesterday, is she introspecting on her intentions? You have no way to tell; all you have is the statement, with some effect on you. When people are asked to introspect on their own processes, experiments show that they are terrible at explaining their own behavior (Nisbett & Wilson, 1977). Yet despite these obvious inadequacies, some behavior analysts still thoughtlessly take introspection as valid. The present paper attempted to explain why we should exclude private events and to suggest that a molar view of behavior renders them irrelevant.

Abstract

Viewing the science of behavior (behavior analysis) to be a natural science, radical behaviorism rejects any form of dualism, including subjective–objective or inner–outer dualism. Yet radical behaviorists often claim that treating private events as covert behavior and internal stimuli is necessary and important to behavior analysis. To the contrary, this paper argues that, compared with the rejection of dualism, private events constitute a trivial idea and are irrelevant to accounts of behavior. Viewed in the framework of evolutionary theory or for any practical purpose, behavior is commerce with the environment. By its very nature, behavior is extended in time. The temptation to posit private events arises when an activity is viewed in too small a timeframe, obscuring what the activity does. When activities are viewed in an appropriately extended timeframe, private events become irrelevant to the account. This insight provides the answer to many philosophical questions about thinking, sensing, and feeling. Confusion about private events arises in large part from failure to appreciate fully the radical implications of replacing mentalistic ideas about language with the concept of verbal behavior. Like other operant behavior, verbal behavior involves no agent and no hidden causes; like all natural events, it is caused by other natural events. In a science of behavior grounded in evolutionary theory, the same set of principles applies to verbal and non-verbal behavior and to human and nonhuman organisms.

Keywords: behavior analysis, behaviorism, dualism, evolution, mental, private, verbal behavior, molar paradigm

Source: Earlier versions of this paper were presented at Association for Behavior Analysis, May 1995, and American Psychological Association, August 1995. The author thanks Howard Rachlin for thoughtful comments on an earlier draft of the paper. Originally published in *The Behavior Analyst*, 34 (2011), pp. 185–200. Reproduced with permission of Springer Nature.

Defining behaviorism, B. F. Skinner (1974, p. 3) wrote, "Behaviorism is not the science of behavior; it is the philosophy of that science." One may define behaviorism by its central proposition—what all behaviorists agree on—that a science of behavior is possible (Skinner 1953, 1974; Watson, 1913; see Baum, 2005 for further discussion). Watson (1913) proposed further that the science of behavior should be a natural science, and Skinner (1945), coining the term "radical behaviorism" similarly asserted that the science of behavior (behavior analysis) is a natural science (Skinner, 1953). One implication is that behavioral events are natural events and, just like the weather or natural selection, involve no agency, but are explained by other natural events (Baum, 1995c). Another implication is that the science leaves nothing important out—i.e., that it is sufficient.

Advocates of radical behaviorism often say that its chief distinguishing characteristic is its treatment of private events. They say it is unlike other versions of behaviorism because it treats private events as well as public events and therefore avoids the accusation that it ignores inner life (Skinner, 1974; Moore, 2008). For example, Skinner (1974) wrote, "A science of behavior must consider the place of private stimuli as physical things, and in doing so it provides an alternative account of mental life. The question, then is this: What is inside the skin, and how do we know about it? The answer is, I believe, the heart of radical behaviorism" (pp. 211–212). Having said this much, however, advocates carefully point out that "private" differs from "mental." In the view of radical behaviorists, mental things and events seem to occur in some inner, imaginary space, usually called the mind. Since this inner, imaginary space and all its contents are nowhere to be found in nature, radical behaviorists see mental events as fictional and deny them any role. Private events, in contrast, are said to be just like public events except that they occur within the skin (Skinner, 1969c; 1974; Zuriff, 1979). For example, Skinner (1969c) wrote:

> An adequate science of behavior must consider events taking place within the skin of the organism, not as physiological mediators of behavior, but as part of behavior itself. It can deal with these events without assuming that they have any special nature or must be known in any special way. The skin is not that important as a boundary. Private and public events have the same kinds of physical dimensions (p. 228).

The radical behaviorists' denial of mental inner space and its contents is a rejection of a dualism that is fundamental to modern, common-sense folk psychology. In the common-sense view, the self dwells in inner space while the body deals with the outer world. Accordingly, it seems obvious that thoughts, feelings, and images remain forever intimate and private while outer actions alone are available for the inspection of others. For example, a cartoon shows a husband saying to his wife, "Nobody's ever understood me, Joyce, not my teachers, not my parents, my boss, my so-called friends—just you, baby—you're the only one who's ever listened. . ." Above the wife is a box saying, "Christ, will he ever put a cork in it?" Between her and her box is a string of circles, which we immediately understand to indicate that the words in her box are private—unknown and unknowable to the man. The rejection of this fundamental inner-outer dualism is one of the features that makes radical behaviorism radical (Baum, 1995c; Baum & Heath, 1992; Catania & Harnad, 1984).

In this paper, I am going to argue that, in comparison with anti-dualism, the role of private events in radical behaviorism is peripheral and inessential. They are brought to the center in a misguided effort to render behaviorism acceptable to laypeople by suggesting that they offer an account of mental life. *I am not saying they do not exist.* Many different types of private events occur within the skin: neural events, events in the retina, events in the inner ear, sub-vocal speech (i.e., thinking), and so on. All of these are possibly measurable and, therefore, possibly public. I will argue that private events are not *useful* in a science of behavior, and, far from being a key defining aspect of radical behaviorism, private events constitute an unnecessary distraction. Private events are irrelevant to understanding the function of behavior—that is, activities in relation to environmental events. Because the origins of behavior always lie in the environment, the origins of behavior are public. Measuring private events might help to understand the mechanisms of behavior, but understanding function is propaedeutic to studying mechanism; one must know what one is trying to explain before one can explain it. Roughly speaking, the distinction between function and mechanism is the difference between understanding why behavior occurs and understanding how it occurs. Understanding function entails relating an activity to environmental events (present and past), whereas understanding mechanism entails tracing the causal chain between environment and behavior. I will argue that the ideas of private stimuli and private behavior, in particular, are irrelevant to understanding behavior in relation to environment. To see why, we must first review the problems with dualism.

Dualism

Most, if not all, of the sciences had to eliminate dualism early in their histories. The habit of supposing an immaterial world or immaterial causes behind or within the material world cannot work for science, because the relationship between the immaterial and the material remains forever a mystery. When we read about Descartes's theory that the soul influenced the flow of animal spirits by moving the pineal gland, we wonder without hope of an answer how the soul could move the pineal gland. The historian Benjamin Farrington (1944/1980), writing about the origins of Greek science, contrasted the Babylonian creation myth, in which the god Marduk created the waters and lands, with Thales's proposal in the 6th century B.C.:

> The general picture Thales had of things was that the earth is a flat disc floating on water, that there is water above our heads as well as all round us (where else could the rain come from?), that the sun and moon and stars are vapour in a state of incandescence, and that they sail over our heads on the watery firmament above and then sail round, on the sea on which the earth itself is afloat, to their appointed stations for rising in the East. It is an admirable beginning, the whole point of which is that it gathers together into a coherent picture a number of observed facts *without letting Marduk in* (Farrington, p. 37; italics in the original).

Farrington's main point was that scientific thinking originated in the rejection of dualism. Science seeks explanations ("coherent pictures") of natural events in other, related, natural events, not in non-natural causes. As the need was for physics then, so it is for a science of behavior now.

Eliminating dualism from a science of behavior, however, presents a formidable problem. English and other languages of the Western World incorporate mind-body dualism so intimately that it is difficult to talk about behavior without using terms that sound dualistic. Skinner (1974) complained of this and warned his readers to resist being misled by phrases such as "I have in mind" and words such as "choose" and "aware." The linguist Benjamin Whorf (1956) wrote eloquently about the inner-outer dualism inherent in what he called the "habitual thought and behavior" of Western culture:

> Now, when **WE** think of a certain actual rosebush, we do not suppose that our thought goes to that actual bush, and engages with it, like a searchlight turned upon it. What then do we suppose our consciousness is dealing with when we are thinking of that rosebush? Probably we think it is dealing with a "mental image" which is not the rosebush but a mental surrogate of it. But why should it be **NATURAL** to think that our thought deals with a surrogate and not with the real rosebush? Quite possibly because we are dimly aware that we carry about with us a whole imaginary space, full of mental surrogates. To us, mental surrogates are old familiar fare. Along with the images of imaginary space, which we perhaps secretly know to be only imaginary, we tuck the thought-of actually existing rosebush, which may be quite another story, perhaps just because we have that very convenient "place" for it (Whorf, 1956, pp. 149–150).

Anticipating behaviorists' objections to mental representations, Whorf notes that "mental surrogates" are hard to escape because they are built into the English language and other aspects of Western culture. Scientific views that run counter to the "habitual thought and behavior" of the culture, such as relativity theory, encounter difficulty getting accepted, Whorf argued, because they must speak "in what amounts to a new language." This must apply with at least as much force to a science of behavior. Indeed, laying stress on "private" instead of "mental" may be seen as an attempt to talk in a new language that still makes contact with ordinary English.

Two Uses of "Private"

The word "private" gets used in two different ways (cf. Baum, 1993b; Lubinski & Thompson, 1993, pp. 667–668; Rachlin, 2003). In the common-sense, folk-psychology view alluded to earlier, a private event can only be known to its possessor. It might seem self-evident, for example, that thinking can only be known to the one who thinks. According to this notion, private events are private in principle, can never be known to another, and therefore are qualitatively different from public events. To try to exorcise this qualitative difference, some behaviorists have claimed that private events are exactly like public events except in the size of the audience—private events always having an audience of one, and public events having an audience greater than one (e.g., Moore, 1995). Such a move fails, however, to erase the dichotomy. For example, how does one distinguish between a potentially public event that happens to have an audience of one (i.e., occurs when the actor is alone) from a private event? If size of audience were the only criterion, then my singing when I am alone would be a private event, but would become a public event if my wife were there to hear it. This would contradict the notion that private events are private in principle, because it is a practical matter—accidental—whether my wife happens to be there or not. Thus, size of audience is insufficient, and if private events are private in principle, they must be so according to some unstated, unanalyzed other criterion. One suspects it is precisely the sort of inaccessibility indicated by circles in cartoons that places them in a world forever inside.

What that other criterion is matters little, however, because, whatever it is, it constitutes a qualitative difference between private and public events. Accepting in-principle private events would reintroduce the inner-outer dualism that was to be avoided. Instead of the mind-body problem we would have the equally intractable problem of how an in-principle private event could serve as a stimulus for public behavior. How would anyone know if it occurred or how it was connected to a public act? If it cannot be made public, even with the help of instruments, it remains a ghostly cause, and its effects remain a mystery.

The second use of "private" makes it *purely* a practical affair. In this view, the privacy of singing when I am alone really is the same as the privacy of a thought or feeling. No private events are private in principle; thoughts and feelings are public in principle, if only we are able to invent apparatus to observe them. This idea depends on the faith that with enough technical advances, even the subtlest thought or feeling in one person could be observed by another. One has to believe, for example, that brain-scanning technology could advance to the point where an arrangement like that in Figure 15.1 would be possible—that a person's head might be put in a machine—say, a helmet—that would be attached to a monitor, and if the person thinks *Who am I?*, the words "Who am I?" appear on the screen. This view at least has the advantage that it truly makes no distinction between private and public events, thereby leaving no mysteries. The idea that private

Figure 15.1 The implication of taking all private events to be public in-principle
Note: To suppose that all private events are only private by accident, not in principle, some sort of arrangement like this would have to be possible. Whenever a private thought or feeling occurred in a person wearing the helmet, the thought or feeling ("Who am I?" here) would be displayed on the monitor.

behavior and private stimuli are only accidentally private, however, encounters at least three problems. The first is that it rests on an article of faith that cannot be disconfirmed. No anti-privacy machine exists at present, and possibly none will ever exist.

Whatever its disadvantages, the notion that private events are public in principle remains the only tenable position for radical behaviorism. Skinner apparently recognized this. In his 1945 discussion of private events, he wrote:

> The response "My tooth aches" is partly under the control of a state of affairs to which the speaker alone is able to react, since no one else can establish the required connection with the tooth in question. There is nothing mysterious or metaphysical about this; the simple fact is that each speaker possesses a small but important private world of stimuli. So far as we know, his reactions to these are quite like his reactions to external events. Nevertheless the privacy gives rise to . . .the . . . difficulty . . .that we cannot, as in the case of public stimuli, account for the verbal response by pointing to a controlling stimulus. . . It is often supposed that a solution is to be found in improved physiological techniques. . . But the problem of privacy cannot be wholly solved by instrumental invasion. No matter how clearly these internal events may be exposed in the laboratory, the fact remains that in the normal verbal episode they are quite private (1945/1961, pp. 275–276).

Skinner here points to a second problem with the anti-privacy machine. From a practical point of view, even if private events might be "exposed" in the laboratory, in everyday life ("the normal verbal episode") private events remain private. Even if the anti-privacy machine existed, it would only be available in the laboratory and not in everyday life, which is most of the time and of primary interest.

Skinner (1945; 1974) took pains to distinguish his view from what he called methodological behaviorism, the view that private events are inaccessible to direct scientific study, but may be studied indirectly in verbal reports. He criticized methodological behaviorism particularly for preserving dualism (Skinner, 1974). He argued instead that "what is felt or introspectively observed is not some nonphysical world of consciousness, mind, or mental life but the observer's own body" (Skinner, 1974, pp. 18–19).

Introspection, however, is notoriously unreliable; that is why Watson (1913) rejected introspection as a method. Skinner presumably would agree, but in the preceding quote he seems to credit introspection with some degree of accuracy. People often express confusion or uncertainty about private events (Is that a pain or an itch? Am I embarrassed or angry?), and also frequently lie in response to questions like, "What are you thinking?" In particular, introspection could never reliably render private events public. The unreliability of introspection brings us to the third problem with accidental privacy.

The third and biggest problem is that, even if an anti-privacy machine were invented, the machine would always be subordinate to the testimony of the person being interrogated. Even if a "solution" to privacy like the anti-privacy machine (Figure 15.1) were to be realized, and the monitor showed all manner of private events, "Who am I?" "pain in foot," "seeing a chicken," or "hearing Beethoven's Ninth Symphony," still nothing would prevent the person being observed from denying that any such event is occurring. Imagine the machine were brought into a court of law, and the monitor showed, "I shot the sheriff," and the person said, "I never thought any such thing; your machine is lying." What could the onlooker do then? Insist the person is lying? The anti-privacy machine still requires the person to corroborate the outcome, presumably on the basis of introspection, which is always unreliable. Thus, even an anti-privacy machine, were it to be invented, would fail to solve the problem of privacy altogether. Its promise proves to be an empty promise, and we cannot assert with certainty that privacy is accidental or that "Private and public events have the same kinds of physical dimensions" (Skinner, 1969c; p. 228).

Private events may be inferred by the verbal community in everyday affairs, but inferred private events can never serve as scientific explanations of public behavior (Skinner, 1974, p. 17–18; "the role of the environment"). If behavior analysis is a natural science, then putting together "coherent pictures," to use Farrington's phrase in the earlier quote, requires observed activities (natural events) to be related to observed events in the environment (past and present natural events). Behavior originates in the environment. Even if we learn much about the physiology of behavior, we only learn about mechanisms and not about the origins in the environment. (See Thompson, 2007, for a review of research on mechanisms.) If we learn that a certain hormone induces nest-building in canaries, we still need to know what, under normal circumstances, stimulates secretion of the hormone (e.g., lengthening day in spring), and, beyond that, we still need to know what history of natural selection brought about this mechanism. Similarly, even if we were able to measure events in the human brain that would permit us to predict behavior, we would still need to study the environmental events, past and present, which led to the brain events and the behavior.

As Heath and I explained in 1992, explanations in behavior analysis are historical. Folk psychology, cognitive psychology, and physiological psychology focus on immediate causes of behavior—e.g., thoughts, information processing, and neurotransmitters. Behavior analysis, like evolutionary biology, finds explanations in the past—in a history of selection. Thus, evolutionary biologists would seek to understand how natural selection, acting on populations of birds over millions of years, resulted in canaries building nests *and* in their hormones being triggered by increasing day. If Tom's car won't start when he needs to get to the airport, and he thinks, "Mary owes me a favor," and calls Mary to give him a ride, behavior analysts need to explain how the calling *and* the thinking came about, considering Tom's history of asking for help, with terms like "favor," and his more specific history with Mary. At best, the thinking is additional behavior to be explained, but usually, as Skinner noted, the thinking goes unobserved. Particularly if it is unobserved, Tom's thinking doesn't cause Tom's calling. Behavior might be caused by environmental events like food, injuries, and people exchanging favors, but, in a natural science, it cannot be caused by unobservable events.

Some confusion has arisen among behaviorists on this score of unobservable causes. For example, Zuriff (1979) identified what he called 10 inner "causes" of overt behavior implied in Skinner's writings. He commented that, in comparison with mental causes, these private events leave no mystery about "their ontological status so that metaphysics does not stand in the way of prediction, control, and interpretation of behavior" (p. 8). A little earlier in the same article, however, he suggested that covert stimuli "are hypothesized to function the same as public stimuli, except that they are located on the other side of the skin" (p. 8). This seems to imply that private events are hypothetical. This impression is strengthened by the further statement that "the properties of covert stimuli and responses are inferred from observations of analogous overt stimuli and responses." Finally, Zuriff notes approvingly that radical behaviorism starts "with the external world of stimuli and responses and then [moves] them inside the skin where necessary," apparently suggesting that private events are inferred whenever one runs out of public explanations. We are left with an ambiguous description, in which private events are hypothesized or inferred, considered internal as opposed to external—a usage that sounds dualistic—and yet are pronounced to have ontological status that is unambiguous. A contradiction arises because inferred private events produce no less specious explanations and have no less mysterious an ontological status than inferred mental events. The possibility of turning private events into public events, and thereby disambiguating their ontological status, remains out of reach in everyday life and is attainable, if at all, only in the laboratory. If behavior analysis is a science, we cannot explain observed behavior by simply making stuff up, even if we insist that the stuff we are making up is "just like" the stuff we observe. Only in folk psychology do private thoughts cause behavior.

Even in the context of laboratory experimentation, some behaviorists have advocated inferring private events. Lubinski and Thompson (1993) claimed that they trained pigeons to report on private events. Their experiment is diagrammed on the right in Figure 15.2, along with a conventional conditional discrimination on the left. In brief, a hungry pigeon was given one of two drugs, A or B, before its daily session. If drug A was given, pecks at the key marked with the corresponding letter (A in Figure 15.2) produced food; if drug B was given, pecks at the other key (marked with a B in Figure 15.2) produced food. When the pigeons pecked correctly, Lubinski and Thompson concluded that the pecks were under stimulus control of the different private feelings produced by the different drugs. In the conditional discrimination diagrammed on the left, a red or green key is first presented as the "sample," and then (sometimes after a delay) the choice keys, with the letters A and B on them are presented. If the sample was green, pecks at the letter A produce food; if the sample was red, pecks at the letter B produce food. In both experiments correct performance may be explained by public events: the colors and the drugs. In the conditional discrimination, particularly if a delay elapses between the sample and the choice, one might be tempted to posit some private event—a trace or representation of the sample—to control the pecking at the choice key. The discrimination, however, is between the red and green circles. No need arises to put copies of the circles inside the pigeon, and keeping the stimuli public—in the environment—avoids confusion over who sees the circles—i.e., the actual pigeon and not an imagined inner pigeon peering into an imagined inner space. Similarly, in the Lubinski-Thompson experiment, instead of inferred private stimuli, the equivalent of an inferred copy of the red circle, one may point to the public drugs. Just as one may omit imagined inner representations of the circles, one may omit imagined inner feelings produced by the drugs and avoid confusion over who feels the feelings and where the feelings reside. The preponderance of correct responses constitutes a discrimination between Drug A and Drug B. That is all. In either experiment, and in any discrimination, the decision about what is a correct response and what is an error depends on what the experimenter knows—the color or drug presented—which is public. A discrimination consists of a change in

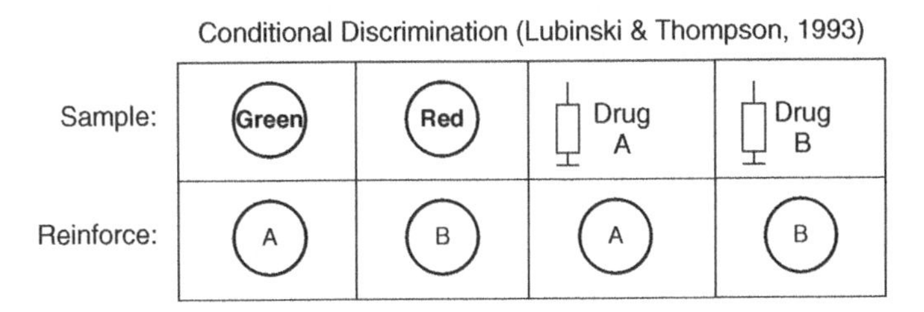

Figure 15.2 The Lubinski-Thompson (1993) experiment compared to a conditional discrimination of color

Note: They claimed that the pigeons' discrimination reflected feelings produced internally by the drugs, but their results are more easily understood as discrimination between the (public) drugs themselves.

behavior with a change in environment, but an onlooker (experimenter) must judge the change in environment (see Herrnstein, Loveland, & Cable,1976 for further discussion). The drug might produce changes in the pigeon's body, but as long as these changes go unmeasured—remain private—they are useless for explaining the pigeon's behavior; the public events of the drugs and the colored circles suffice.

Philosophers who regard behaviorism as incomplete pose the following challenge (e.g., Dennett, 1978). Imagine that Tom rides the Number 4 bus home every day. We see him riding the bus, but no cause for this behavior is evident. He must be riding the bus because he wants to go home and believes this bus will take him there. Thus, behavioral accounts are incomplete, because one cannot explain behavior without reference to mental causes. Behaviorists respond that such explanations are circular, because the only way we know that Tom wants or believes is that he behaves (e.g., rides the bus). The causes are not evident because they lie in the past—in Tom's history with home and buses.

Including private events in behavioral accounts undermines the behaviorists' response. The philosopher may reply that private events hardly differ from wants and beliefs. Tom might be sitting on the bus and reciting to himself that he needs to get out at 79th Street. How different is that?

The Dilemma of Private Events

Radical behaviorists who consider private events to be useful additions to explanations of behavior sit on the horns of a dilemma. Should private events be included or should they be excluded? On the one hand, to exclude private events would be to deny what almost everyone says—that his or her private thoughts and feelings determine public behavior; to deny this would seem to open behaviorists to the philosophers' accusation that behaviorism is incomplete because it neglects an important part of behavior, the very accusation that Skinner strove to avoid. On the other hand, to concede the importance of private events is to introduce hypothetical private events that appear to be—and perhaps actually are; see Zuriff (1979), discussed above—hidden causes and to undermine the behaviorists' claim to a true natural science of behavior. Either way, the mentalists seem to win.

If explanations are sought in public events and all privacy is assumed to be accidental—and there is no other consistent position for behaviorists—then the position is the same as that of Watson (1930/1970), who argued, for example, that thought is sub-vocal speech. Instead of "sub-vocal," Skinner used the word "covert." Neither term solves the problem that private events remain hidden when one is explaining another creature's behavior.

Behaviorists should be careful about the claim that radical behaviorism deals with thoughts and feelings at all, because laypeople are likely to conclude that radical behaviorism incorporates the conventional notion of thoughts and feelings—as things or events in mind-space. Radical behaviorism admits to no such inner space. That denial makes the verbal behavior of behaviorists unconventional (Hineline, 1995), and that unconventionality poses the same dilemma: Should radical behaviorism be presented as if it deals with conventional concepts, making it seem acceptable on false grounds, or should it be presented as the truly radical position it is—the complete denial of dualism—risking its seeming inadequate and implausible? What is the way out? How to preserve the science of behavior and yet have the science be complete and plausible? I argue that the answer lies in adopting a molar view of behavior.

The Molar View of Behavior

Organisms fill the seas, land, and air because they carry genetic material and because that genetic material reproduces more often when in organisms than when not. Otherwise the genetic material would have remained in the original soup. (See Dawkins, 1989b, for a book-length discussion.) Why did selection favor organisms? What is the advantage? In a word, it is *behavior*. To be an organism, to be alive, is to behave. Organisms interact with their environment, and that commerce with the environment is behavior, and its importance lies in its effects on reproductive success via the environment. Organisms produce offspring, feed themselves and offspring, build shelters, avoid predators, and change the world around them in myriad ways. All of these advantageous effects occur through time—on average and in the long run. Behavior is, by its very nature, extended in time. Just as any one individual in a population may fail—may die without leaving progeny—so any individual action may fail. Advantage and success occur over time—on average and in the long run. Just as natural selection operates on populations and cannot be understood by looking at individuals, so behavioral selection operates on extended patterns of activity and cannot be understood by looking at moments. At any particular moment, for example, we might see a pigeon poised with its back parallel to the ground and its beak extended, but when we see an extended sample in which it marches along pecking at seeds on the ground, only then do we understand that it is foraging. The insight that behavior is commerce with the environment tells us both that behavior is extended in time and that behavior and its effects are concrete and measurable. In other words, *all the behavior and effects that matter are public.*

In the molar view of behavior, activities are more extended or less extended in time, which means they have the property of *scale*; more extended activities are defined on a longer time scale than less extended, more local, activities (Baum, 2002; 2004). A canary building a nest gathers material, puts it in the nest, and works it in with its feet. Building the nest is a more extended activity, defined on a longer time scale, and its parts—less extended activities—are defined on a shorter time scale.

The philosophers' challenge, "Tom is riding the Number 4 Bus because he wants to go home and believes that this bus will take him there," leads behaviorists to respond that this explanation is circular, because the only evidence for the wanting or believing is Tom's behavior of riding the bus, getting off at the correct stop, and getting home. That response overlooks a problem with the philosophers' argument itself. The challenge begins with a false premise: that Tom can ride the bus at a moment. "Momentary behavior" is an oxymoron. By its very nature, behavior is extended in time. If Tom is sitting on the bus, we cannot tell if he is going home, to the store, or somewhere else. A momentary snapshot is subject to maximal uncertainty; only with a larger time sample do we become certain about what Tom is doing.

The temptation to posit private events arises when an activity is viewed on too small a time scale. If we view a snapshot of a moment, we see, for example, Tom with a shovel in the garden, but we have little idea what activity is occurring. Viewing over a slightly longer timeframe, we see that Tom is digging a hole. Viewing on a scale longer than that, we see that Tom is digging a ditch. Longer still, and we see he is laying a pipeline. Longer than that, and we see he is installing a waterfall in his garden. And so on. At each time scale, we see public activity, and no problem arises. But, let Tom stop for a while and lean on his shovel, looking at the ground, and then the temptation arises to suppose he is thinking privately about his project. However, we don't know what he is doing at that moment; he might be resting or thinking about getting something to eat. In a larger timeframe, we might see that he resumes digging after a while, and even though he took a short break, he is still working on his project. Whatever covert speech may have occurred hardly matters, because Tom is engaged during the period of observation in the activity of digging a ditch, laying a pipeline, or installing a waterfall. Seen on a longer time scale, the activity is continuous, and any private events that occur may be ignored (Baum, 2002).

Suppose that after pausing, Tom resumes digging in a different direction, and we ask why. Tom says that he encountered a buried electric line and had to dig around to avoid it. We might say that Tom encountered a problem that he solved by changing direction. Whatever sub-vocal or overt verbal behavior may have occurred, it was part of an extended activity—solving the problem. Any private actions or stimuli were neither causal nor essential. The verbal behavior and the change in direction were both due to encountering the buried electric line, a public event. Dealing with the electric line was a less extended part of digging the ditch and laying the pipeline.

The molar view also allows us to avoid hypothesizing the private events that are called "feelings" or sensory events. The temptation to view seeing, hearing, and being in pain as private events arises when we look at behavior over too short a time span. Does the zebra see the lions stalking it? At a moment, we cannot say. We have to watch for a while, until the zebra takes evasive action, before we conclude that the zebra sees the lions. A police officer asks a motorist, "Didn't you see that stop sign?" If the motorist says no, the officer might be tempted to suppose some private seeing occurred, but would have no basis to conclude the motorist is lying, because the officer has seen only the subsequent driving past the sign.

Indeed, the point may be extended to all inferred events, private or mental. Carrying on from the point that in everyday life and in the laboratory most of the time we have access only to public stimuli and public behavior, Rachlin (1994, 2003) argued that mental and private events may be identified with the public activities from which they are inferred. Drawing on the writings of Aristotle and Gilbert Ryle (1949), Rachlin identified mental events like believe, want, intend, know, hear, see, be in pain, and so forth with extended patterns of (public) behavior. For Jane to believe that the death penalty is wrong, for example, means Jane speaks out against it whenever the subject comes up, gives money to organizations that work to oppose it, joins in demonstrations against it, and so on. If enough of these activities occur, over a period of time, people around Jane will assert that she believes the death penalty is wrong. Jane herself will assert her belief on the same grounds. No private or mental event need come into the account.

Following Rachlin, we may go a step further and assert that, seen in the context of her other overt activities, Jane's activities about the death penalty *are* her belief in its wrongness. Anyone who watches the extended patterns of Jane's activities could know as well as Jane what Jane's desires and beliefs are. Indeed, such an observer might know better than Jane, because another person's actions are easier to observe than one's own; people pay money to psychotherapists for exactly this reason.

Are Sensations Private?

Events that might be considered private sensations or private stimuli may be treated the same way as beliefs and desires. Philosophers pose the following problem for behaviorists

(Rachlin, 2003). Suppose that two persons are seated in a room where music is playing, and neither is moving but one of them is deaf. How could the two be distinguished except by their private experience of the music? This challenge is really just another version of Tom riding the bus. If one is restricted to observing them at a moment, one cannot say which person is deaf and which can hear. Afterwards, however, one of them will talk about the music and enjoyment of it, whereas the other will have nothing to say about it. In a more extended timeframe, the distinction between deafness and hearing is readily made; the extended patterns of public behavior of the two persons make the difference (Baum, 2011a). Suggesting that one person is enjoying the music privately would be the wrong answer, because it would concede the mentalists' point by referring to a hidden mental criterion.

A more challenging example is pain, because pain is usually taken to be the quintessential private event. As we saw earlier, Skinner considered pain to be a private stimulus. To understand why this is an error, Figure 15.2 may help, because it made the point that inferred inner feelings were unnecessary to understanding discrimination. Although some insult to the body stimulates nerve endings that may be involved in pain, the cut, burn, pressure, blow, or tear is the origin of the pain and always observable. The stimulating of the nerve endings is like light stimulating receptors in the retina. If Jane stops her car at a red light, the stimulus controlling her stopping is the red light, not an inner representation or sensation of the red light. Similarly, if Jane has a pinched nerve in her spine, the pinched nerve is the event contributing to her pain, not an inner representation or sensation of pain. When she complains, "I am in pain," she is not complaining about an inner sensation or private stimulus, but about the pinched nerve (assuming she is not faking). The pinched nerve may be regarded as a stimulus, but it is not private, except perhaps in the trivial sense that no one has taken the necessary x-rays.

Laypeople and philosophers often claim that one may be in pain but not show it. On that basis, they insist that pain must be private. Rachlin (1985) argued that this is a logical impossibility, because to be in pain *is* to show it. If a soccer player flops to the ground, clutching his leg, rolling about, grimacing, and groaning, we are likely to say he is in pain. If thereby he stops the game to his team's advantage, we are tempted to conclude he is faking. We will only decide later, if ever, in a longer timeframe, on the basis of his ability to continue playing or his limping, whether he was faking or not. Conversely, if someone actually succeeds in not showing any pain behavior at all, we conclude that person was not in pain; regardless of what the person might claim later, for all practical purposes, he or she was not in pain.

Similar to Skinner's (1945) treatment of such utterances, another approach to understanding the claim, "I was in pain but didn't show it," is to ask what conditions might occasion such speech. If Tom makes the claim, one possibility is that he shut himself away in a separate room, say, and thereby rendered all his behavior necessarily private. People usually mean by the claim, however, that others were present but saw no pain behavior. The claim is based on the possibility that some conditions (e.g., an injury or a pinched nerve) might be present that would ordinarily result in public pain behavior, but that some other conditions (e.g., being at a wedding) might override the usual activity. If Tom succeeds in arranging that no one sees any of his pain behavior, then everyone around him concludes he is not in pain. In contrast, if he shows pain behavior, and no one sees any circumstance to conclude he is faking, then usually onlookers will conclude he is in pain and will act sympathetically—try to soothe him, offer palliatives, call an ambulance, and so on. Whether or not the person is in pain resides in the onlookers' behavior, particularly the onlookers' behavior in an extended timeframe.

A football player who is hit by an opposing player but goes on to receive a pass might after the game complain and groan while x-rays show that he has a broken rib. The immediate causes of his pain behavior are the broken rib and the presence of sympathetic onlookers. If he is asked whether he was in pain while making that great catch, he might say he was in pain but ignoring it at that moment. But, how could he know that? Even if the broken rib was affecting nerve endings that could in turn affect his brain, his nervous system was responding only to the broken rib. If he was ignoring anything, he was ignoring the broken rib—the injury now made public—not some inner pain *thing*, not a private stimulus. To onlookers, he was not in pain then, even if the x-rays combined with his pain behavior lead present onlookers to conclude he is in pain now.

The conclusion that one's being in pain depends on the judgment of onlookers, rather than on one's own judgment, might seem counterintuitive. A layperson might still insist that he or she has been in pain but not shown it. More accurately, we might reply, you succeeded in engaging in so little pain behavior that no one around noticed. You were faking not being in pain, so to speak, and people around you saw no reason to behave as if you were in pain. They would have behaved so, too, if you exhibited pain behavior but you seemed to be faking. Ultimately, we still decide about what a person is or is not doing on the basis of prior and subsequent behavior in an extended time span.

The real solution to the problem of privacy is to see that private events are unnecessary to understanding behavior. They might or might not exist; they are irrelevant. A complete account of behavior can be had without them. Recalling that behavior exists only as commerce with the environment and consists of activities more extended or more local in time, we need not talk about any private events to understand the function of behavior. Mechanisms inside the skin, particularly in the nervous system, but also in glands and muscles, are important to understanding how behavior is accomplished, but understanding how the environment causes an organism to behave one way rather than another depends on a larger timeframe—the history of the individual and the species to which the individual belongs (Baum, 2002, 2005). If behaviorists wish to understand why people talk about private and mental things and events or to avoid the accusation that behaviorists fail to address people's inner life of thoughts and feelings, they may follow Rachlin's suggestion that private and mental terms are verbal behavior occasioned by extended patterns of behavior.

From an evolutionary perspective or a therapeutic perspective, only public behavior matters. Whatever a human or nonhuman animal may think or feel privately, the private thinking and feeling cannot affect reproductive success; only commerce with the environment—moving about, gaining resources, interacting with conspecifics, avoiding predators, and the like—events that are observable, measurable—i.e., public—can advance reproductive success. Natural selection cannot affect inner events—whether they are labeled mind, psychology, philosophy, thinking, or feeling—but natural selection can favor advantageous behavioral tendencies and patterns, as long as they are influenced to some extent by genes. If a therapist were to change a client's private thoughts and feelings without changing any public behavior (were such a thing possible), the therapist would have failed, because the aim of therapy, even psychoanalysis, is to help the client live more effectively. If Jane asserts that she feels better about her life but continues her addiction, stays in an abusive relationship, cringes from her boss, and continues to attempt suicide, no one should believe her. Indeed, for any practical purpose, only public behavior matters. A safety engineer doesn't want people only to think privately that wearing a seat belt is good; the actual buckling-up is what matters. If we can predict, control, and understand public behavior, our understanding will not be incomplete due to the omission of private events, because private events are irrelevant; only public behavior matters to evolution and for all practical purposes.

The Mistake of Private Events

Whorf's point about the need to "speak in another language" is well illustrated by the concept of *verbal behavior*, which amounts to speaking about lay concepts like language, reference, and meaning in an entirely different vocabulary (speaking about language in a different "language," Whorf might say). Skinner (1957) defined verbal behavior as operant behavior of a "speaker" reinforced by the behavior of another organism present (the "listener") and acquired as a result of membership in a verbal community of speakers and listeners. The definition covers not only speech, but also gestures (e.g., signing). Skinner aimed, however, not to establish a distinct category but exactly the opposite: to liken verbal behavior to other operant behavior and to overcome the seeming difference (Baum, 2005). Much of the confusion about private events derives from failure to grasp fully the implications of replacing mentalistic notions about language with verbal behavior. If a dog limps, whines, and whimpers, we may unhesitatingly say that it is in pain, our utterance being occasioned by its pain behavior. If a pre-verbal infant cries, grimaces, whines, whimpers, and swipes at its ear, we may say it is in pain or has an earache, our utterance being occasioned by its pain behavior. If Jane, an adult human, grimaces, groans, and holds her face, we may say she is in pain or has a toothache, our utterance being occasioned by her pain behavior. If, in addition, she says, "I have a toothache," that utterance is *just more pain behavior*; it only makes our utterances about her pain more likely and more sympathetic (Baum, 2011a).

Many philosophers and other mentalists, committed as they are to inner-outer dualism, would insist that first-person statements like, "I am in pain," differ fundamentally from third-person statements like, "She is in pain." They do so because they assume that first-person statements are based on private events, whereas third-person statements are based on public events. Usually, they assert also that first-person statements are "incorrigible." They mean by this that no one can question what Jane says about herself, because she alone is privy to the private events that underlie her statement. Even if we set aside the possibility that Jane is lying or faking, we know that first-person statements can be unreliable—people change what they say—for example, an athlete may report no pain from an injury in the heat of play, but complain of the pain later.

From the viewpoint of radical behaviorism, first-person utterances and third-person utterances are instances of verbal behavior, and they are controlled by similar, if not identical, conditions in the environment. We look at the dog's paw for a thorn, in the child's ear for a swollen ear drum, and Jane's dentist will find the decay that explains all of her pain behavior, including her saying she is in pain. Injuries, pinched nerves, excessive blood flow to the brain, and other afflictions all are potentially made public and when made public make our responses to pain behavior more sympathetic and less suspicious of faking. When Jane complains of a toothache, she is not peering at some inner pain thing (or a private stimulus) and reporting on it; she is responding to the injury in her tooth (Baum, 2011a). When Skinner (1945/1961) wrote famously, "...my toothache is just as physical as my typewriter" (p. 285), one wonders just what he meant. He treated the toothache as a "private stimulus," but the statement remains cryptic. Is the private stimulus the injury to the tooth? That would be physical. But he says "toothache," not "tooth." The private stimulus cannot be some inner pain thing; that would not be physical. In the molar view, the toothache is the pain behavior ("...hand to jaw, facial expressions, groans, and so on" p. 277), which Skinner called "collateral responses," plus the person's verbal complaints and assertions—that behavior *is* just as physical as a typewriter.

Much confusion arises from the notion that Jane "reports on" or "observes" some inner private event when she says she is in pain. The mentalistic way of looking at observing is to suppose that it is a single activity directed at different objects. Observing a cow differs from observing a flower, in the mentalistic view, because inner attention is directed toward two different objects in the external world. The weakness of this view appears when we ask questions like, "Who does the inner attending?" and "Is the observer in the external world with the objects?" (See Baum, 2011a, for additional discussion.)

In radical behaviorism, which rejects mentalism and dualism in favor of monism, the observer or reporter is the whole organism, and the behavior of observing or reporting is public verbal and nonverbal behavior. Observing a cow and observing a flower are not the same activity directed at two different objects, but are two qualitatively different activities. One pattern consists of orienting toward the cow, saying that it looks like a Holstein, that it seems skinny, and so on; the other pattern consists of orienting toward the flower, smelling it, saying that it is lovely, perhaps picking it, and so on. When we see such behavior, we say the person sees (observes) the cow or the flower. The presence of the cow or flower alone cannot suffice to produce the behavior of "observing" or "reporting"—other conditions usually have to be met, such as the presence of other people who might respond to the utterances and a history of interactions with cows or flowers. The activities are occasioned by all of these circumstances, but not by any inner copy of a cow or flower (Skinner, 1969c). Moreover, if one "imagines" a cow or flower—sees it in the absence of the thing seen (Skinner, 1969c)—still the imagining involves no inner copy or private event. The person behaves more or less as he or she did when the thing was seen (with eyes open in good light; see Rachlin, 2003, for additional discussion of imagination).

As it is with cows and flowers, so it is with pain and other so-called "private events." When one reports on the oboe playing in a piece of music, one is engaging in verbal behavior that includes words like "oboe," "plaintive," "surprising," and so on. It is occasioned by the music. No inner oboe enters the picture. Similarly, when one reports on pain, one is engaging in verbal behavior that includes words like "hurts," "sharp," "excruciating," and so on. No inner pain thing enters the picture, and if the person is not faking, the pain behavior is occasioned, in part, by an injury or other condition that is at least potentially public.

Conclusion

In the mentalistic view of verbal behavior, which relies on phrases like "using language" and "symbolic communication," a speaker is said to "produce" speech—that is, to act as an agent who talks for his or her self. A natural science includes no place for hidden, unobservable causes—not spirits, not essences, not an inner self (Baum, 1995c, 2005; Ryle, 1949; Skinner, 1969c). Radical behaviorism views all behavioral events as natural events, like earthquakes, rain, sunsets, cell division, birth, death, and taxes, including verbal behavior. Utterances are episodes of verbal activity, like running a race or walking home. Speech, like bird song, comes down to sounds that affect the behavior of conspecifics (humans) who hear them. Thus, when someone speaks of thoughts or feelings, we need not imagine private events as causing the utterance, but rather we must seek the determinants in environmental events present and past. The past events are invisible in the present, but they were public and observable, and all inferences about them are testable, unlike inferences about private events. When a person says, "I hear music," "I see cows," or "My foot hurts," a science explains those utterances with other natural (environmental) events, such as music, cows, injury, and the presence of listeners. The same holds for utterances like, "I feel like going home" and "I thought about the problem." Viewing these utterances on a time scale broader than the moment renders hidden events irrelevant, and these utterances require no private events to explain them. In a science grounded in evolutionary theory, verbal behavior requires no new principles to explain it, and the same set of principles applies to the behavior of verbal and non-verbal organisms.

Afterword

This article appeared along with five commentaries and my response to the commentaries. In his introduction, the editor of the journal, Schlinger (2011), raised a couple of points of his own in a commentary which I had not seen. This is my response to his commentary.

Schlinger (2011) claimed that: (a) I wrote that private events occur at too small a scale to be observed; (b) "self-talk" can contribute to solving problems like a fax machine not working; and (c) *private* may mean just "unobserved" rather than unobservable.

Schlinger (2011) misinterprets my saying that the temptation to posit private events arises from viewing an activity on too small a time scale. He incorrectly takes this to mean that the objection to imagining private events to explain activities is that private events occur on too small a scale to be measured. Possibly he is confusing two different usages of "scale." He might be saying that private events are too faint to be measured, but the usage here is *time* scale. When people imagine private events to try to explain an action, the events they imagine generally are brief, because of the brief time span being examined, but their brevity is not the problem, their imagining is the problem. If I see Schlinger fix the fax machine by pressing the reset button, my imagining that he thought *I should press the reset button* is no explanation, it is only fantasy and has no place in a science.

Suppose, Schlinger says, "I thought I should press the reset button, and then I pressed the reset button." First, we have no reason to believe him; he could be mistaken or he could be lying. A more interesting possibility is that he is imagining the thought just as another person might. Asked how he fixed the fax machine, he might himself imagine that he thought, "I should press the reset button," simply because that seems to make sense. His having fixed the machine, combined with the presence of an inquiring listener, occasions the narrative, "I thought I should press the reset button, and then I pressed the reset button." The role of "then" is critical, for insertion of that little word suggests the thinking caused the acting.

Convention alone prescribes that thoughts must precede actions. The explanation of actions as the result of preceding thought is only a cultural recept. Having grown up in Western culture, Schlinger imagines that the thought *I should press the reset button* preceded his pressing the button, but he has no basis for this. Assuming the thought actually occurred (we have no reason to believe him, but we don't wish to contradict him), it might have accompanied the button pressing or might even have followed it. He reconstructs the events according to the culturally prescribed form, inserting "then" to make the narrative acceptable. Even if the thought did apparently (to Schlinger) precede the action, he has no reason to suppose that the action wasn't initiated before the thought. The most Schlinger or anyone else can say is that he fixed the fax machine by pressing the reset button and *says* that he thought, "I should press the reset button," (omitting "then"). Perhaps someday neuroscience will progress to the point where thoughts might be measurable, and then the order of thinking and acting might be researchable.

Schlinger (2011) displays confusion over the meaning of "molar," maybe because he is unfamiliar with my previous explanations (Baum, 1995a, 2002, 2004). He quotes Catania (2011) displaying similar confusion, probably for the same reason. Activities are extended in time and are made up of parts that are themselves activities on a smaller time scale. Those parts are made up of activities on a still smaller time scale. Thus, we may say that activities are always nested within more extended activities (except perhaps for the activity called "living," which would be nested only in the context of reincarnation.) Hineline (2011) complained that "molar" may connote a long time scale. That is why he suggests that "multiscale" might help to avoid confusion. Whether we call it "molar" or "multiscale," it is the same view. A pigeon's peck may be part of an extended activity of pecking, but even the peck has parts: forward head moving, opening of the beak, closing of the eyes, closing of the beak, backward head moving, and opening of the eyes (Smith, 1974). In the molar view, all behavior takes time, and the pigeon's peck is extended in time, even if it takes less than a second.

Schlinger's (2011) discussion of the meaning of "private" raises some issues critical to thinking about the status of private events. I agree that the distinction between accidental privacy versus in-principle privacy corresponds to the distinction between "unobserved" versus "unobservable." If I sing alone in my room, that action is accidentally private in the sense that it is unobserved but observable. Similarly, neural events, muscle contractions, and glandular secretions may be unobserved but observable. Thoughts, feelings, and sensations, however, when taken as private events, are not only unobserved but are unobservable. Schlinger asks how we tell the difference.

The difference between unobserved and unobservable has nothing to do with access to the senses, because we measure constructs like electron flow even if we cannot see or hear electron flow; the senses come in finally when we read a display or some state of an instrument. Thus, one might argue that just because inner thoughts, feelings, and sensations are inaccessible to the senses, still they might be measurable by some to-be-invented technology like the antiprivacy machine (Baum, 2011c).

When we say an event is unobserved but observable, we are saying that events like that event have been observed in the past. The planet Neptune was speculated to be unobserved but observable because other planets had been observed and their effects measured in the past. Inner thoughts, feelings, and sensations are probably unobservable, because no one has ever succeeded in measuring them in the past. A pigeon's discrimination between red and green lights depends on the lights, not on sensations of red and green. The participant in a psychophysical experiment who is asked to estimate the brightness of a light reports on the light, not on the sensation of the light, as Boring and Stevens supposed. When Skinner (1945) wrote of a "private world of stimuli," he inadvertently reintroduced the dualism he meant to reject. No such world can be found inside the skin or anywhere else. Private stimuli either do not exist (unobservable) or are hypothetical (unobserved but observable). Either way, they cannot offer a valid explanation of (observable, public) behavior. An unobserved condition of the body might be measurable, and, if measured, would no longer be private, but would have become public and, therefore, must have been observable. Behavior, however, would depend on the formerly-private bodily condition, not some hidden private stimulus—unless one were to equate the bodily condition with a "stimulus." Schlinger (2011) asks whether this transition from unobserved to observed could ever happen. It could only happen if one were to bend the meaning of "stimulus" to include any bodily condition that was private in the limited sense that it was unobserved but observable (presumably with instruments). Such a broadening of the meaning of "stimulus," however, would raise problems. If an organism is dehydrated, should dehydration be considered a stimulus? Is a dog scratching itself responding to a private stimulus? Skinner (personal communication) said, "No." Why should a person scratching an itch be treated any differently?

16

Ontology for Behavior Analysis: Not Realism, Classes, or Objects, but Individuals and Processes

Foreword

This paper might be thought of as an update to my thinking about ontology. It summarizes the problems with realism and dualism for behavior analysis and explains how behavior may be viewed as process. Behavior analysts must reject dualism to have a coherent science and must reject realism because it leads directly to dualism. One implication, touched on in the previous paper, is that explanations that rely on private events need to be avoided if the notion of privacy implies dualism. Recognizing that behavior is process opens the way to a conception of behavior beyond discrete responses and compatible with evolutionary theory.

Abstract

Realism, defined as belief in a real world separate from perception, is incompatible with a science of behavior. Alternatives to it include Eastern philosophy, which holds that the world is only perception, and pragmatism, which dismisses the belief as irrelevant. The reason realism is incompatible with a science of behavior is that separating perception of objects from real objects leads directly to subjective–objective or inner–outer dualism. This dualism, in turn, leads directly to mentalism, the practice of offering inner entities as explanations of behavior. Positing unobservable causes renders a science incoherent. Ontology for behavior requires two distinctions: (a) between classes and individuals; and (b) between objects and processes. These distinctions allow a workable ontology in which behavior consists of activities that are extended in time (i.e. processes) and are ontological individuals—functional wholes with parts that also are activities. Such an ontology provides coherence to a science of behavior.

Keywords: realism, dualism, process, activity, individual, pragmatism

Source: Originally published in *Behavior and Philosophy*, 45 (2017), pp. 63–78. Reproduced with permission of Cambridge Center for Behavioral Studies.

Science and Philosophy of Behavior: Selected Papers, First Edition. William M. Baum.

> *It is indeed an opinion strangely prevailing amongst men, that houses, mountains, rivers, and in a word all sensible objects, have an existence, natural or real, distinct from their being perceived by the understanding... yet whoever shall find in his heart to call it in question may, if I mistake not, perceive it to involve a manifest contradiction. For what are the forementioned objects but the things we perceive by sense? and what do we perceive besides our own ideas or sensations?*
>
> (George Berkeley, 1717/1939, p. 524; italics in the original.)

The prevalence of the belief in a world of things existing independently of our perceptions is "strange" because it has no basis in logic. All we have is our perceptions of objects or our experience of objects. Say, I have a tree near my house; I see it, feel it, climb it, and so on. If the tree I experience has some existence as a real tree distinct from my experience of it, I have no access to that real tree and could never prove it does actually exist separately.

James Boswell (1791/2007), in *Life of Samuel Johnson*, relates that Samuel Johnson (1709–1784), upon hearing about Berkeley's argument, kicked a stone and said, "I refute him *thus*" (p. 310). The error involved is so obvious that scholars doubt Boswell's account (Womersley, 2007). Johnson's foot, the stone, the kick, and the vocalization are all just Johnson's (and Boswell's) perceptions. The foot's engaging the stone need have no reality distinct from those perceptions. Indeed, from Boswell's viewpoint, Johnson himself is just Boswell's perceptions.

The physicist Erwin Schrödinger (1887–1961), one of the founders of quantum theory, added to Berkeley's argument by pointing out that a material existence apart from our experience is superfluous:

> . . .if, without involving ourselves in obvious nonsense, we are going to be able to think in a natural way about what goes on in a living, feeling, thinking being (that is, to see it in the same way as we see what takes place in inanimate bodies)—without any directing demons, without offending against, say, the principle of the increase of entropy, without entelechy or *vis viva* or any other such rubbish—then the condition for our doing so is that we think of *everything* that happens as taking place in our *experience* of the world, without ascribing to it any material substratum as the object *of which* it is an experience; a substratum which . . .would in fact be wholly and entirely superfluous (Schrödinger, E. (1961/1983), pp. 66–67; italics in the original).

The present paper aims to discuss three topics: (a) realism and its alternatives; (b) the trouble with realism; and (c) a workable ontology for behavior analysis. By realism, I mean the belief in a world of things that exists independently of our perceptions or experience. Philosophers would likely call it "naïve realism," because they have proposed other types. Schoneberger (2016), in a discussion of realism and pragmatism, called it "metaphysical realism." Hereafter, I use "realism."

Realism and Its Alternatives

One alternative to realism may be found in Eastern philosophy, which anticipated Berkeley's skepticism by thousands of years: that our experience seems to tell us of a physical world, but our experience is illusory, like a dream. For example, in the Hindu text, the *Bhagavad Gitā*, we find:

> Never is this born, nor does it die, nor having been does it ever cease to be; unborn, eternal, un-decaying, ancient; this is not disintegrated by the disintegration of the body. (Chatterji, M. M. (1960), Ch. 2, par. 20.)

A similar view may be found in Zen:

> When all these deep things are searched out, there is after all no 'self'. Where you descend, there is no 'spirit,' no 'God' whose depths are to be fathomed. Why? Because Zen is a bottomless abyss (Suzuki, 1964, p. 43).

The spiritual master, Meher Baba (1894–1969) expressed the view:

> God alone is real; He is infinite, one without a second. The existence of the finite is only apparent; it is false; it is not real. (Meher Baba, 1973, p. 384).

All of these quotes, though differing in choice of words, express the oneness of Reality and the illusoriness of the world of things. The illusion is characterized by the seeming manyness of things, whereas Reality is one.

Probably the principal challenge to Berkeley and Eastern thinking, the observation that seems to compel realism, is the ability of two observers to report about the same object. You and I may both agree that a tree stands before us. How is this possible if the tree is just our perceptions or is part of an illusory world? Wouldn't the tree have to be really there? The question presupposes a separation between you and me that the Eastern view does not. Schrödinger, who was a student of Vedanta, put the matter this way:

> For philosophy, then, the real difficulty lies in the spatial and temporal multiplicity of observing and thinking individuals. . . .I do not think that this difficulty can be logically resolved, by consistent thought, within our intellects. But it is quite easy to express the solution in words, thus: the plurality that we perceive is only *an appearance; it is not real.* Vedantic philosophy, in which this is a fundamental dogma, has sought to clarify it by a number of analogies, one of the most attractive being the many-faceted crystal which, while showing hundreds of little pictures of what is in reality a single existent object, does not really multiply that object (Schrödinger, E. (1961/1983), p. 18; italics in the original).

Thus, the answer denies that multiple individuals' similar reports necessitate an independent reality apart from the individuals' perceptions and affirms instead that the multiplicity of individuals itself is part of the illusion; in Reality all is one. The world of things is illusory and has no independent real world of manyness behind it. Reality is completely different: one, indivisible, and beyond the illusory world. (For a book-length presentation of the many manifestations of this view in different times and places, see Aldous Huxley's (1945) *The Perennial Philosophy*.)

Schoneberger (2016) urges behavior analysts to adopt a version of realism that he attributes to the philosopher Richard Rorty (e.g., 1979, 1989). Although still maintaining a real substance that exists independently of our experience, Schoneberger considers this reality to have no intrinsic structure of its own. Instead, it is shaped by what we say and do about it—a sort of blank canvas on which our speech and actions write. It seems to be similar to what Zen calls the "Nameless"—what is real before any thinking (Suzuki, 1964).

Schoneberger's (2016) proposal might actually resemble Eastern philosophy, because the reality he envisions would be a oneness that contrasts with the manyness of our experience. For example, the existence of giraffes (Schoneberger's example) is contingent on our having a term "giraffe" with which we carve out giraffes from the non-differentiated (one) reality. The difference is that Schoneberger (and perhaps Rorty) doesn't suppose that the one reality exists beyond our limited, worldly experience.

Another alternative to realism derives from pragmatism. William James (1907/1974) presented pragmatism as having dual aspects: as a method for settling disputes and as a theory of truth. He pointed out that some questions lead only to endless disputes back and forth, with no satisfactory resolution:

> Is the world one or many?—fated or free?—material or spiritual?—here are notions either of which may or may not hold good of the world; and disputes over such notions are unending. The pragmatic method in such cases is to try to interpret each notion by tracing its respective practical consequences. What difference would it practically make to any one if this notion rather than that notion were true? If no practical difference whatever can be traced, then the alternatives mean practically the same thing, and all dispute is idle. Whenever a dispute is serious, we ought to be able to show some practical difference that must follow from one side or the other's being right (pp. 42–43).

In other words, if the answer to a question would in no way change the way science would proceed, then the question itself is at fault and merits no attention. The question of whether there really is a real, independent, objective world out there apart from our experience qualifies as one of those questions about which dispute is idle. James wrote that our conception of an object consists of nothing beyond its practical effects: "—what sensations we are to expect from it, and what reactions we must prepare" (p. 43). What matters about a bicycle is that I see it, call it by its name, may lend it to a friend, may ride it myself. Pragmatism remains agnostic about whether a *real* bicycle exists behind these effects. One alternative to realism, then, is pragmatic agnosticism (Barnes-Holmes, 2000). One would hold that we need not bother about a reality we cannot know and is useless to science; we may concentrate on what we can know and use—our experience itself.

The Trouble with Realism

Should any of this philosophical discussion matter to behavior analysts? I think it should, because realism creates incoherence, and if any behavior analysts subscribe to realism, they should beware. In the physical sciences—physics, chemistry, astronomy, and geology—realism may raise few problems, because the physical sciences deal with non-living things that may be treated separately from the observer without confusion; whether they exist independently may be of little note. To an astronomer, whether the real universe is really expanding or whether the perceived universe is perceived to be expanding may not matter, because the data are the data.

Realism may present problems in some areas of biology, however. What is true of the physical sciences might also be true of physiology and evolutionary biology, because physiologists can treat living cells as mechanical things, and evolutionary biologists can treat populations of organisms without discussing their individual interactions with the environment. When biology treats behavior, however, problems arise. For behavioral ecology, which deals with individual organisms interacting with the world around, realism creates complications, because questions about consciousness may intrude. For behavior analysis, the matter becomes crucial, because an account of consciousness lies at the heart of its mission, even if that only means explicating why the term is useless.

Realism is disastrous for behavior analysis because it implies dualism with its incoherence. If things have a real existence independent of our perceptions of them, then two worlds must exist: (a) the world of real things; and (b) the world of our perceptions. This follows from the separation of a tree from our perceptions of the tree. According to realism, the tree is "out there," in the real, independent world, whereas our perceptions are somewhere else, a second world that becomes "in here." In other words, realism leads immediately to subject-object dualism or inner-outer dualism, in which perceptions are subjective or inner, and things are objective or outer. Once we suppose an inner world of subjective perceptions, we may populate it with all kinds of other subjective things—an inner self with intentions and so on. Once this division exists, behavior seems to be part of an outer world while perceptions and the like are part of an inner world. How does this happen?

Dualism becomes inevitable when I apply realism to myself. Just as I perceive a tree before me, I perceive my own body. I see I have arms and legs. When I look in a mirror I see a body that resembles other people's bodies. According to realism, my body, like the tree, belongs to the real world. It is made of the same sort of stuff as the tree. I perceive the world of things, and I perceive my body as part of that world. But, where am I in that picture? Who does this perceiving? The existence of that real, objective world, independent of me, if that is what we study, requires someone separate from it to observe it. That separate observer must be me, my self, but not being in the real world, my self must be somewhere else. That somewhere else is the inner world, the world of perceptions and other mental things. Thus, realism, separating objects from their perceptions, leads inevitably to the inner self and all its problems—that is, to incoherence, because we know of no way that an invisible, non-material thing could cause behavior (Baum, 2016b).

Is subject-object dualism benign? Burgos (2016) argued that dualism in itself is no threat to the understanding of behavior; it only becomes a threat when it passes into mentalism, the practice of invoking inner entities as causes of outer events. Burgos may be correct in principle, but in practice, dualism leads inevitably to mentalism, because Burgos's argument neglects the reason for positing an inner world in the first place. Once we suppose dualism, we are likely to explain our saying, "There is a tree" by asserting that the verbal behavior is caused by the perception of the tree. Talk of the inner world exists precisely to "explain" our own behavior and the behavior of others. For everyday discourse, this talk may be benign. For a science of behavior, it is disastrous, because it results in incoherence. Hidden causes become acceptable—even necessary—to explain behavior, not only one's own behavior, but that of others, too. If this holds for me, it must also hold for other people. The inner self, unseen, also called ego or personality, takes in information, processes it, makes decisions according to its intentions, desires, and beliefs, and causes concordant behavior, and a science of behavior becomes impossible. Drawing on Eastern philosophy, Schrödinger put the matter this way:

> 'There's another one like you sitting over there, thinking and feeling go on in him too.' And now everything depends on how we go on: whether with 'I am over there too, Self is over there, that is myself'; or with 'There is a self over there, like yours, a second one.' It is the word 'a' which differentiates the two ideas, the indefinite article, degrading 'self' to a common noun. It is only this 'a' which . . .fills the world with ghosts and drives us helplessly into the arms of animism (Schrödinger, E. (1961/1983), p. 35).

Supposing each person to have an inner self separate from the objective world leads inevitably to mentalism ("animism"), to the view that Ryle (1949) called the "ghost in the machine." That is the trouble with realism. If behavior analysts eschew realism, however, what alternative can they pursue? A view rooted in Eastern philosophy is compatible with science, but the philosophical stance of pragmatism may offer the best approach (Baum, 2017a).

A Workable Ontology for Scientific Study of Behavior

Ontology is the branch of philosophy that is concerned with being and existence. It specifies the things that exist to know about and goes hand-in-hand with epistemology, which is the branch concerned with what and how we know about those things. For example, Newton's law, F = ma, makes the ontological assertion that forces, masses, and acceleration exist, and epistemological considerations would focus on how to study those things.

If one thinks of ontology as based on realism, one might be led to suppose that ontology requires faith, because belief in a real world independent of experience and inaccessible to our senses is an article of faith (Barnes-Holmes, 2000). Thus, in a realism-based ontology, if I see a tree I should have faith that a real tree exists behind my experience of the tree.

Ontology need not be realism-based, however, and need not require faith. Pragmatist ontology is possible (Barnes-Holmes, 2000). Such ontology proposes things available for study that might be useful to making sense of our experience. For example, when the heliocentric view of the solar system challenged the geocentric view of the universe, it introduced new things to be known about: the solar system and satellites. Seen in the light of the verbal behavior of scientists, pragmatist ontology specifies terms that might be useful in understanding our experience, such as "solar system" and "satellite." Such terms are occasioned by invariances and bring together observations that might otherwise seem disparate. In this way, terms like "operant," "reinforcer," and "stimulus control" organize and make sense of our experience of behavior, whether in the laboratory, in applied settings, or in everyday life. The physicist Ernst Mach (1960/1933), who adhered to pragmatism, explained:

> To find, then, what remains unaltered in the phenomena of nature, to discover the elements thereof and the mode of their interconnection and interdependence — this is the business of physical science. It endeavors, by comprehensive and thorough description, to make the waiting for new experiences unnecessary; it seeks to save us the trouble of experimentation, by making use, for example, of the known interdependence of phenomena, according to which, if one kind of event occurs, we may be sure beforehand that a certain other event will occur (pp. 7–8).

For Mach, the business of scientists was to describe phenomena in terms that bring phenomena together and reduce our puzzlement over the events in our experience. Accordingly, Mach had a pragmatist view of explanation:

> When once we have reached the point where we are everywhere able to detect the *same* few simple elements, combining in the ordinary manner, then they appear to us as things that are familiar; we are no longer surprised, there is nothing new or strange to us in the phenomena, we feel at home with them, they no longer perplex us, they are *explained* (p. 7; italics in the original).

The pragmatist view of truth follows from this approach. James, for example, denied absolute truth of the sort suggested by realism. Instead of supposing that theories approximate some ultimate real world, James regarded truth as comparative. A theory is more or less true insofar as it is useful in making sense of our experience. Kuhn (1970), whose views coincided with pragmatism, maintained that a theory or paradigm that explained more phenomena or explained them more elegantly would gain more adherents among scientists and eventually become dominant for a while, until a superior theory or paradigm appears. Among the aspects of a paradigm, as conceived by Kuhn, are proposals of terms, which may be viewed as ontological claims. "Solar system" and "satellite" are examples.

Two fundamental distinctions have been useful in ontology: (a) the distinction between class and individual; and (b) the distinction between object and process. These may help to understand ontological considerations that apply to the study of behavior.

Class versus individual

A book-length explanation of the difference between a class and an individual may be found in biologist Michael T. Ghiselin's (1997) *Metaphysics and the Origin of Species*. The present discussion draws on that book. A class is an ontological type that is defined by a set of properties. Classes are characterized by having instances that conform to the properties. For example, "pieces of furniture with four legs" would be a class with instances like chairs and tables, and "table" would be a class also, but defined by function instead of structure. The word "individual" is often taken to be synonymous with "organism," but ontologically speaking organisms are only one type of individual. More generally, an individual is a concrete thing that is situated in time and space and functions as an integral whole. Instead of instances, an individual is made of parts, and these parts are themselves individuals. An organism is made up of parts like appendages and organs. A species of organisms may be thought of in two ways. Thinking of "human being" as the name of a class, we would say that B. F. Skinner and Isaac Newton are instances. Thinking of the species *Homo sapiens* as an individual, we would say that Barack Obama, as a member of the species, is a part of a whole population, which, in evolutionary biology, is defined as a reproductive unit (Mayr, 1970). When he dies, he is no longer a part of the species, but the species goes on; a salient property of individuals is that, in contrast with classes, which are fixed by their properties, individuals can change while still retaining their identity (Ghiselin, 1997). Classes are defined by their properties, are fixed forever, and may even have no instances (e.g., "mental cause" or "person more than 10 feet tall"). For a long time, the class "living coelacanth" was considered to have no instances, but when a live one was discovered, that class came to have instances. If a defining property of the class had been "fossil," the class would have changed with the change in properties. Individuals, in contrast, have no defining properties, can be defined only ostensively (e.g., "That is my dog Fido"), and can only be described. Individuals have a beginning and an end and occupy a certain geographical location. Organisms are born and die. Species result from speciation events, have a certain geographical range, evolve and change, and may ultimately go extinct. Classes themselves cannot do anything, although their instances may be individuals that have functions.

This distinction between class and individual is important to behavior analysis because it applies to behavior. When Skinner defined the operant as a class, he specified the properties a response must have to possibly produce a reinforcer, and responses having those properties were instances of the class. The actually occurring responses that result from requiring those properties, however, are another matter. Those responses do not necessarily even have the required properties. A rat interacts with a lever in a variety of ways, some of which operate the lever, and some of which do not. Catania (1973) distinguished between the "descriptive" operant and the "functional" operant for this reason. The "descriptive" operant is Skinner's definitional operant. As Glenn, Ellis, and Greenspoon (1992) explained, the "functional operant" is an individual, not a class, because it consists of actually occurring responses, not responses that might occur. Whether one retains the concept of "the operant" (supposedly composed of discrete responses) or not, the distinction between class and individual greatly affects our understanding of behavior, as we will see after we take up the distinction between object and process.

Objects versus processes

Firstly, instead of dualism, we may embrace monism or, as Skinner (1945) put it, "the 'one' world." To do this, we must put aside the notion of a real existence independent of our experience. In the spirit of pragmatism, we can aim to make sense of our experience, be it populations of organisms or the behavior of individual organisms.

Secondly, we need to recognize that sciences in general, and behavior analysis, in particular, focus on process—that is, change through time. When we examine all the arguments for and against realism, we discover that almost all have to do with objects. We see propositions like, "I perceive that tree because a real, independent tree is there." This focus on objects makes a lot of the trouble, because it leads naturally to that separation of the object from its perception.

When people speak of objects, they seem to be discriminating something that has boundaries and remains stable in a changing world. In ordinary speech, a tree seems to be an object. To a botanist, an ecologist, or an evolutionary biologist, however, the tree is a process. It develops from a seed, grows to maturity, reproduces, interferes with other plants, acquires symbionts and parasites, eventually dies, and is finally reabsorbed into the earth. A focus on process is fundamental to the sciences.

The physical sciences focus on process less obviously, because they often examine the structure of things—rocks, stars, plants, DNA, and atoms. The study of structure, however, is not usually an end in itself, because scientists try to understand the way things work or function and how they change or evolve. Accordingly, the physical sciences aim at understanding volcanism, the life-cycles of stars, growth of plants, replication of genetic material, motion, decay, and transfer of energy—all processes. Whether on a short time scale (atomic physics) or a long time scale (geology or astronomy), sciences aim to understand processes.

The study of living things focuses on process more obviously. Beyond interest in molecular structure and anatomy, physiology concerns itself with processes like metabolism, circulation, cell division, secretion, and uptake. More obvious still, evolutionary biology focuses on process when it concerns itself with changes in populations and speciation. History becomes important, because, ontologically speaking, species are individuals—entities that can change while still retaining their identity (Ghiselin, 1997). Thus, a species may be thought of as a lineage—a population with a history that contains the process of its origin.

When we come to a science of behavior, all ambiguity about process vanishes, because behavior itself is process. The most basic datum, response rate, is a process. Even if one thinks of behavior as composed of discrete responses, a response may be taken as an event, which is a process seen in a small time frame. A response rate is a process seen in a longer time frame. Choice is a process in which behavior is divided among two or more activities. Studying the structure of behavior, often called its "topography," only illuminates its process, its function. Thus, a pigeon's pecking at a key that produces food is a different process from its pecking at a key that produces water (Jenkins & Moore, 1973).

The trouble with taking behavior to be composed of discrete responses is that discrete responses are easily confused with objects. The discrete response, borrowed from reflexology, originally consisted of little more than a muscle twitch. Even a muscle twitch is a process in a muscle, but when taken as a discrete event with no variance other than its occurrence or non-occurrence—as in the operation of a switch or not—then it becomes the behavioral analog to an object. Such a punctate unit cannot capture the continuity of behavior, but it was historically embodied in Skinner's (1938) concept of "the operant." At best, it seems to be a population of discrete responses. The noun "operant" suggests something like an object—a tree, for example—whereas "operant" as an adjective may be combined with a process, such as "operant activity." Even when behavior is acknowledged to be continuous, commitment to discrete responses makes the behavioral stream seem more like a series of beads on a string than an actually continuous flow (Schoenfeld & Cole, 1972). When we speak of everyday life, discrete events make little sense. What is the discrete event in the activity of watching television?

One must be careful to distinguish between discrete responses and small-scale processes. As noted already, a muscle twitch is a process, and a pigeon's pecking constitutes a process with small-scale parts that occupy fractions of a second, like moving the head back and forth, opening and closing the eyes, and opening and closing the beak (Smith, 1974). When behavior is equated to the operations of a microswitch, confusion may arise, because each switch operation might incorrectly be thought of as a discrete response. Attaching a microswitch to a lever or key, however, turns out to be a fairly reliable way to measure an organism's interaction with the lever or key (Baum, 1976, 2013).

Two earlier papers argued that behavior consists of activities that are temporally extended and have parts that are themselves activities in a smaller timeframe than the more extended activity to which they belong (Baum, 2002, 2013). An activity is a process. Like a species, an activity may be thought of as a lineage—that is, a process with a history of origin and change. A history of reinforcement is such a history. One's playing tennis, for example, may begin with parts that change over time with practice to the point where the beginning activity hardly resembles the mature activity.

Explicit focus on process avoids much of the temptation to realism. Whereas objects might seem to call for explanation because of their seeming permanence, processes more readily can be taken at face value, because they convey change or a lack of permanence. When we measure a response rate, we need not imagine a real response rate apart from the one we measure. Why not study the process we measure and just that?

Ontology for Behavior Analysis

A workable ontology for studying behavior focuses on processes in the form of activities. Variation and change are inherent in activities. Just as I cannot repeat the same act exactly the same way twice, so the allocation of my behavior among the activities of my life changes from time to time. The resemblance to species in evolutionary theory is no accident. Just as variation is inherent in a population of organisms, so variation is inherent in activities. If I drive to work every day, still my drives to work vary, and my activity of driving to work over the course of a year constitutes a population of varying episodes of driving to work. As the parts change, the activity not only varies, but may evolve. If I sometimes drive on Road A and sometimes on Road B to get to Road C, driving to work may in time include less driving on Road A than Road B—perhaps Road B is smoother than Road A—and my driving to work changes through time. The parts of driving to work may be as detailed as analysis requires, including stopping and starting, slowing and speeding, and so on (Wallace, 1965).

When we take behavior to consist of activities, the key measure of an activity is the time it takes up. For example, instead of considering a rat's interactions with a lever as consisting of discrete presses, we may take the number of operations of a switch attached to the lever to be an indicator of the time taken up by "lever pressing" (Baum, 1976). Time is limited, however, because an experimental session lasts for a certain duration, and a day contains only 24 hours. Since a living organism's activities take up all its time, activities compete with one another for time. Every organism has a time budget that describes the allocation of its time to its activities (e.g., Barnard, 1980). When a rat is not lever pressing or a pigeon is not key pecking, it is engaged in other activities, often called "background" activities (e.g., Baum & Davison, 2014). When I am not working, other activities, such as spending time with my family or exercising, take up the time (Baum, 2010, 2013).

One can represent time allocation with a pie chart. Figure 16.1 shows average time allocation of American civilians 15 years or older in 2015, based on data supplied by the United States Department of Labor Bureau of Labor Statistics (www.bls.gov). The 5 activities in the left-hand pie chart, measured in hours, add up to 24 hours. The most time-consuming activity is Health Maintenance ("HeM"; my label), which combines Personal Care Activities (BLS label; 9.64 h) and Leisure and Sports (BLS label; 5.21 h). The catch-all activity called "Other Activities Not Elsewhere Classified" (BLS label; "OTH") takes up only 0.19 h. The activity which I label "Gaining and Using Resources" ("RES"; 5.92 h) is unpacked into its parts in the right-hand pie chart (BLS labels): Eating and Drinking ("Eat"; 1.18 h), Purchasing Goods and Services ("Buy"; 0.75 h), Working and Work-Related Activities ("Wrk"; 3.53 h), and Educational Activities ("Edu"; 0.46 h). All of the other activities represented on the left also have parts, and the BLS parts have sub-parts, and many of those have sub-sub-parts.

Figure 16.1 illustrates two general points. First, the definition of an activity includes its function. For example, buying groceries is an activity that serves the function of bringing home food and other desired items. It cannot be defined only by its topography or structure. Thus, walking by itself is not an activity, because one cannot walk without

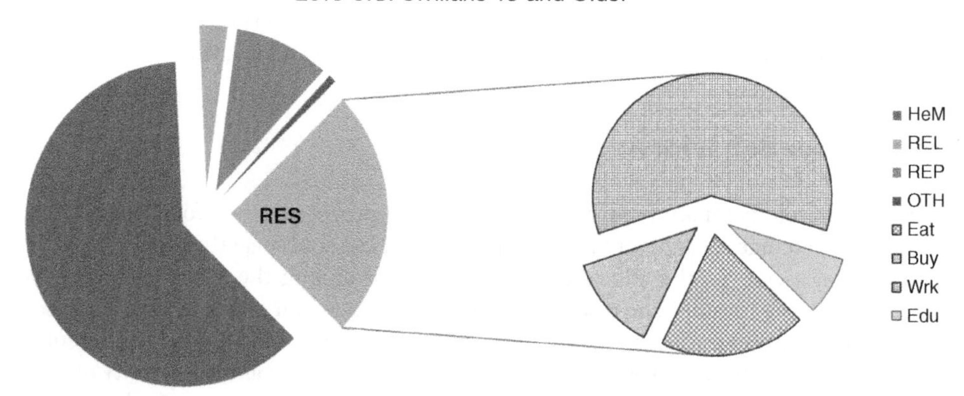

Figure 16.1 Average time allocation in hours of Americans 15 years and older in 2015.
Note: The left-hand pie chart was constructed by combining activities among the 11 most extended activities measured. "Health Maintenance" ("HeM") combines (BLS labels) Personal care activities and Leisure and sports. "Forming and Maintaining Relationships" ("REL") combines (BLS labels) Caring for and helping nonhousehold members, Organizational, civic, and religious activities, and Telephone calls, mail, and e-mail. "Reproductive Activities" ("REP") combines (BLS labels) Household activities and Caring for and helping household members. ("OTH" is a catch-all for activities not otherwise classified.) The right-hand pie chart shows the time taken by activities composing "Gaining and Using Resources" ("RES") and consisting of its parts (BLS labels): Eating and drinking, Purchasing goods and services, Work and work-related activities, and Educational activities. See text for more explanation. Data are from the United States Department of Labor Bureau of Labor Statistics (www.bls.gov).

walking somewhere, whereas walking for exercise or walking to the bank would count as activities, because specifying their functions situates them in time and space. Second, every activity has parts that are themselves also activities. Figure 16.1 shows the parts of "Gaining and Using Resources." Buying groceries would have parts like driving to the store, collecting items in the store, paying for the items, and driving home with the items. Conversely, except for the most extended activity, "Living," every activity is a part of some more extended activity. A rat's food-maintained interactions with a lever consist of lever pressing and eating, less extended activities that are parts of the more extended activity "feeding" (Baum, 2012a, 2013).

I called this way of viewing behavior the "molar" view in the past. The word "molar," however, connotes for most people a highly extended time scale. A more proper label would be the *multiscale* view of behavior, because activities are measured at different time scales.

A more extended activity is composed of parts that are less extended—on a smaller time scale. The parts of buying groceries are activities at a smaller time scale. Any one of those parts has parts at a still smaller time scale—collecting items in the store is composed of parts like going down the cereal aisle and going through the produce section. Indeed, the time scale can be as brief as one might need it to be for research or discourse, as in a pigeon's key pecking, discussed above, in which the parts require less than a second (Smith, 1974).

That an activity is defined by its function and is a functioning whole with parts that are also functioning wholes, tells us that activities are individuals. Just as an organism or a species is a concrete functioning whole situated in time and space, so too an activity is a concrete functioning whole situated in time and space. Just as an organism or species can change and still retain its identity, so too an activity can change and still retain its identity. Skilled activities are good examples; someone's playing tennis can change over time but remains that person's playing tennis (see Baum 2002, 2013 for more explanation).

Conclusion

No matter whether many behavior analysts espouse realism, as Schoenberger (2016) claims, realism is an unworkable ontology for behavior analysis, because it leads at once to subject-object or inner-outer dualism, which leads inevitably to mentalism, and behavior analysts should disavow it. Even if Burgos (2016) is correct that one could logically hold dualism without falling into mentalism, nothing is gained by such a position, because the inner, subjective world would be superfluous. In practice, the only reason for holding onto dualism is to engage in mentalism. In contrast with realism, either Eastern mysticism or pragmatism is compatible with the ontology suggested here (Barnes-Holmes, 2000; Baum, 1995c, 2017a). The molecular view, derived from reflexes (Skinner, 1938), based on discrete responses and contiguity between events has outlived its usefulness. Viewing behavior as composed of activities, instead of discrete responses, allows us to study them in the laboratory and speak of them in everyday life in a coherent manner. The multiscale view makes ontological sense.

17

Berkeley, Realism, and Dualism

Foreword

In the late Middle Ages, science split off from philosophy, and soon after theology did the same. These splits occurred because the practitioners pursued different interests. Today, philosophy aims for analytical truth, whereas science aims for empirical understanding. Philosophers point out the problems with common terms or concepts, and their critiques are often useful for scientists because they help scientists avoid pitfalls that might render theories meaningless. For example, Gilbert Ryle's (1949) book, *The Concept of Mind,* explains how dualism in the form of the mind–body problem gives rise to meaningless statements about mind and behavior. In Rachlin's books, *Behavior and Mind* (1994) and *The Escape of the Mind* (2014), Ryle's critique translates to a more direct non-dualistic understanding of mental terms. When philosophers attempt to solve the problems with concepts like free will and dualism, however, their "solutions" are less helpful, because they usually involve endless distinctions and hair-splitting. Scientists trying to clarify the ontology and epistemology of science, like the physicist Ernst Mach and the psychologist William James, come under criticism from philosophers, who claim that they are naïve and their views oversimplistic. In a sense, the philosophers are correct, if the writings of these scientists are judged by the philosophers' standards. Those standards, however, do not apply, because of the different aim of science—to achieve adequate understanding of our experience.

The philosopher Max Hocutt (2018) wrote a critique of the previous paper about ontology (Baum, 2017b). As a philosopher, he was offended by my failure to engage in the wordplay that is the business of philosophy. My interest is in clarifying the conceptual requirements of a science of behavior. As a scientist, I soon lose patience with the ruminations of philosophers, but I still find that their careful analysis of terms is sometimes useful. For example, Hocutt (2013) convincingly clarified that "values" are actually preferences, not beliefs, because beliefs are subject to evidence, whereas preferences are matters of taste, and "*De gustibus non est disputandum.*"

In his critique of my paper, Hocutt (2018) had much to say derogating Berkeley, and claimed that some theories—like the heliocentric view of the solar system—are true "beyond a reasonable doubt." He alluded to the complexities that philosophers have cooked up, but failed to come to grips with the practical problems that behavior analysts

Science and Philosophy of Behavior: Selected Papers, First Edition. William M. Baum.

face when trying to construct valid theories and explanations of behavior. How could he? That is not his training. In this response to Hocutt, I try to point out the ways in which he fails to appreciate what scientists do—that is, the business of science.

Abstract

Using Hocutt's vocabulary, I repeat that realism leads inevitably to an unacceptable dualism, because realism distinguishes two categories: real things (material) and perceived things (immaterial). This dichotomy is unacceptable because it creates an unsolvable mystery: we have no way to understand how a thing in one category could affect a thing in the other category. By dividing the subject matter, dualism renders the science incoherent. Someone who asserts that only material things exist (Hobbes, according to Hocutt) is not espousing realism, but monism, and in monism, which assumes only one type of stuff or world, the terms material and immaterial have no meaning. Behavior analysis can advance as the study of behavior in relation to environment, past and present, without having to wait for advances in neurophysiology. Neurophysiology has not yet advanced to the point where behavior analysis can benefit from what it says about the nervous system. Someday neurophysiology may help to understand how the brain participates in behavior, and behavior analysis will be able to tell neurophysiologists what phenomena need to be explained.

Keywords: realism, dualism, perception, monism, George Berkeley

Source: Originally published as "Berkeley, realism, and dualism: Reply to Hocutt's 'George Berkeley resurrected: A commentary on "Baum's Ontology for Behavior Analysis"'," *Behavior and Philosophy*, 46 (2018), pp. 58–62. Reproduced with permission of Cambridge Center for Behavioral Studies.

Sometimes a scientist, a behavior analyst, can learn something from philosophers. Reading philosophy is sometimes rewarding. I learned something from Max Hocutt's (2018) comments on my paper. First, I learned that in speaking of dualism, one does better to substitute "distinct categories" for "separate worlds." I stand corrected. So, dualism consists of distinguishing two categories: material and immaterial or physical and mental. Secondly, I learned that the history of the term "realism" is a muddle, and one must be careful using the word.

I will be brief, because I have discussed these matters elsewhere: in the original article that Hocutt criticizes (Baum, 2017b), in another article on perception (Baum, 2019), and in a book, *Understanding Behaviorism* (Baum, 2017a).

The important issue that Hocutt fails to address about dualism is how the things in these disparate categories could possibly affect one another. Dualism, however one defines it, creates a mystery that is inimical to science, because no known means exists for the mental or immaterial things to affect the physical or material things—or the other way around.

In discourse with philosophers, I often find that, when discussing a problem, different philosophers propose different solutions. For example, if you look at what has been written about free will, you will find that philosophers distinguish several kinds of free will and several kinds of determinism. Dennett (1984), in his book, *Elbow Room*, for example, defines free choice as choice preceded by deliberation. Maybe, but that is not what non-philosophers mean; indeed, it is not even close, because deliberation is behavior that may be determined. If you try to discuss what everybody else believes about free will and determinism, philosophers, having defined all these varieties, criticize you for not distinguishing all the meanings they have attached to the terms. Clearly, philosophers variously define dualism and realism in the same way.

Realism is the strange belief, as Berkeley put it, that the objects we perceive have a real existence apart from their being perceived. It is a belief with no basis, however passionately one may believe it. Try an experiment. Pick an object—say, an apple—you can see it, feel its hardness and roundness, smell it, and bite it and taste it. Can anyone prove that a real apple is really there apart from these perceptions? The answer is categorically "no." That is what I meant when I said that perception is all we have.

What about Hocutt's "unperceived objects"? In discussing the problem of error, he asserts that certain theories are true "beyond reasonable doubt." Atoms, for example, would seem to fall in this category of "unperceived objects." Atomic theory, however, is not so established as Hocutt might think. The recent super collider experiment produced some puzzling results, which physicists are still struggling to reconcile with atomic theory. Atoms are not "unperceived objects." They are theoretical constructs, heuristics that allow us to integrate and make sense of many phenomena in our experience with relative ease. They are useful, but they are not objects and probably are unperceivable, not just unperceived.

The phrase "beyond reasonable doubt" applies in a courtroom, where a jury must make a determination of guilt, but the phrase has no place in science. Hocutt's usage amounts just to name calling, saying that anyone who disagrees is unreasonable. Society and the scientific community have historically believed in theories that we today reject; think of essences, phrenology, and inheritance of acquired characteristics. Search on the internet for "flat earth," and you will find a substantial community exists that believes the earth is flat. The reason Hocutt would find their views unreasonable is just that we have a model—the solar system—that, like atomic theory, allows us to make better sense of more of our experience than the flat-earth model. If we had no alternative with which to compare, we would have to say that the flat-earth model is the best we have at this time. Presumably, Ptolemy would have said, assuming the earth is the center of the universe, his theory was the best that could be had at that time.

Ptolemy probably assumed and at least some flat-earthers assume that the Bible is the literal truth. They reject the theory that the Bible was written and edited by humans, a theory espoused by many biblical scholars. True believers are apt to reject the theory of evolution by natural selection for similar reasons. According to surveys, they make up a substantial minority of Americans. Hocutt, apparently, would say they are unreasonable, but they start with different assumptions from Hocutt, and his name calling won't persuade anyone. Only exposure to the sense-making effects of evolutionary theory will persuade someone who might be willing to examine basic assumptions.

Hocutt's basic assumption is that the world is real, but reasoning from this assumption doesn't help integrate experience as does atomic theory or evolutionary theory. Assuming an idea to be false and then reasoning to the conclusion that the idea is false constitutes a logical error. When Johnson claimed to refute Berkeley by kicking a stone, he committed this logical error. The circularity was Johnson's, not Berkeley's.

Instead of clarity, realism creates unsolvable mysteries like the mind-body problem. This is a pseudo-question, a question that entails a nonsensical assumption. "How many angels can dance on the head of a pin?" is a famous pseudo-question, because it assumes that angels are such things as could dance on the head of a pin. "What happens when an irresistible force meets an immovable object?" is another, because the simultaneous existence of irresistible forces and immovable objects is a logical impossibility. Similarly, "How does the mind affect the body?" is a pseudo-question, because we have no way to understand how things in the immaterial category could possibly affect things in the material category. Like angels on a pin, the mind is not such a thing as could affect behavior. Yet, dualism and realism raise this pseudo-question and create other confusions too, splitting up phenomena into those involving the mind and those involving the body, with the result that the science cannot cohere.

Hocutt objects to my assertion that realism leads inevitably to metaphysical dualism. He cites Hobbes as someone who held to realism but was not a dualist. If someone asserts that only the material is real, the statement has no meaning except in contrast to the immaterial. If Hobbes believed as Hocutt says, then Hobbes held to monism. In monism, only one world or one type of stuff exists, but no warrant exists to call it material or immaterial. It just is.

If we embrace monism, then what should we say about the brain and behavior? Hocutt seems to think that neurophysiology has advanced to the point where it has something significant to say about behavior. I disagree, because as far as I can tell, fMRI studies have only confirmed that when a person behaves one way in one situation and another way in another situation, a difference exists in the brain. Just the other day, I listened on the radio to a prominent neurophysiologist talking about creativity and saying that the brain invents ideas, stores them, evaluates them in the frontal cortex, and puts them into action. This seems to me no different from what would have been said 50 years ago. Such talk is both nonsensical and misleading, because only the whole person can do things like invent, remember, evaluate, and act (Baum, 2018). How the brain participates in behavior remains to be seen.

If, as Hocutt says, materialists hold that mental activity is a "function" of events in a material brain, then he is describing an incoherent view. The word "function" papers over the question of how brain activity could possibly cause mental activity. If mental activity is immaterial, we find ourselves back in the mire of dualistic nonsense. Rachlin (2014), in his book, *Escape of the Mind,* offers a coherent alternative: mental activity is publicly observable behavior extended in time. One thing is sure: consciousness will never be found in the brain. Whatever consciousness is, it cannot be identical to neuronal activity in the brain. A view more like Berkeley's, and compatible with Eastern mysticism is that consciousness *is,* and what we need to understand is what all *this* is about. Put another way, our experiences are a given, and science aims to see what sense we can make of them.

I do not say the brain is irrelevant. A complete understanding of behavior requires us to know how the brain participates—what are the mechanisms of action. Until neurophysiologists have something helpful to say about those mechanisms, however, making stuff up about the workings of the brain is worse than useless, because it gives a phony sense of understanding that discourages the search for valid understanding. I only ignore the brain because we understand so little about how it participates. Behavior analysis need not wait for neurophysiology to proceed, no more than Darwin needed genetics to invent the theory of evolution by natural selection. Knowing genetics is helpful to evolutionary theory now, and someday knowing about the brain will be helpful to behavioral theory, but the study of behavior in relation to the environment will show neurophysiologists just what phenomena they should be trying to explain.

18

What is Suicide?

Foreword

The editor of this online journal, *Animal Sentience*, invited me to comment on an article asking about suicide in nonhuman animals. I took a behaviorist stance toward the question and questioned the empirical grounds for calling any death "suicide." The upshot is that preventing suicide depends on addressing the environmental causes of self-injurious behavior, whether in humans or nonhumans.

Abstract

Whether a person committed suicide is often difficult to determine, and intent particularly so. If it's difficult for humans, how much more so for nonhuman animals. A nonhuman observer would remark that humans usually avoid self-harm, but sometimes engage in self-injurious behavior. If instead of speculating about suicide we focus on self-injurious behavior that is sometimes lethal, we recognize continuity of species and also understand and possibly remedy self-injurious behavior. No need exists to impute doubtful capacities on animals for one to be kind and compassionate toward animals, because kindness and compassion toward humans and other animals benefits the one who practices them.

Keywords: suicide, self-harm, self-injurious behavior, behaviorism

Source: Originally published as "What is Suicide? Comments on 'Can Nonhuman Animals Commit Suicide?' by David Peña-Guzmán" in *Animal Sentience*, 20 (2018).

In an organized society, when a person dies, a coroner or equivalent official determines the cause of death and issues a certificate. Often the coroner easily decides on "natural causes," because the person died in a hospital of disease. Otherwise, the coroner has to decide among accident, murder, and suicide. The decision may be difficult, particularly

Science and Philosophy of Behavior: Selected Papers, First Edition. William M. Baum.

with suicide. A person is found hanging by the neck: is this suicide or murder? A person is found drowned: suicide or accident? A person overdoses on heroin: accident or suicide?

The traditional view of suicide focuses on intent, but questions about intent often find no clear answer. Even when a person leaves a suicide note, if it is posted where someone might see it in time (on a neighbor's door, on the internet), the self-harm might constitute a "cry for help." The Buddhist monk or the Tunisian street vendor who immolates himself may be acting politically, rather than intending to die. The same may be true of "suicide by police." On top of all this ambiguity, suicide is often impulsive, triggered by some momentary event; people who survive jumping off the Golden Gate Bridge report regret on the way down. Finally, complicating the coroner's job will be any stigma attached to suicide, social pressure against calling the cause of death to be suicide. Suicide, it seems, is much in the eye of the observer.

Neither free will nor a concept of death withstands scrutiny in trying to define suicide. Free will is an unsustainable concept that retreats the more we learn about the causes of behavior. Creatures that respond to the loss associated with death cannot be said to have a concept of death rather than a response to loss.

The traditional view of suicide defines it in such a way that it would be a peculiarly human phenomenon. It would imply a discontinuity between humans and other animals that contradicts one of the implications of evolutionary theory, which holds that all traits of a species derive from antecedents in ancestral species. The author correctly brings this so-called "continuity of species" to bear on suicide.

Continuity of species, however, cuts two ways. Just as one may ask whether nonhuman animals commit suicide, one may turn the question around and ask, "What behavior do nonhuman animals engage in that would correspond to what is called 'suicide' in our society?" The author correctly points to self-injurious behavior in animals. Both humans and nonhuman animals engage in self-injurious behavior under some circumstances.

What if, instead of looking at dolphins' self-injurious behavior and likening it to human behavior, we imagine what a dolphin might say when looking at human self-injurious behavior. Better yet, suppose a Martian were looking at both human and dolphin self-injurious behavior. This nonhuman would observe that human creatures and dolphins usually avoid harm to themselves but sometimes engage in self-injurious behavior, sometimes even to the point of death. Lethal self-injurious behavior would be seen as a subcategory of the more general category.

If we are to understand and prevent lethal self-injurious behavior in humans, we need to stop moralizing about suicide and identify the causes of self-injurious behavior in general. Some of the causes are genetic or physiological, but many are environmental, sometimes historical events and sometimes ongoing. Categories include: isolation and idleness; pain and illness; loss and disappointment; torture and bullying; availability of guns and drugs. The list could be extended, but the point is that these are factors that can be remedied and which people can be taught to cope with.

Finally, the question about animal suicide begs the question of compassion and kindness to other beings. As the Buddhists often point out, compassion for others benefits the one who practices it. One's own mental and physical health are elevated by practicing kindness and compassion towards other humans and also to other animals. By doing so, one also avoids the trauma to oneself of committing acts of cruelty, as the effects of participating in warfare amply demonstrate. No need exists to rationalize kindness and compassion by trying to impute abstract notions about "personhood" to animals.

19

Relativity in Hearing and Stimulus Discrimination

Foreword

I wrote this paper in response to Ghiselin's (2018) disagreement with a point I made in my book, *Understanding Behaviorism* (2017a). I cited the Zen *koan*, "If a tree falls in the forest and no one is there to hear it, does it make a sound?" I wrote that the behaviorist's answer has to be "no," because the meaning of "hear a sound" requires someone to act in response to the sound. Ghiselin took issue with this—admittedly counter-cultural—idea and asserted instead that sound consists of vibrations. To be clear not only about Ghiselin's error here but also about notions of a "real" world that invite realism and dualism, I wrote this paper. The deeper point is that all discriminations depend on comparison between an observer's behavior and another organism's behavior.

Abstract

What does it mean to hear a sound? What does it mean to perceive anything? Sound has no objective reality, such as "vibration." Of two people together, one may hear a sound and one may not. We know only that their actions—their judgments—differ. Such comparison underlies all discriminations. In experiments on concept learning, for example, pigeons peck when they are shown a slide containing human beings and don't peck when the slide contains no humans. The experimenters judge beforehand whether the slides contain human beings or not, and the pigeons' concept of human being is determined by the comparison between the experimenters' judgments and the pigeons' pecking or not. Similarly, to tell which of two people is hearing or deaf, an observer that can hear must judge whether their behavior corresponds with the observer's judgments. In an experiment by Lubinski and Thompson (1993), in which pigeons pecked at different keys depending on which of two different drugs they had received beforehand, the experimenters judged which drug had been injected, and the pigeons' pecking corresponded to the experimenters' judgments. If two persons' judgments differ, they can only resolve the difference by deciding that one of them is mistaken. If no one is there to hear a tree fall in the forest, from the point of view of a science of behavior, it made no sound.

Keywords: realism, stimulus discrimination, sound, hearing, behaviorism

Science and Philosophy of Behavior: Selected Papers, First Edition. William M. Baum.

Source: Originally published in *Perspectives on Behavior Science*, 42 (2019), pp. 283–289. Reproduced with permission of Springer Nature.

If a tree falls in the forest, and no one is there to hear it, does it make a sound? (Traditional Zen *koan*.)

Does the tree falling make a sound? Can sound exist apart from hearing? What does it mean to hear a sound? What does it mean to perceive anything? The standard behavior-analytic answer is that an auditory discrimination or any other discrimination consists only of a change in behavior concomitant with a change in stimulation. We find this account in textbooks and the writings of behavior analysts. How adequate is this definition? This paper aims to show that the statement omits something rarely made explicit and often omitted altogether: the relativity in stimulus discrimination.

Appealing to common sense or the culturally received view of the world, the layperson says, "If a tree falls in the forest, of course it makes a sound. Trees falling always make a sound." If, however, no one is there to hear it (including animals and recording devices), how could anyone know the tree falling made a sound? Inductive reasoning is always logically flawed, because no matter how many times an event occurs, you never know if it will happen again. On one hand, a layperson says that the tree falling makes a sound even if no one is there to hear it, but, on the other hand, someone must be there to hear the sound to know for sure. The conundrum comes down to the question, "Does sound exist apart from its being perceived, or does sound exist only in our perceiving it?" Put another way, "Is sound an objective reality—'vibration,' according to Ghiselin (2018)—and our perception of it subjective, or is sound only a perception?" The *koan* aims to question the existence of any such thing as objective reality.

To advance our understanding, we may ask, "What does it mean to hear a sound?" If you hear a sound, the way anyone knows of it is that your behavior changes. For example, Ted and Sue are together, and Ted says, "Did you hear that?" Ted discriminates in the sense that his behavior before hearing the sound differs from his behavior after hearing it. Sue might make no such discrimination and say, "No, I heard nothing," only hearing what Ted said. If Ted heard it, and Sue didn't, how can we know if the sound is real? If sound is vibration, did it not vibrate in Sue's ear too? It may be real for Ted, but it is not real for Sue.

How is it possible for Ted to hear a sound and Sue not to hear it? The layperson might respond that Ted must have been attending to the sound, whereas Sue wasn't. This "explanation," however, is an example of mentalism. It does no more than restate the original observation that Ted's behavior differs from Sue's, because the only evidence for attention is Ted's behavior.

A valid alternative relies directly on the difference in their behavior. Ted says, "It sounded like a crash, like a tree falling in the forest," and he may go into the forest to see if any trees may have fallen lately. His questioning, describing, and investigating *is* his hearing the tree falling. More technically, Ted's behavior induces an observer to say that Ted heard a sound. Thus, to hear a sound means to behave in ways that an observer will recognize as hearing a sound.

The layperson might challenge such relativity and ask, "What is the sound itself?" The question itself is misguided. Just as one cannot speak without saying something or walk without walking somewhere, one cannot hear without hearing a sound. The sound is not something separate—e.g., vibration—but is intrinsic to the hearing. No essence defines sound apart from hearing. Stimuli and the actions they induce have no ontological status apart from one another. To hear a tree falling differs from hearing a bell ring; the behavior—what the hearer might say or do—differs. Even if sometimes instruments measure vibrations that may or may not accompany hearing, on many occasions—e.g., dreaming and hallucinating—no vibration is present. Ted heard the sound, and Sue did not. All we know is that Ted's behavior changed from one time to the next, whereas Sue's behavior did not change.

In a report on concept learning in pigeons, Herrnstein and Loveland (1964; see also Herrnstein, Loveland, & Cable, 1976) made a general point about perception that applies also to hearing. The experiment entailed showing pigeons slides from *National Geographic* magazine and requiring a (hungry) pigeon to peck at a key next to the slide in order to produce a bit of food. Some slides contained one or more human beings, and other slides contained no humans, but both types of slide included the same sorts of settings: scenes from jungles, deserts, farms, and cities. Pecking the key only produced food when a slide contained a human, and the pigeons came to peck almost exclusively when the slide contained a human. Moreover, they did this even with slides they had never seen before. Thus, the pigeons exhibited the concept "human being." Among the questions that may arise is, "Is the pigeons' concept of human the same as a human's concept of human?"

In explaining their procedure, Herrnstein and Loveland said, "For any one session, approximately half the photographs contained at least one human being: the remainder contained no human beings—*in the experimenter's best judgment*" (p. 549–550; emphasis added). The statement reveals that the experiment was, in fact, comparing two sets of judgments: those of a human being and those of a pigeon. If the pigeons classified the slides in the same way as the human, the pigeons had the concept "human being." Herrnstein et al. (1976) followed up with experiments on the concepts tree, water, and a specific person and obtained similar results to the experiment with human being. They conducted exhaustive tests comparing pigeons' difficulties and errors with humans' (the experimenters') and found no notable differences. They concluded, "In summary, from aggregate measures of performance and from an inspection of the pictures themselves, it is clear that the pigeons used principles of classification that approximate those we use ourselves, at least in complexity" p. 290). They considered their results to indicate that, for at least some pigeons, ". . .the pigeons' stimulus classes were essentially isomorphic with ours" (p.296).

These experiments drive home a basic point about stimulus discrimination. In each experiment, the human researchers selected slides: with human beings or without, with trees or without, with water or without, and with the specific person or without. The evidence for the concept arose from comparing the humans' judgments with the pigeons' judgments. This observation, however, holds for any discrimination, no matter whether complex, as in the concept experiments and everyday life, or simple, as in typical experiments on stimulus control.

As a classroom demonstration, I used to bring in a pigeon in an apparatus with a response key and feeder. The key could be transilluminated red or green for a minute or two at a time. The pigeon was trained to peck at the red key and not peck when the key was green. For me and most students, the change of behavior with the change of color was clear, but suppose one student was color blind. The color blind student would be at a loss to explain why the pigeon sometimes pecked and sometimes did not, because that student could not compare the pigeon's changing behavior with judgments of color. Even in such simple discriminations, the experimenter's perceptions, perhaps inadvertently, lay down the standard by which performance is assessed. If a pigeon pecks when the experimenter thinks it should not, the experimenter might call it an "error," but calling it so would depend on the experimenter's comparing his own perception with the pigeon's performance. The relativity of performance to perception never goes away.

Concept learning and color discrimination entail judgments by the researcher and the pigeon in the same sense modality—vision—but nothing requires that the two judgments be based on the same sensory input. Researchers on bird song or echolocation in bats and dolphins compare visual displays with the animals' changing behavior, which discriminates among auditory signals. A researcher studying behavioral thermoregulation in a lizard compares (visual) temperature readings with the lizard's moving in and out of the sun and shade; at that point, how the lizard discriminates temperature remains to be determined.

An experiment by Lubinski and Thompson (1993) offers an instructive example. These researchers trained pigeons to peck at two response keys. Before each session, a pigeon received an injection of one of two different drugs. Pecks at the left key produced food if the injection was Drug A, and pecks at the right key produced food if the injection was Drug B. After a number of sessions, the discrimination formed: pecks went to the left key following injection of Drug A, and pecks went to the right key following injection of Drug B. When they administered a third drug, Drug C, before the session, pecks went to the left key; Drug C produced an effect on behavior similar to Drug A.

Lubinski and Thompson explained their results by asserting that the pigeons were discriminating on the basis of private stimuli produced by the drugs. They supposed that the private stimuli produced by Drug A differed from those produced by Drug B, and that the pigeons' pecking was directed to the left key or the right key by the different private stimuli. To explain the pecking going to the left key with Drug C, they supposed that Drug C must produce private stimuli more like those produced by Drug A than by Drug B, and that the pecking was directed by those similar private stimuli.

Relying as it does on unobserved stimuli within the pigeon, the explanation seems misguided, at least in the context of a science. Speculation about unobserved stimuli adds nothing to the account, because the basis of the discrimination remains unknown. To explain observations, one must not invoke invisible causes. The analogy with a color discrimination would say that the pigeon discriminates red from green on the basis of private stimuli generated inside it by the different colors. Instead, we say that the discrimination is between the red key and the green key, with no appeal to private stimuli necessary. What was the observable basis for the discrimination between the drugs?

Just as the experimenters in the concept experiments sorted the slides, the experimenters in this experiment sorted the drugs. They must have looked at the label on the bottle and judged it to be Drug A, Drug B, or Drug C before injecting it. As in the concept experiments, they compared their judgments with the pigeons' pecking. The similarity between Drug C and Drug A, lay solely in the similarity of the pecking. One need not infer similar private stimuli to conclude that Drug C has effects similar to Drug A, because the effect is the pecking. We may conclude only that the experimenters' discrimination between the drugs corresponded to the pigeons' discrimination between the keys.

The situation of Lubinski and Thompson (1993) resembles that of the researcher studying thermoregulation. The researchers compare changes in their visual signals with changes in the subject's behavior and, finding a match, report a discrimination. The physiology underlying both discriminations remains to be understood.

This is not to say that private stimuli never become public when means are discovered to measure them or that speculation about physiology can never be fruitful. Postulating unobserved private stimuli as causes of observed behavior, however, is little different from postulating mental states that could never be observed; the only difference is the faith that private stimuli, with the right instrumentation, might become public. If, for example, measurement of a pigeon's heart rate indicated that Drug A increased heart rate, whereas Drug B did not, one would begin to have some evidence for the basis of the discrimination.

To return to hearing, Rachlin (2014) discussed the following situation. Two people, Adam and Eve are sitting still in a room while a recording of a Mozart string quartet is playing. Adam can hear, but Eve is deaf. How can one tell the difference? A layperson might say that one of them is hearing the music privately, but private hearing is no answer, because it appeals to an unobservable essence, and, besides, it doesn't tell which one is deaf. The answer is that in this narrow time frame, no one can tell which person is deaf, one can only tell in a wider time frame. Afterwards, Adam may say that he enjoyed the music, whereas Eve may say, "What music?" To tell that Adam can hear and Eve cannot, we need to see adequate samples of their behavior over a period of time. The patterns of their behavior will differ. Rachlin (2014) concluded, "The crucial difference between Adam and Eve is that Adam may do different things when sounds are present than when they are absent, while Eve generally does not do different things in the presence and absence of sounds" (p. 20)

Rachlin omitted from his account one important element: A third person (by implicit assumption, hearing) must be observing Adam and Eve to judge whether they can hear or not. Let's call this observer Mary. Mary witnesses occasions like those shown schematically in Figure 19.1. On one occasion, she hears no sound and observes that Adam and Eve behave the same (X). On another occasion, Mary hears a sound and observes that Adam and Eve behave differently: Adam's behavior changes to Y, but Eve's behavior remains X. After enough such different occasions, Mary will conclude, based on comparing her hearing with Adam's and Eve's across several such occasions, that Adam can hear and Eve is deaf. As in the concept experiments and the experiment by Lubinski and Thompson, the observer's (Mary's) observations are the judging standard.

Taking this line a bit further, we may now suppose that the observer is not Mary, but Jane, who is deaf. What does Jane witness in Adam's and Eve's behavior? As observer, Jane hears no difference between the columns in Figure 19.1; both occasions are equivalent. She, like the color blind student with the pigeon pecking at the red and green key, can only be puzzled by the changes in Adam's behavior from X to Y; they would appear inexplicable.

Now let's suppose we give Jane a microphone attached to an oscilloscope. Even though she cannot hear the sounds on the different occasions, she can now see them. Now she can discriminate silence from sound, but visually. Now she has a judging standard and

	Judged: Silence	**Judged: Sound**
Adam's Behavior	X	Y
Eve's Behavior	X	X

Figure 19.1 The difference between Adam's hearing and Eve's deafness
Note: When the situation is judged by a third-party observer to be silence, Adam's actions and Eve's actions are similar (X), but when the observer judges that sound is present, Adam's actions change (Y) while Eve's do not.

can, like Mary, conclude that Adam can hear and Eve is deaf. Although Jane's standard now is visual instead of auditory, she can compare her judgments with Adam's and Eve's behavior on multiple occasions.

Jane's situation with the visual sounds resembles the situation of Lubinski and Thompson (1993) with their pigeons' discrimination of the drugs. Instead of visual sounds, Lubinski and Thompson had visual cues in the form of the labels on the different bottles of the different drugs. Just as Jane could not hear whatever events accompanied changes in Adam's behavior, Lubinski and Thompson could not feel the pigeons' private feelings, but they did have visual cues that allowed them to compare their judgments to the pigeons' key pecking.

To return to our first example with Ted and Sue, we see that their behavior differs, because Ted behaves in ways that induce Sue or another observer to say that Ted hears the sound but Sue does not. They can resolve the difference in their judgments in one of two ways. Either Ted was mistaken or Sue was mistaken. The resolution depends on their relationship and the different consequences of the different resolutions. Sue may decide that she must have been mistaken because Ted is usually right about such matters, and they may go together into the forest to investigate. Ted may decide that he must have been mistaken, because he doesn't want to bother going into the forest, and whether a tree fell or not doesn't matter. The first resolution is like Ted claiming to have a pain and Sue responding sympathetically (Baum, 2011c; Rachlin, 1985). The second resolution is Ted retracting his claim to have heard a sound and saying he didn't hear anything after all.

Why does this relativity of judgments matter? The short answer is that dualism renders a science of behavior incoherent, because dualism leads to positing unobservable causes (Baum, 2016b). Explicitly acknowledging the comparison of judgments steers us clear of dualism, which is the basis for mentalism. The differences between the judgments of Ted and Sue, Adam and Eve, Mary and Jane, or of a researcher and a pigeon all occur in the same (one) world. No need arises to distinguish material from nonmaterial or objective from subjective. Whether Herrnstein and Loveland (1964) or Lubinski and Thompson (1993), the researchers' judgments were not objective and the pigeons' judgments subjective or the other way around. The judgments exist in the one world, neither subjective not objective.

Conclusion

The Zen *koan* with which we began is a didactic device; it is not intended to have a yes or no answer. In the larger context of Zen teaching, the *koan* is meant to take the student beyond dualistic thinking. Science, too, seeks to go beyond the dualistic thinking of common sense built into a language like English (Baum, 2016b, 2017a, 2017b). It focuses on the one world, the world of our experience. Whereas Zen seeks to go beyond our worldly experience, science seeks to make sense of our worldly experience. Thus a science of behavior seeks to make sense of our experience of behavior—i.e., our observations of organisms' behavior. To understand what it means to hear a sound, we focus on the behavior that we describe as "hearing a sound." To the behavior analyst, the answer to the *koan* has to be "no," because no one was there to hear the tree fall and no one could behave in the way that we would call "hearing a tree fall." That is the account of "hearing a sound." It leaves out nothing necessary. In particular, it leaves out the misleading duality of sound and hearing that common sense offers, but that is a good thing.

Part III

Culture and Evolution

20

Rules, Culture, and Fitness

Foreword

This paper had two aims. First, it expressed my frustration and alarm that so few behavior analysts are familiar with evolutionary theory and that theory and practice in behavior analysis is so rarely informed by evolutionary theory. Second, the paper aimed to set the concept of rule-governed behavior on a solid foundation, both logically and in relation to evolutionary theory, culture, and cultural evolution. This second aim was ambitious; it received a longer treatment in my book, *Understanding Behaviorism.*

Although I would change nothing in the basic arguments, if I were writing the paper in 2022, I would use a different vocabulary and lay more emphasis on the concept of induction as outlined in Papers 7, 10, and 11 of the present book. In 1995 I had only begun to appreciate the integrating potential of induction for theorizing about behavior. I was still using the vocabulary familiar to behavior analysts at the time, the vocabulary of reinforcement. Today, for example, I would say that rule making is induced by correlations in the environment and that rule following is induced by the rule from the rule-giver. The criticisms I directed at current conceptions of rule-governance still seem valid. Although the futility of structural definitions of behavior and stimulus was outlined by Skinner from the 1930s (e.g. 1938, *Behavior of Organisms*), theories often failed to rely on functional definitions. One of the benefits of integrating with evolutionary theory is that all definitions of behavior and environment may be seen as functional, because all behavior serves functions, either proximate or ultimate.

Abstract

Behavior analysis risks intellectual isolation unless it integrates its explanations with evolutionary theory. Rule-governed behavior is an example of a topic that requires an evolutionary perspective for a full understanding. A rule may be defined as a verbal discriminative stimulus produced by the behavior of a speaker under the stimulus control of a long-term contingency between the behavior and fitness. As a discriminative stimulus, the rule strengthens listener behavior that is reinforced in the short run by socially mediated contingencies, but which also enters into the long-term contingency that enhances the listener's fitness. The long-term contingency constitutes the global context for the speaker's giving the rule. When a rule is said to be "internalized," the listener's behavior has switched from short- to long-term control. The fitness-enhancing consequences of

Science and Philosophy of Behavior: Selected Papers, First Edition. William M. Baum.

long-term contingencies are health, resources, relationships, or reproduction. This view ties rules both to evolutionary theory and to culture. Stating a rule is a cultural practice. The practice strengthens, with short-term reinforcement, behavior that usually enhances fitness in the long run. The practice evolves because of its effect on fitness. The standard definition of a rule as a verbal statement that points to a contingency fails to distinguish between a rule and a *bargain* ("If you'll do X, then I'll do Y"), which signifies only a single short-term contingency that provides mutual reinforcement for speaker and listener. In contrast, the giving and following of a rule ("Dress warmly; it's cold outside") can be understood only by reference also to a contingency providing long-term enhancement of the listener's fitness or the fitness of the listener's genes. Such a perspective may change the way both behavior analysts and evolutionary biologists think about rule-governed behavior.

Keywords: rule, rule-governed behavior, culture, fitness, evolutionary theory, rule giving, rule making, rule following, bargain

Source: This paper is gratefully dedicated to my teacher, Richard J. Herrnstein. A version was presented at the Association for Behavior Analysis meeting in Atlanta, May 1994. The author thanks P.N. Hineline, A.S. Kupfer, J.A. Nevin, H. Rachlin, M.E. Vaughan, and G.E. Zuriff for helpful comments on earlier drafts. Originally published in *The Behavior Analyst*, 18 (1995), pp. 1–21. Reproduced with permission of Springer Nature.

It took about 70 years before the biological community completely accepted Darwin's theory of natural selection. From the outset, evolutionists discussed species-typical behavior, but it was only a matter of time before discussion advanced beyond fixed-action patterns. After the synthesis of Darwinian evolution with genetics in the 1930s, biologists talked increasingly about individual organisms' interactions with the environment. In the 1960s, with the rise of behavioral ecology and sociobiology, they began to talk about learned (i.e., operant) behavior and culture.

This progression should impress psychologists in general and behavior analysts in particular. As it continues, the interests of evolutionary biologists overlap more and more with those of anthropologists, sociologists, psychologists, and behavior analysts. What will happen to behavior analysis if the evolutionists never hear about it? Works like Richard Dawkins's *The Selfish Gene* (1989a), widely read and acclaimed by biologists and laypeople, discuss operant behavior and culture with no reference to behavior analysis. It is inevitable that behavior analysis will be integrated into evolutionary biology; the question is how this will occur. One possibility is that evolutionists will raid our treasury of concepts and take what they find useful while inventing or reinventing other concepts for talking about human behavior and culture. If that happens, behavior analysis will become a historical footnote, a short-lived movement within psychology just before the Darwinian revolution took over. Another possibility is that behavior analysts will explicitly build bridges to evolutionary biology and convince the evolutionists that they may profit by listening to us. I have tried to make a start on this in my book, *Understanding Behaviorism: Science, Behavior, and Culture* (1994b).

When Watson founded behaviorism with his 1913 manifesto, "Psychology As the Behaviorist Views It," he tied the study of behavior explicitly to evolutionary theory

(Baum, 1994a). Within 10 years, however, he reverted to the earlier anthropocentric viewpoint that had characterized psychology since its beginnings (Logue, 1978, 1994). Instead of continuing to relate human and animal behavior to evolutionary history as well as to individual experience, he focused on learning and individual experience to the exclusion of phylogeny. As a result, behaviorism is described in introductory psychology textbooks today as radical environmentalism (Todd, 1994). The most often quoted excerpt from Watson is his boast about what he could do with a dozen healthy infants. As long as evolutionists get their impressions of behaviorism from such descriptions as these, they will continue to regard behaviorism and behavior analysis as irrelevant. It is time for behavior analysis to come full circle and recognize that Watson was right the first time. Our choices are to reach out, get on the field and play the game, or remain forever on the sidelines.

Why Tie Behavior Analysis to Evolutionary Theory?

Logically, there is every reason for behavior analysis to align with evolutionary biology. The two fields' interests increasingly focus on the same subject, their common interest in behavior concerns its function in exchange with the environment, and they share the same historical, selectionist mode of explanation. Yet, with a few exceptions (e.g., Glenn, 1991; Petrovich & Gewirtz, 1991; Staddon, 1983), behavior analysts have been slow to integrate evolutionary theory into their discussions of behavior. Perhaps this is due to a lingering anthropocentrism, left over from behavior analysis's roots in psychology. Whatever the reason, the absence of an evolutionary context represents a weakness and a danger for behavior analysis.

I discussed the problem at greater length earlier (Baum, 1994b). In brief, evolutionary biology is making inroads into subject matter traditionally assigned to psychology, topics such as decision making, learning, and, in particular, human social behavior and human culture. The biologists' progress derives from the logic of the Darwinian paradigm, which considers humans and human culture to be just particular instances of more general concerns about how natural selection works. Boyd and Richerson (1985)—a biologist and an anthropologist—adapting a quote from Darwin, argued that "Trying to comprehend human nature without an understanding of human evolution is 'like puzzling at astronomy without mechanics'" (p. 1). It was inevitable that evolutionists would come to be interested in the same subjects as behavior analysts. The only question is how behavior analysts will relate to evolutionary biology.

The danger here is not of being wrong, but of becoming irrelevant. Evolutionary biology will push ahead because it has the weight of the Darwinian revolution behind it. To remain relevant, behavior analysis must discard anthropocentrism and embrace evolutionary biology.

Rules As an Example

In this paper, I shall discuss rules and rule-governed behavior as an example of a topic that requires an evolutionary context for full understanding. Perhaps nothing is so uniquely human as rule-governed behavior. This explains much of its fascination, but also raises the danger of anthropocentrism. When we see rules as uniquely human, we may forget to relate them to principles that apply to behavior in general, principles that derive from evolutionary theory.

Since Skinner's (1969a, 1974) discussions of rule-governed behavior, behavior analysts have written a lot about rules. There is one entire book (S. Hayes, 1989), large portions of other books (e.g., L. Hayes & Chase, 1991) are devoted to the topic, and experimental work appears regularly in the *Journal of the Experimental Analysis of Behavior*. None of this discussion, however, has placed rules in the larger context that Skinner's ideas logically implied: the context of culture and evolution. Glenn's (1988, 1991) discussion of the relations among behavioral, cultural, and biological evolution represents a step in this direction. It makes sense to discuss rules in this context because the rules of a culture are an important part of its practices. The discussion of rules cannot be complete without an account of their place in culture and their origin, like other practices, in a history of evolution, both cultural and genetic.

Some of what I shall say will seem familiar. I shall begin by reviewing some basic concepts about evolutionary theory. Then I shall move to defining rules in a way that places them in a cultural and evolutionary context. In a sense, the whole point will be to define rules in a new way, consonant with evolutionary theory. I shall define a rule according to three criteria. First, a rule is a verbal discriminative stimulus, given by a speaker because of its likely influence on the behavior of a listener. Second, the rule signals a short-term contingency in which some target behavior of the listener may be reinforced by the speaker. Third, the rule is occasioned by a long-term contingency on the same behavior, a contingency with its own discriminative stimuli and long-term consequences that affect the listener's fitness or the fitness of the listener's genes.

Behavior Analysis and Evolutionary Theory

Evolutionary history affects behavior in a variety of ways, acting on the structure of the body by way of genetic evolution (Baum, 1994b). By this means, it provides species-typical patterns of behavior that emerge as a result of normal development. These are important in their own right, but they also make up the raw material of operant behavior (e.g., Segal, 1972). A striking example in human beings is children's early babbling and approximations to speech. Natural selection assures sensory sensitivity to important environmental events, that certain behavior is easily acquired or easily comes under the control of certain stimuli, and so on. To fully understand rule-governed behavior, probably all of these influences need to be considered, because rules are environmental events associated with conspecifics and come easily to control a variety of operant behavior.

To understand the acquisition of rule-governed behavior, we need at the least to remember that natural selection ensures that certain events function as reinforcers and punishers. Reinforcers and punishers that are shared by all the members of a species may be called *species-typical* reinforcers and punishers. They may result simply from normal development, with no need for experience—the smile, for example, is both a fixed-action pattern and a reinforcer—or they may just be so readily acquired that everyone acquires them in the normal course of events; just as ducklings normally imprint on their mother, so human beings normally acquire reinforcers as diverse as candy, physical affection, and vocalizations of approval.

Species-typical reinforcers and punishers, like any traits, are selected by their long-term effects on reproductive success (i.e., fitness). Given that individuals vary and that some of the variation depends on differences in genetic inheritance, the genotypes of reproductively more successful individuals will be represented more frequently in the gene pool of the species. It is often convenient to speak of fitter individuals, but strictly speaking, and whenever there is any conflict between the individual and the genes, fitness is considered to be a property of genes. Genes reproduce more or less successfully by virtue of the individuals in which they occur.

Relations to fitness are nearly always molar, in the sense that they apply on the average and in the long run, rather than in individual instances. This applies to reinforcers and punishers: A reinforcer usually increases fitness, and a punisher usually decreases fitness; these relations hold on the average and in the long run. Usually eating enhances fitness, but an individual organism that eats well but dies without reproducing cannot be said to have had its fitness increased by eating. Like any molar relation, the relation between eating and fitness cannot be observed in individual instances. The genes that provide the basis for eating to function as a reinforcer are selected despite such unlucky individuals, because those individuals possessing those genes, *as a type,* reproduce more often than other types. The same holds for genes that make a smile a reinforcer and for genes that make a frown a punisher. This line of reasoning will be important to our discussion of the long-term consequences of rule-governed behavior and their impact on fitness.

Rules and Fitness

Evolutionary theory offers a way to understand both why people follow rules and why people give rules. We shall take up rule following first, and then tum to rule giving. We begin with an analysis of rule following in familiar terms, in which a rule is seen as a discriminative stimulus (S^D) generated by the (verbal) behavior of the rule-giver (speaker) that increases the likelihood of some operant (rule-governed) behavior on the part of the rule-follower (listener).

Two Contingencies

A rule relates to two contingencies (cf. Cerutti, 1989; Zettle & Hayes, 1982). In one, which is short-term and social, it participates as the discriminative stimulus. I shall call this the *proximate* contingency (cf. Alcock, 1993). The other, which is long-term and ultimately related to fitness, occasions the giving of the rule and involves the same behavior as the proximate contingency. I shall call it the *ultimate* contingency (cf. Alcock, 1993). In technical terms, relative to the proximate contingency, the speaker's giving the rule is a mand; relative to the ultimate contingency, it is a tact (Skinner, 1957).

Figure 20.1 diagrams the two contingencies in symbols and gives an example. Each contingency is represented in the basic form S^D:B→S^R, where S^D represents a discriminative stimulus, B represents operant behavior, S^R represents reinforcement, the colon represents the occasion-setting function of S^D and its effect of increasing the likelihood of B, and the arrow indicates that B makes S^R more likely. The two contingencies share the same B, but each has its own S^D and its own S^R. (Strictly speaking, B is not exactly the same for the two contingencies; it overlaps between them—a point we shall address later.)

The proximate contingency. The rule, the S^D of the proximate contingency, appears at the upper left of the diagrams in Figure 20.1. It is produced by the verbal behavior of the speaker (B_V in Figure 20.3). The listener hears or sees the rule, and if the proximate reinforcement, shown at the upper right, is effective, then the listener's behavior B will be strengthened or maintained—that is, the listener will "follow" the rule. In the example

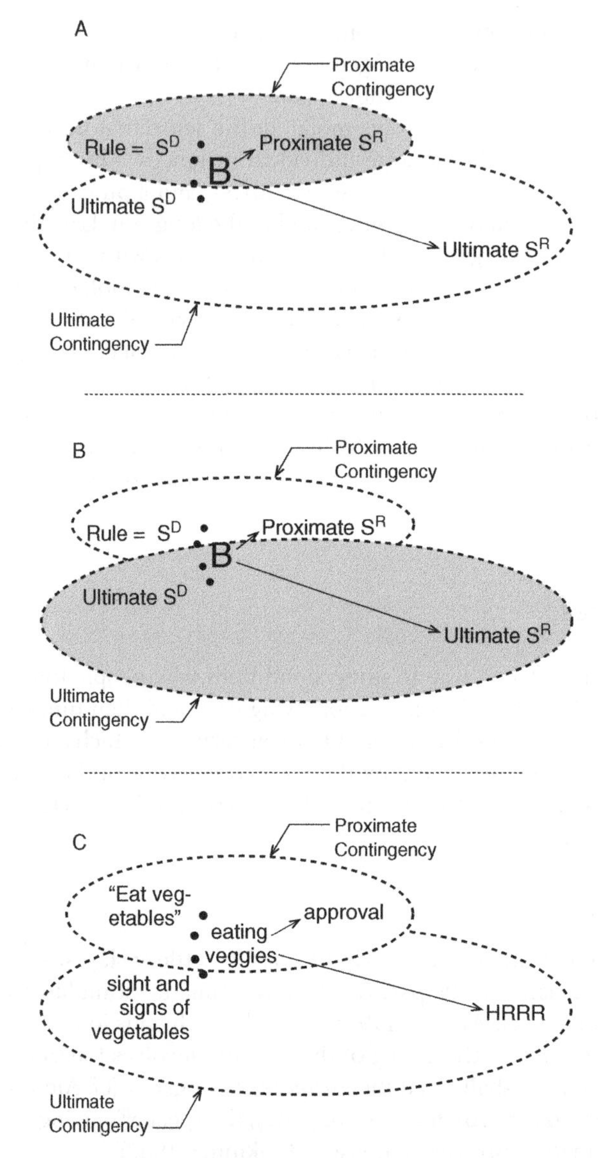

Figure 20.1 The two contingencies of rule-governed behavior
Note: A: The shaded ellipse indicates the proximate contingency. B: The shaded ellipse indicates the ultimate contingency. C: An example, in which the rule (proximate S^D), "eat vegetables," occasions the behavior of eating vegetables; this behavior is reinforced in the short run by approval. Availability of vegetables might ultimately occasion the same behavior, which is ultimately reinforced by increased fitness as a result of improved health, resources, relationships, and reproduction (HRRR). Both contingencies conform to the standard pattern S^D:B $\rightarrow S^R$.

(Figure 20.1C), the speaker (e.g., a parent) says something like, "Eat your vegetables," which is equivalent to "If you eat vegetables, then you will develop properly and remain healthy." If the listener (e.g., a child) eats vegetables, the speaker provides the proximate reinforcer, which may be approval or simply the withholding of disapproval. The effect of this social contingency is to strengthen the eating of vegetables.

The ultimate contingency. The S^D of the ultimate contingency, shown at the lower left of the diagrams in Figure 20.1, may be any of a class of environmental conditions. In the example, this class might be called "sights and signs of vegetables." It includes advertisements, vegetables in the store, and vegetables on the table. The ultimate S^R, shown at the lower right of the diagrams, is a result that usually enhances fitness (reproductive success) in the long run. In the example, avoidance of disease or preservation of health in the long run makes it more likely that the listener will survive long enough and remain healthy enough to reproduce successfully. Eating vegetables enhances the listener's fitness. If the speaker is a parent or relative of the listener, it also enhances the speaker's fitness. The long-term nature of the ultimate contingency is symbolized by the longer arrow from B to the ultimate S^R.

"Internalizing." Figure 20.1 suggests a distinction between control by the proximate contingency and control by the ultimate contingency. When the proximate contingency controls behavior (i.e., when the rule exerts stimulus control and the short-term social reinforcement maintains the behavior), then the behavior constitutes rule following or may be said to be rule governed. If control shifted to the long-term contingency, then the behavior would no longer be rule governed, because the ultimate discriminative stimuli would control (govern) the behavior and the ultimate reinforcement would maintain it. In the example in Figure 20.1C, eating vegetables would come under control of the environmental conditions that indicate their availability and would be maintained by good health.

When one speaks of a rule being "internalized," it means that control over the behavior has shifted from the proximate contingency to the ultimate contingency. Although the mentalistic view has the rule move inside, in the behavior-analytic view the change is from one form of environmental control to another. The behavior remains regular, but has ceased to be rule governed and has switched to long-term control.

Much human behavior starts out as rule-governed behavior and switches to long-term control. This usually occurs when one person instructs another, for example. The shift is likely if the ultimate reinforcer is powerful and relatively conspicuous, like money or opportunities to reproduce. A novice salesperson is told, "Always ask for more than you expect to get." The behavior of negotiating initially is reinforced by supervisors but soon comes under the control of the extra money earned in the long run. A young woman might be instructed, "Don't trust a man unless he is willing to make a commitment to you." After some experience with men, her behavior might come under stimulus control by signs that are correlated with reliability and might be reinforced by actual reliability. Still, nothing has been internalized; control has just shifted to the ultimate contingency.

When control switches from the short to the long term, the behavior may also change. The rule-governed behavior of the novice may resemble later skilled performances, but the two may be far from identical. Indeed, the surest sign of the switch may be the change from more variable, deliberate performances to more coordinated, automatic performances. A new contingency is likely to select new behavior. Although Figure 20.1 suggests that the behavior involved in the proximate and ultimate contingencies is the same, that is a simplification. At least with the shift from novice to master, it would be more accurate to show two overlapping categories of behavior.

The shift to long-term control might be common and often desirable, but it may never occur. Particularly with morals and social conventions, behavior may remain rule governed. This may be because the ultimate contingencies involved are particularly subtle. Someone who was taught to tell the truth early in life may continue to do so until death just because it "seems wrong" to lie. Truth telling might never come under the control of its long-term relation to stable social relationships; one might say it remains *conventional* or *principled* truth telling. Principled (rule-governed) truth telling might overlap substantially with ultimately controlled truth telling—it might be good enough for most purposes—but the two would be distinguishable in some circumstances. For example, a principled truth-teller might tell a painful truth in a situation in which the ultimate truth-teller would prevaricate for the sake of friendship.

Rules in a Cultural Context

When we turn from rule following to rule giving, explanation requires two additional steps. First, we must explain how rule giving and rule following fit into culture. Second, we must explain how the giving of any particular rule becomes a practice of a culture.

Culture and Fitness

I have discussed culture and its relation to fitness elsewhere (Baum, 1994b, chap. 12 and 13), but a brief summary here will help to establish perspective.

From an evolutionary viewpoint, culture consists of behavior that is learned as a result of group membership and is transmitted from member to member within a group. As with other forms of learned behavior, the existence of culture depends on the selection of genes that make it possible and likely. Those genetic variants that make for culture are selected if the existence of something like culture resulted in their enjoying higher fitness than their competitors—that is, if their presence in an individual usually results in that individual's behaving so as to increase the frequency of those genetic variants in future generations. Such a view includes not only individual fitness—an individual's usually producing more surviving offspring as a result of culture—but also genetic fitness that results from altruism toward kin—an individual's usually producing more copies of genes shared with kin by helping kin to produce more surviving offspring. In other words, it includes a person's enjoying better health and more resources—factors that improve that person's reproductive success on the average—and a gene's enjoying better reproductive success on the average (e.g., because a maiden aunt assists her nieces and nephews). Culture exists because it enhances the fitness of the genes that promote it.

Practices. A culture is composed of practices. These include all sorts of manufacture, social conventions, rituals, songs, stories, and sayings. Only as a result of group membership do people learn how to assemble television sets, how to greet a friend, or how to sing the national anthem. Just as genes are transmitted from parent to child, practices are transmitted from one member of a culture to another, though often from people other than parents and to people other than children. We may learn to brush our teeth in childhood from our parents, but we also learn to play games from friends, and as adults we learn skilled performances from instructors.

Only some people within a group need to engage in a practice for it to be part of the group's culture. Even if only some people in the United States tell their children the story of Hansel and Gretel, telling the story is still a practice of our culture. Even if only some people use chopsticks to eat, eating with chopsticks may still be a practice of our culture; it just occurs with a lower frequency than eating with cutlery.

Many of the practices of human culture consist of giving rules. This is particularly true of conventional morals and instructions like "Thou shalt not kill" or "A penny saved is a penny earned." Sommerville (1982) argued that fairy stories play a similar instructive role. Giving advice and giving instructions tailored to specific situations constitute practices. We shall return to rule giving as a practice after we discuss rules in relation to fitness.

Cultural evolution. The practices of a culture change as time goes by. Skinner (1981) and others (Baum, 1994b; Boyd & Richerson, 1985; Dawkins, 1989a) have pointed out that cultural evolution can be explained by a process of selection analogous to natural selection and operant conditioning. Various practices compete with one another within the culture, and variants may coexist for a time or one variant may completely replace another. Watching television at home competes with going to the movie theater, but the two practices may coexist indefinitely. The frequency of driving a horse and buggy, however, has dropped almost to zero in competition with driving automobiles.

Skinner reasoned that cultural variants are selected by their reinforcing and punishing consequences. Reinforcement and punishment, in turn, reflect effects on fitness (Baum, 1994b; Boyd & Richerson, 1985). Yet cultural practices like smoking tobacco often seem to run counter to fitness. How can this be?

Genetic evolution moves at a snail's pace in comparison with cultural evolution. The genetic bases for reinforcers like nicotine, sweets, and aggression were selected in a different environment from the one in which we live now. The environment has changed far too rapidly for any significant genetic evolution to have occurred. As a result, practices arise that afford short-term reinforcement that conflicts with long-term fitness. Cultural evolution continues, however, and new practices evolve that are selected by their long-term effects on fitness. When the ill effects of tobacco smoke become apparent, laws are passed limiting smoking, and anti-smoking educational campaigns are launched. Similar activities occur for aggression and eating sweets.

Educational campaigns illustrate well where rules fit into culture. Exhortations and explanations to "Just say no" and "Eat right for health" are rules. The practice of giving a rule like this exists (is selected) precisely to offset relatively immediate reinforcement and to bring behavior into contact with long-term contingencies that involve fitness more directly.

Comparison with cultural materialism. The views of the anthropologist Marvin Harris (1980), called *cultural materialism,* resemble the behavior-analytic view of culture in defining culture as behavior. In this, they both contrast with traditional anthropological definitions that rely on abstractions like values, ideas, and beliefs. Glenn (1988, 1991) discussed at length the affinity between Harris's view and the behavior-analytic view (see also Malagodi & Jackson, 1989).

Harris (1987; Harris & Ross, 1987) appeals to evolutionary theory to explain cultural practices, just as a behavior analyst might (Baum, 1994b, chap. 13). Where a behavior analyst would refer to contingencies of reinforcement, Harris refers to economic considerations. Although his explanations are couched in different terms, Harris (1980) tries to accomplish within anthropology the same task that behavior analysts try to accomplish within psychology—to offer concrete explanations of behavior instead of explanatory fictions or abstractions that on examination explain nothing.

The correspondence between Harris's views and behavior analysis is strong, but not perfect. Anthropologists like Harris often focus on the institutions of a culture rather than on the behavior of individuals. Whereas anthropologists assign institutional practices to groups, behavior analysts have steadfastly assigned behavior only to individuals. Behavior analysts might argue that, although an institutional practice may be characteristic of a group, its occurrence in the behavior of any individual is explained by contingencies that affect that individual's behavior. Glenn (1988), however, developed the concept of a *metacontingency* to bridge the gap between individual practices and institutions, attempting to capture the "interlocking" contingencies among the members of a group. Some evolutionary biologists have argued that, when groups are well defined and compete with one another, selection among individuals may be overwhelmed by group selection (Wilson & Sober, 1994). The analogous phenomenon for culture would require that different institutional practices compete either within a group or between groups. It remains to be seen whether a concept like metacontingency is necessary. We may speak of manufacturing cars as a cultural practice, but the practice of holding a job (which might be in a car factory) may prove to be a more useful unit. In this spirit, rather than calling a rule a practice, I have been calling *giving* a rule a practice.

An Evolution-Based Definition

According to the view presented here, a *rule* is a verbal discriminative stimulus (a) that sets the occasion for a listener's behavior to be reinforced by short-term consequences arranged by the speaker and (b) that is given by a speaker's behavior under the stimulus control of a long-term contingency connecting that same listener behavior ultimately to the fitness of the listener's genes. For example, the Golden Rule, "Do unto others as you would have them do unto you," is equivalent to "If you are nice to people you meet, then they will tend to be nice to you." This is still abbreviated, because history determines what actions are nice. The two contingencies involved are: (a) If the listener acts nicely, the speaker reinforces that behavior in the short run; and (b) if the listener acts nicely toward others, those others will sometimes reinforce that behavior with helpful responses that lead directly or indirectly to better health, more resources, and opportunities to reproduce. The remainder of this paper will explain this definition further.

One qualification needs to be made right away. If a person's behavior remains rule governed and never shifts to control by the long-term contingency (i.e., remains conventional or principled) one cannot say that person's rule giving is directly under the control of experiences with the ultimate contingency. The speaker may only be repeating an utterance spoken by someone else. Such secondhand conventional rule giving is under stimulus control, not of the original ultimate contingency but of a contingency between obeying the rules of a community and acceptance by the community (vs. disobeying the rules and ostracism). If group acceptance enhances fitness, this qualifies as an ultimate contingency, even if its connection to fitness is indirect (a point we shall take up in the next section). To be perfectly accurate, conventional rule giving should begin with something like, "I have heard... "or "They say...."

Even if a rule spreads like this from person to person, remaining secondhand, still somewhere along the line someone must have given the rule firsthand, under the control of experiences with the original ultimate contingency. Because, in the long run, only rules that enhance fitness are selected, rule giving is ultimately reined in by natural selection (Boyd & Richerson, 1985). Moreover, any listener hearing the rule given secondhand might rediscover its original basis, particularly if he or she fails to comply and suffers dire consequences. Having defied convention by going barefoot in town and having received a nasty cut, the speaker now gives the wearing-shoes-health rule from firsthand experience. Someone who quits smoking because of the disapproval of family and coworkers may subsequently discover an improvement in health and give the smoking-illness rule firsthand.

Rules and Fitness: HRRR

The ultimate contingency affects fitness in the long run because ultimate contingencies involve four sorts of consequences: health, resources, relationships, and reproduction (symbolized in Figure 20.1 by the mnemonic HRRR). An ultimate contingency usually involves more than one of these. Indeed, according to evolutionary theory, in the final analysis all consequences bear ultimately on reproduction. Genes that make for health-promoting mechanisms and practices that promote health are selected because good health usually (on the average and in the long run) makes reproduction more likely.

The connection between reproduction and fitness is only the most direct of the four. Rule giving about mating and mate selection is bound to be common in any culture. We take for granted the reasons behind "It takes real caring to get through the hard times" or "If a couple doesn't do well in bed, they won't do well at all" or "Choose someone who shares your interests." Successful reproduction, including good child care (e.g., "Stop at two"), is nearly always valued.

The connection between resources and fitness, like the connection between health and fitness, is indirect. Having more resources, or having the right resources, usually leads to more successful reproduction. Money, for example, makes one more attractive to a potential mate and makes successful child rearing more likely. When we come to examples like following directions to get to Boston or to assemble a table, the connection to fitness is less direct, but it is still there. Why do you want to get to Boston? Why assemble the table? Chances are, getting to Boston or having the table is connected with achieving status, fostering a relationship, or making money.

The connection between a relationship and fitness can be either direct or indirect. It is most direct in a relationship with a mating partner. Rules in the form of marital counseling are common because of ultimate contingencies between successful marriages and fitness. The connection between a good relationship with a child and fitness is also direct. Reproduction is only successful if one's children are successful; hence the prevalence of rules about parenting.

Rules that promote benevolent behavior toward relatives derive from the connection between helping relatives and the fitness of shared genes. Genes that make for altruism toward kin are selected by the increase in frequency of those genes that results from the reproductive success of kin (Dawkins, 1989b). Caring for one's child helps the fitness of the genes shared by parent and child twice as much as caring for one's niece or nephew. As a result, relatives are favored over nonrelatives in all cultures. The rule "Charity begins at home" reflects this, as do exhortations like "Take care of your brother" and "Be good to your cousins."

The connection to fitness is least direct in a relationship with someone who is neither a mate nor a relative. Because I have discussed such relationships in detail elsewhere (Baum, 1994b, chap. 11), I shall be brief here.

The link to fitness depends on reciprocity. Relationships persist over time because of mutual reinforcement, each person providing reinforcement for the other's actions. This reinforcement takes the form of help with a variety of problems that involve preserving health, gaining resources, and reproducing. A friend may alert you to a new medicine, lend you money, or introduce you to a potential mate. Group membership may be thought of as a relationship between an individual and a group in which conventional behavior, including making sacrifices, is reinforced by other group members who may have no personal relationship with the recipient. The benefits are the same as in more personal relationships.

In summary, every rule is given directly or indirectly under the control of a long-term contingency by which the rule-governed behavior preserves health, gains resources, builds relationships, or affords opportunities to reproduce. The ultimate contingency ultimately enhances the fitness of the listener or the listener's genes.

Fitness and Reinforcement

The relation between rules and fitness explains why the listener's compliance to the rule serves to reinforce the speaker's verbal behavior of giving the rule. A rule is given, as Skinner (1971) wrote, for the "good of others"—that is, their good is the speaker's good in the long run. The long-term benefits to the speaker, as for the listener, are in health, resources, relationships, and reproduction. This holds even if the rule giving itself is initially rule governed. An inexperienced manager may instruct salespeople because some supervisor told him to. An inexperienced parent initially may tell a child to wear a coat because that is what good parents do. After some experience, however, the rule giving shifts to control by the long-term contingency between giving instruction and higher profits or between giving the coat-health rule and the child's continued health.

The long-term reinforcement provides the basis for the listener's compliance to function as a (presumably conditional) reinforcer of the speaker's rule giving. The manager instructs a new salesperson because, in the manager's experience, the salesperson's compliance leads to higher profits for the manager in the long run. Without a long-term payoff, however, the manager would offer no instruction. A child's putting on a coat when it is cold outside affects both the fitness of the child and the fitness of the parent. Unless the child grows into an adult who will reproduce, the parent's investment is lost. A child's illness is aversive both to the child and to the parent. If the child's coat donning shifts to long-term control, it will be maintained by avoidance of chill and illness (Herrnstein, 1969). Why the stimuli corresponding to chill and illness should be aversive, however, must be traced to fitness. Those of our ancestors who treated chill and illness with indifference are no longer represented in the population. Similarly, those of our ancestors who treated the signs of illness in their offspring with indifference are no longer represented in the population. Some signs of illness in a child may require experience on the part of the parent, but obvious symptoms like bleeding, crying, coughing, and lethargy probably require no special experience to render them aversive. If giving the coat-health rule is occasioned by the parent's experience of a contingency between children wearing coats and children showing no signs of illness, then giving the rule is maintained by avoidance of signs of illness.

Rule Giving, Rule Following, and Rule Making

Every culture includes rule giving on the part of speakers and rule following on the part of listeners. Another term, *rule making,* the activity of giving a rule the first time, which is strictly under the control of experience with the ultimate contingency, might be useful as a means for talking about the origins of rules.

Rule giving as a practice. There are two senses in which rule giving may be a cultural practice. First, the giving of a particular rule may be a practice. Every culture includes customary sayings and admonitions like "Blood is thicker than water" and "Fools rush in where angels fear to tread." Like the Golden Rule, these sayings are highly abstract but, with appropriate history, become applicable to specific concrete situations. "Blood is thicker than water," for example, may alter a listener's behavior in seeking a promotion if the competition is the boss's nephew or may strengthen the likelihood of helping out a cousin.

The second sense in which rule giving is a cultural practice is that a culture may include a general practice of instruction, in which relatively specific rules are given in response to concrete situations. Examples include the vegetables- and coat-wearing-health rules discussed earlier and the giving of rules like "If you want to take good photos, you'll have to get a camera with a good lens" or "The best way for you to go into Boston is Route 93." Instead of being customary sayings, these are specific instances of customary helpfulness. Social intercourse is peppered with instances of this tendency to share experiences; it costs the speaker little and may strengthen highly reinforced behavior in the listener. Low cost to the giver and high benefit to the receiver are the conditions under which altruism toward nonrelatives becomes likely and under which people may be said to behave "for the good of others" (Alcock, 1993; Skinner, 1971). The speaker's rule giving is reinforced, but only some of the time on the average and in the long run.

Rule following as a skill. If rule giving may be a generalized cultural practice, rule following is an even more essential cultural skill. Children are taught to follow commands even before they learn to speak. Later they learn to follow instructions and advice, generalizing from parents to teachers and friends. Lost in an unfamiliar area, people unhesitatingly inquire directions of total strangers. These well-brought-up strangers, of course, generally give directions worthy of being followed.

Although the importance of skillful rule following is essential to culture, it tends to be taken for granted. Its importance becomes apparent when it fails. Neglected children miss the training and grow up into adults who fail to do what they are told (i.e., commit crimes and resist instruction).

Origins of rule making. Like any other operant behavior, rule giving requires a special account of its first occurrence. After the giving of a particular rule has been reinforced, its repetition is understood, but where did the rule come from in the first place? To explain the first occurrence of nonverbal behavior, we appeal to induction and response generalization (Segal, 1972). Operant behavior originates either in nonoperant behavior (induced or elicited; e.g., autoshaping) or in operant behavior that is under other control that transfers to a new situation.

Rule making—giving a rule for the first time—arises from other operant behavior. It depends on a history of reinforcement for verbal behavior under the control of regularities in behavior or environment. A statement like "Don't ask Liz questions in the morning; she's always grouchy," a rule in the present sense, occurred for the first time after several morning encounters with Liz, but the very first time it occurs, it depends on a history of reinforcement for such generalizations. This history may go back to a child's early training in naming objects, then in naming events, then in talking about simple relations, and so on. The discriminative stimuli get progressively more complex with further training.

The discriminative stimuli that occasion rule making may be firsthand encounters or stimuli generated as a result of firsthand encounters—i.e., data. For the surgeon general to declare that smoking is bad for health, it may have been unnecessary to have known anyone ill from smoking. It may only have been necessary to have looked at some tables of correlation coefficients. When one is taught to examine data and draw conclusions, one is being trained in rule making. It is good for a culture to include the practice of training citizens in rule making, because from thence arise new practices and more rapid adjustment for the sake of long-term impact on fitness. Long-term effects on fitness offer an evolutionary explanation of the existence of rule giving, rule following, and rule making.

This sort of explanation has been rare in behavior analysis (Glenn, 1991). One may wonder whether it is essential, because evolutionary explanations may seem to affect practice and experimentation little or not at all. Such a view is unlikely to be correct, because both practice and experimentation are affected by the concepts brought to bear in inventing new methods and experiments. Without the concept of genetic inheritance, one would never arrive at discussions of genetic diseases and genetic engineering. Evolutionary thinking has already affected behavior analysis by sparking recognition of species-specific behavioral tendencies that interact in various ways with operant behavior (Segal, 1972; Staddon, 1977). In addition, there are more general reasons to unite behavior analysis with evolutionary theory: intellectual completeness and survival as a discipline.

Verbal Behavior

Our concern with rules requires us to focus on verbal behavior, because we are concerned in part with discriminative stimuli and consequences provided by a speaker for the sake of changing the likelihood of some type of action on the part of a listener.

Speaker and Listener

A rule is generated by the verbal behavior of a speaker. Not all verbal behavior is rule giving, but rule giving is a type of verbal behavior, because it is reinforced by its effects on the listener. We are primarily concerned with auditory rules, but there is no reason a rule cannot be signed, gestured, or written. To say that a rule is a verbal discriminative stimulus is to say only that it is generated by the speaker's verbal behavior and that it exerts stimulus control over the listener's behavior.

Insofar as we are concerned with rule following, we are concerned with the behavior of the listener. The listener behaves in response to the rule, in ways that have been reinforced in the presence of such stimuli in the past. In behaving in the way that we call following the rule, the listener may reinforce the speaker's verbal behavior, but more importantly for present discussion, the listener's rule following is reinforced either by some action or nonaction on the part of the rule-giver. The rule-giver might either show signs of approval or withhold disapproval. When nothing happens as a result of rule following, it usually means that following the rule is avoidance behavior (e.g., of disapproval)—that is, it is behavior maintained by negative reinforcement. Much rule following is enjoined by the threat of punishment, as when the speaker says, "Do this or else."

Rules and Rule-Governed Behavior

The word *rule* has had different meanings in different contexts (Reese, 1989). The vernacular usage differs from the behavior-analytic usage, and the standard behavior-analytic usage differs from the one that evolutionary theory would suggest.

The Standard Behavior-Analytic Definition

Skinner (1969a, 1974) and others (e.g., Blakely & Schlinger, 1987; S. Hayes, 1989) define a rule as a verbal description or specification of a contingency or a verbal statement that points to a contingency. Trouble arises because of the terms "description," "specification," and "points to." Calling a rule a description or specification of a contingency is inaccurate and misleading. It is inaccurate because rules, as they are actually given, often fail to describe or specify a contingency. It is misleading because it implies the concepts of meaning and reference used in traditional mentalistic accounts of language. I shall discuss each of these points in turn.

Rule giving, like any verbal behavior—like any operant behavior—is defined by its function rather than by its structure. A parent might say to a child, "Put on a warm coat so you won't get sick," which would count as a description of a contingency, but the parent might just as well say, "Put on a coat, it's cold outside," or simply, "It's cold out," or might even just open the closet door or gesture in its direction. Because all of these variants achieve the same effect (the child puts on a coat), they are all members of the same operant, the same functional class. Few members of this functional class actually are descriptions of the contingency between wearing a coat and health. Yet they all function as a rule. Saying that they are implicit descriptions of the contingency might solve this problem, but would require an unusual meaning of *implicit,* such as "belonging to the same functional category as explicit descriptions."

Focusing on variants of a rule that have the structure of descriptions may be misleading, however, by suggesting that the meaning of the rule resides somehow in its structure, rather than in the conditions governing the behavior of giving it. In his 1945 paper on psychological terms, Skinner pointed out that the meaning of a term should be understood as the conditions of its occurrence. The point applies to all verbal behavior, including rule giving. Calling a rule a description implies that the rule refers to the contingency it describes and that the speaker has a meaning, intention, or idea in mind when stating the rule. Strictly speaking, rule giving, like any verbal behavior, has no meaning and refers to nothing, because the explanation of its occurrence implies only a certain history, certain establishing conditions, and certain discriminative stimuli (Baum, 1994b, chap. 7). The notions of meaning and reference are simply foreign to behavior analysis.

Stimulus Control of Rule Giving

In the vernacular, rules are often spoken of as things that are learned and known. Mentalistic explanations place the rules inside the organism (Hineline & Wanchisen, 1989). Behavioral explanations instead look to past and present environment. When not placed inside the person, rules in the vernacular are often summaries of regular features of the environment or someone's behavior. Statements like, "As a rule, it starts snowing in December" and "As a rule, I get up at 6:00 in the morning" are based on environmental events, but on no one particular event. The stimulus conditions that control their occurrence extend through time over many particular instances.

Similarly, many instances over a period of time control a statement like, "As a rule, if I leave the trash cans uncovered, raccoons get into them." This example is occasioned not simply by one event being relatively frequent, but by certain combinations of events being relatively frequent, in particular: (a) trash cans uncovered and trash disturbed; and (b) trash cans covered and trash undisturbed. Figure 20.2A illustrates the events involved. Of the four possible behavior-consequence combinations, only two occur frequently. The check marks suggest relative frequency. To the extent that I usually remember to cover the trash cans, the trash is usually undisturbed; those few times when I leave the cans uncovered almost always result in disturbed trash, but occasionally I am lucky and the raccoons don't get into them even though they are uncovered (one check). Technically

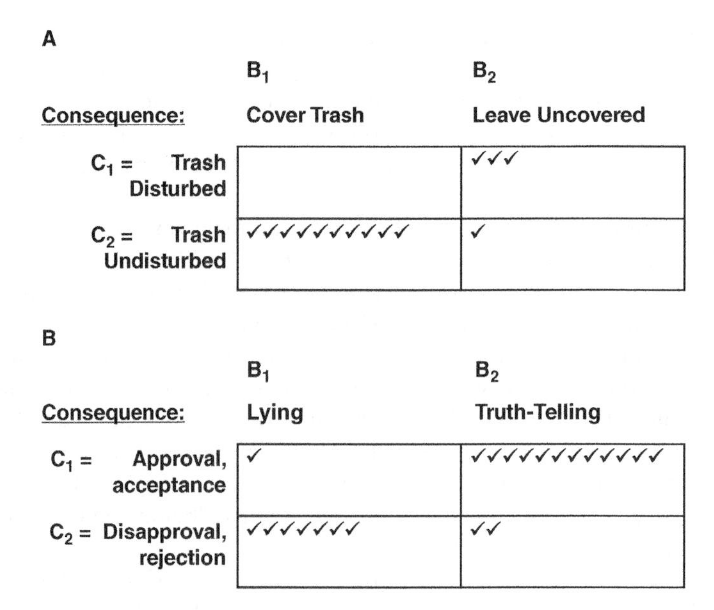

Figure 20.2 Diagrams of experience with two contingencies
Note: Check marks indicate relative frequency of occurrence of four possible events: $B_1 + C_1$, $B_1 + C_2$, $B_2 + C_1$, and $B_2 + C_2$. A: The particular contingency illustrated concerns covering trash cans and raccoons disturbing the trash. B: The particular contingency concerns telling the truth, lying, approval, and disapproval.

speaking, the utterance, "If I leave the trash uncovered, then raccoons get into it," is a tact under the stimulus control of a relation between covered and uncovered trash cans and undisturbed and disturbed trash (i.e., a contingency).

Figure 20.2B shows another, social, example. The contingency is between lying and disapproval, on the one hand, and honesty and approval, on the other. The checks suggest that all four conjunctions may occur, but only two are relatively frequent. Giving the rule, "If you tell the truth, people will generally like you, but if you lie, they will generally dislike you," is under stimulus control of this contingency.

Like any discriminative stimulus, a contingency that serves as a discriminative stimulus must be defined by its function. Just as the effective stimulus may differ from an observer's guess, so an effective contingency, as a stimulus, may differ from "reality" or the ostensible reason given by the speaker. The functional contingency that controls rule giving may be neither veridical nor as described by the rule.

Veridicality. Two sorts of reasons may prevent a contingency from being taken as "reality." First, a contingency consists of a limited set of events. Giving a rule sometimes resembles induction on the basis of a small sample. Early in one's life, it might seem that lying rarely results in punishment, but with more experience, one may conclude just the opposite. On the basis of different (limited) experience, one speaker may say buying stocks is good, whereas another speaker may say it is foolhardy. Second, events in a contingency are unequal in their effects. Instances that include a definite outcome appear to have more of an effect than instances in which behavior occurs with no outcome. In experiments on contingency judgment, people judge contingencies to be stronger when the frequency of outcomes is high, even when there is no behavior-outcome contingency at all (Allan & Jenkins, 1983; Alloy & Abramson, 1979). If they are exposed more to the contingency, however, their judgments tend to become more veridical (Dickinson, Shanks, & Evenden, 1984). Although the rules of a culture may be based on long and reliable experience with contingencies, when experience is limited and open to bias, there is no reason to expect everyday rules to be veridical.

Ostensive contingency versus functional contingency. Even when a rule appears to be a description of a contingency, the functional contingency may differ from that described. The contingency in Figure 20.2B, for example, might occasion a rule like, "If you lie you will go to Hell, but if you tell the truth you will go to Heaven." The functional contingency concerns approval and disapproval by members of the community in the here-and-now, not in the afterlife. In the Middle Ages, a physician might give the rule, "If there is fever, then we must let blood." This fever-blood-letting rule probably had little to do with the health of the patients; more likely it was controlled by a contingency that related blood letting to gaining wealth and status.

A much-discussed example is the Jewish dietary laws (e.g., Harris, 1987). The list of forbidden foods is long and varied: camel, swine, shellfish, birds of prey, animals that creep on their bellies, all insects except grasshoppers, crickets, and locusts, and so on. Individual items on the list have been attributed to considerations of health or economics (Harris, 1987). Another explanation is culturally based: The prohibitions were designed to separate the Israelites from the surrounding tribes. The functional contingency might be that if Jews cannot eat with non-Jews they are less likely to marry non-Jews. Whatever the functional contingency, it differs from the ostensive contingency—that keeping the laws prevents God's anger.

Verbal behavior that is occasioned not by immediate circumstances but by past experience with a contingency, verbal behavior for which experienced contingencies serve as a discriminative stimulus, is particularly important for arriving at a sound behavior-analytic definition of rules. We avoid the problems with the standard definition by recognizing that all rule giving in the behavior-analytic sense is verbal behavior under the control of experienced contingencies. The reverse, however, would be false; although all contingency-controlled verbal behavior might be rule giving in the vernacular sense, not all such verbal behavior is rule giving in the behavior-analytic sense, because something more is required; rule giving is characteristically reinforced by the listener's behavior.

Reinforcement of Rule Giving

The stimulus control of rule giving tells more about the speaker's behavior than the listener's. Discussion of rules, however, invariably revolves around the idea of rule-governed behavior, which is behavior of the listener, and to which the rule relates as a discriminative stimulus. When a listener hears instructions or advice, the listener's behavior changes. "Tum left at the corner to get to the bank" may be a report of a contingency, but it also affects the listener's behavior, and it is given for the sake of its effect on the listener's behavior. That the listener's following the rule is essential may be seen by speakers' behavior when listeners fail to follow advice or instructions. Speakers then behave as if accustomed reinforcement for their behavior had been withheld.

Technically speaking, therefore, the giving of a rule is as much a mand as a tact (Skinner, 1957). It not only is occasioned by the speaker's experience with a contingency, but it has a characteristic reinforcer in the rule following of the listener. The giving of instructions or advice is reinforced by signs of compliance on the part of the listener.

Figure 20.3 illustrates the relations governing the speaker's rule giving. The contingency, indicated as a box, sets the context (S^D) for the rule giving (B_v). This produces

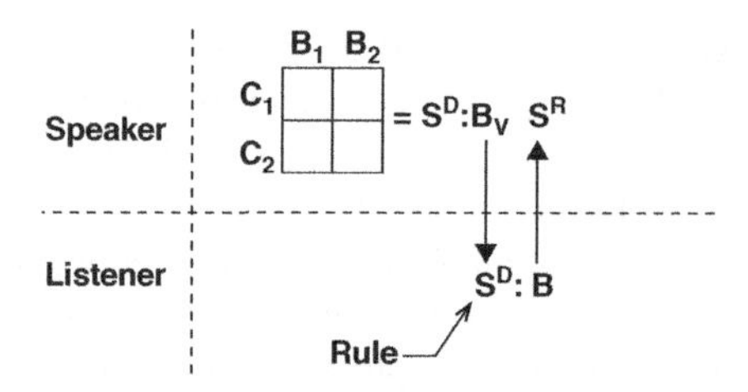

Figure 20.3 Diagram of an interaction in which a speaker's verbal behavior (B_V) generates a rule (S^D) governing the listener's behavior
Note: Experience with a contingency, as in Figure 20.1, serves as the discriminative stimulus for the speaker's rule giving (B_V). The listener's behavior (B) provides reinforcement for the speaker's rule giving.

a discriminative stimulus for the listener's compliance (B), which produces reinforcement for the speaker's rule giving. The listener's compliance also serves as a discriminative stimulus for the speaker to supply reinforcement for it, a point to which we shall return shortly.

We might amend the standard definition of a rule to avoid the problems with "description" and "specification" by saying that a rule is a verbal discriminative stimulus that is functionally equivalent to a contingency statement in the paradigmatic form of "if B, then C," where B is behavior and C is a consequence. We also need to add that the giving of the rule is reinforced by signs that the listener will engage in B and that this common reinforcement defines the phrase *functionally equivalent.* This is good as far as it goes, but it is too inclusive, because we restrict the label *rule* only to some of these if-then statements.

Bargain versus Rule

Some if-then statements govern behavior only on a specific occasion and only in the short run. A price posted in a store constitutes an implicit if-then statement like, "If you will give $1.69, then you may take away this loaf of bread." Another example would be, "If you'll let me look at your comic book, then I'll give you a piece of my donut." To distinguish such verbal discriminative stimuli from rules, I shall call them *bargains.*

In everyday terms, giving a bargain is an offer of short-term mutual benefit. In behavior-analytic terms, it is speaker behavior with characteristic reinforcement—the listener's compliance (i.e., a mand; Skinner, 1957). It has an additional property, however, in that the reinforcement for the listener's compliance also is characteristic. Seen this way, bargains would usually be called promises or threats in everyday usage. For example, "Your money or your life" is a bargain, because it is equivalent to the if-then statement, "If you give me your money, then I will spare your life."

In contrast with giving a bargain, giving a rule is occasioned by a long-term contingency that benefits the listener on no particular instance of compliance, but on the average and in the long run. Wearing a coat in cold weather may afford no immediate benefit to a child, but the parent gives the coat-health rule and reinforces the behavior because wearing a coat is generally good for health. The behavior that is strengthened by the rule may never be directly reinforced; it may remain forever under the control of social contingencies. It may also shift to control by the long-term contingency (i.e., it may become internalized in the sense described earlier).

Although the distinction between bargains and rules might seem clear-cut, it is easy to conceive of ambiguous examples. When a teacher tells a student, "If you do this page of arithmetic, then I will give you a sticker," that might appear to be a bargain. We might also construe it as a rule if we ask why the teacher would say this in the first place. The teacher's behavior is reinforced with money and other social reinforcers, but other people arrange reinforcement for the teacher's behavior because, on the average and in the long run, it is good for the child to complete arithmetic problems and other educational tasks. All the bargains we have considered allow the addition of some long-term benefit to the listener; bread is a resource, and sharing a comic book may cement a friendship.

However fuzzy the distinction between bargain and rule, it remains useful anyway, for two reasons. First, it reminds us of the importance of long-term contingencies to understanding rules. Second, it reminds us of the need to explain the behavior of both the listener and the speaker. The difference between rules and bargains is a matter of degree, rather than an absolute dichotomy, partly because it lies in the degree of importance we attach to long-term contingencies, and partly because it lies in the extent to which there is clear benefit to the speaker. A bargain implies obvious benefit both for the listener's accepting the bargain and for the speaker's offering it, whereas a rule implies benefits that are obvious only for the listener's rule following. That the benefits of the speaker's rule giving are less obvious means only that, when we are done explaining rule following, we still need to explain rule giving.

Comparison with Zettle and Hayes (1982)

Zettle and Hayes (1982) criticized the definition of rules as "contingency-specifying stimuli," arguing that it is "excessively narrow in some respects, and excessively broad in others" (p. 77). They defined rule-governed behavior as "behavior in contact with two sets of contingencies, one of which includes a verbal antecedent" (1982, p. 78). This resembles the present analysis if Figure 20.1 captures the meaning of "in contact." Zettle and Hayes distinguish what they consider to be two types of rule-governed behavior, *pliance* and *tracking*. They define pliance as "rule-governed behavior primarily under the control of apparent speaker-mediated consequences" (1982, p. 80). They offer as an example handing over one's wallet in response to a thief's saying, "Your wallet or your life."

This example and their discussion suggest that Zettle and Hayes (1982) explain pliance entirely by history and short-term contingencies; it requires no reference to long-term contingencies. Their definition remains ambiguous because of the word "primarily." If we focus only on the short-term speaker-mediated contingency (the proximate contingency), their idea of pliance fails to conform to the definition of rule-governed behavior as involving two distinct sets of contingencies. The thief's utterance is only under short-term control. Blakely and Schlinger (1987) interpreted pliance this way, arguing that a listener's response to a verbal discriminative stimulus under short-term control is simply a discriminated operant. On such a view, pliance (or compliance) would fail to qualify as rule-governed behavior. Ambiguity remains, however, because Zettle and Hayes offer other examples (e.g., a parent says, "It's cold outside," and the child puts on a jacket) that would qualify. The distinction Zettle and Hayes sought to make remains useful, however. In present terms, if we consider pliance to be behavior entirely under short-term control, then what they called a *ply* is what I have called a bargain. It is a verbal discriminative stimulus equivalent to an if-then statement involving only speaker-mediated consequences. Requiring no long-term contingency for its explanation, it would not be called a rule.

Zettle and Hayes define tracking as "rule-governed behavior under the control of the apparent correspondence between the rule and the way the world is arranged" (1982,

p. 81). As an example, they offer following directions. Their discussion suggests that they probably intended something like what I am calling simply rule-governed behavior. Ambiguity arises from attributing behavior to an apparent correspondence. In present terms, rule-governed behavior is under the control of the rule given by the speaker. The *speaker's* verbal behavior is under the control of past experiences with the long-term contingency of Figure 20.1, as shown in Figures 20.2 and 20.3. In this sense, the *speaker's* behavior might be said to be under control of an apparent correspondence, but as indicated earlier such a locution raises the problems of reference and intentionality.

Further ambiguity arises because Zettle and Hayes emphasize that in tracking "the speaker does not mediate compliance" (1982, p. 81). True, when a listener follows directions like, "The way to get to Greensboro is to follow I-85," the speaker usually cannot supply consequences, and getting to Greensboro is a consequence itself. These considerations, however, give no basis for distinguishing direction following from other forms of rule following. Direction following originates in a history of speaker-mediated (proximate) consequences; no one would follow directions unless trained to do so, and children usually receive extensive training to follow all sorts of rules, including advice and directions. The origins would make following a particular set of directions rule-governed behavior like any other. That consequences like getting to Greensboro may maintain direction following suggests that direction following, as a general functional class transcending particular sets of instructions, is internalized in the sense discussed earlier. It is possible that Zettle and Hayes intended tracking to refer to a listener's tendency to follow rules even in the absence of speaker-mediated consequences because, in the listener's history, following rules generally paid off.

The Need to Explain Rule Giving

According to the standard behavior-analytic definition, the giving of a rule is explained by its being occasioned by the long-term contingency and reinforced by the listener's compliance. This, however, leaves open the questions of why the long-term contingency should occasion rule giving and why compliance should reinforce it. In addition to its other shortcomings, the standard definition allows no straightforward answer to these questions. What about the coat-health contingency should induce a parent to exhort a child to wear a coat? What good accrues to the parent if the child does wear a coat? The definition suggested here permits an explanation in a larger perspective—the cultural and evolutionary context.

Conclusion

The virtue of defining rules and rule-governed behavior in terms of long-term contingencies of fitness is that it integrates the behavioral analysis with evolutionary thinking. It makes the strength of the behavioral analysis available to evolutionists. At the very least, it offers a richer, more detailed account of how culture and individual human behavior work, but potentially it might change the way evolutionists talk about behavior. For behavior analysts, integration with evolutionary theory affords a large and powerful context within which to understand individual behavior and culture.

21

Being Concrete about Culture and Cultural Evolution

Foreword

This paper attempted to lay out a behavioral approach to culture. It sought to overcome the problem that culture is often defined abstractly as the values and beliefs of a society. The scientific study of culture requires that it be measurable. Behavior is measurable, whereas the usual vague understanding of values and beliefs would prevent measurement. Any attempt to measure values and beliefs requires translation into behavior, even if only to answers questions in a survey.

A key question about culture is, "What are the units?" Many different answers have been offered by biologists and anthropologists, but generally unsuitable for scientific study—unless translated into behavior. Dawkins's (1989a) invented term *meme* supposedly denoted a structure somewhere in the brain. Since the meme is inferred from the behavior it supposedly represents and produces, the notion puts one no further ahead in understanding the units of culture. A meme is nothing if not behavior (see Simon & Baum, 2011, for a full discussion). When the term "meme" entered popular culture, it did so with a telling change: it denoted something concrete—an image, a recording, or a piece of text. This popular usage of "meme" translates to a behavioral definition if we regard the *posting* of the meme online as the activity that would be a practice in a behavioral account of culture.

The present paper seeks to anticipate possible criticisms of taking the practice as the unit of culture. One needs to address how practices vary and how one practice may be selected over another. One needs to distinguish idiosyncratic activities of individuals from practices of the group. One needs to explain how practices are transmitted from one person to another. The paper attempts to address all of these challenges.

Abstract

Culture consists of behavior. The units are practices shared by members of a group and acquired as a result of membership in the group. Although it is common to define the units of culture as abstractions, such as culturgens or memes, these abstractions in no way help to explain a group's practices. Instead, they only direct research away from the context and consequences that result in transmission of a practice from one group member to another. If culture is to be described as an evolutionary process, one must have a sufficiently general definition of "evolutionary process" to allow genetic

Science and Philosophy of Behavior: Selected Papers, First Edition. William M. Baum.

evolution and cultural evolution to be examples. This may be accomplished by defining an evolutionary process as composed of variation (within a pool of replicators), transmission (by copying), and selection (by differential transmission). The replicators of cultural evolution are practices, which are units of operant behavior (i.e. behavior under control of consequences and context). Practices are transmitted from individual to individual by imitation and instruction. Instruction may be understood with the concept of a rule, which is a verbal discriminative stimulus (i.e. a verbally created context). Instruction consists of rule-giving, which results in rule-following on the part of the instructee. New rules come into a culture frequently as a result of rule-making, the generation of new rules on the basis of non-social stimuli. Selection in cultural evolution occurs because imitation and instruction of competing practices have differential consequences in the long run, consequences that impact reproductive potential. Since such long-term consequences have little effect on behavior, an adaptive practice (i.e. one that pays in the long run, called "self-control") is strengthened in the short-term by social reinforcers delivered by rule-givers. A behavioral analysis focuses on environmental events that are observable and, so, susceptible to research. The postulating of fictitious inner entities, whether they be memes, "devices," or "modules," only impedes understanding of culture and cultural evolution. Natural selection, because it is short-sighted and opportunistic, may be expected to produce mechanisms that aid and abet valuable behavioral functions. The key to understanding cultural evolution lies in understanding practices in the light of their environmental contexts and short- and long-term consequences.

Keywords: culture, meme, culturgen, practice, cultural evolution, rule, rule governance, imitation, self-control

Source: The author thanks F. Tonneau, N. Thompson, and R. Hinde for many helpful comments. Originally published in N.S. Thompson and F. Tonneau (eds.), *Perspectives in Ethology: Evolution, Culture, and Behavior* (Vol. 13), pp. 181–212. New York: Kluwer Academic/Plenum, 2000. Reproduced with permission of Springer Nature.

Introduction

The main thesis of this chapter is that culture consists of behavior and that cultural change constitutes an evolutionary process. I will argue that certain notions often advanced about culture are incorrect and misleading, because they divert attention away from cultural behavior. The units that make up a culture and pass from one individual to another within a cultural group or society are behavioral, and defining them as behavioral gives us a better picture of how cultural evolution works.

To understand cultural change as an evolutionary process, we need to understand cultural variation, transmission, and selection. I shall first take up cultural units, how they vary, and how to define them in a way consistent with behavioral definitions of culture. I shall then discuss the means of cultural transmission—imitation and instruction—pursuing a behavioral interpretation of instruction as transmission based on rules. Finally, I shall propose that operant reinforcement (Honig & Staddon, 1977) constitutes the means of selection in cultural evolution.

Cultural Units

Several authors have written about the units of culture. Dawkins (1989a) called them "memes." Lumsden and Wilson (1981) and Pulliam (1983) preferred the name "culturgen." Boyd and Richerson (1985) used no new term, sticking to words like "phenotype," "variant," or "behavior," while also asserting, consistent with traditional social-science approaches, that culture consists of "information capable of affecting individuals' phenotypes" (p. 33).

Each of these definitions identifies the cultural unit with an abstraction possessing no specific properties. Boyd and Richerson's definition raises the question of just what "information" might be or where it might reside and how it might be "capable of affecting individuals' phenotypes." Pulliam (1983) writes, "A representation of each culturgen is stored in each individual's long-term memory. . ." (p. 428), apparently oblivious to questions about how and why such storage would occur or how and why something stored might be retrieved. Dawkins's (1989a) meme, likewise, is a representation, "a unit of information residing in a brain" (Dawkins, 1982, p. 109) and transmitted from brain to brain.

Memes and other abstractions

These conceptions of cultural units raise at least four interrelated problems. The first one is superfluity. For example, Dawkins's (1989a) meme is, for all practical purposes, redundant with the behavior it is supposed to cause. One observes that a behavioral pattern is imitated or instructed, and then that it repeats. The notion that the organism has acquired a meme from the imitation or instruction adds nothing to what has been observed, except the vague implication that the brain is involved. If Richard occasionally sings *Auld Lang Syne*, the meme is inferred from the observation that the song is sung, which is also the only evidence for the meme. It contrasts with the concept of the gene, which, although equally hypothetical at its inception, still summarized facts about heredity, such as: that the transfer occurs at conception, that some types dominate others, and that the recessive types nevertheless reappear in the next generation. The meme possesses no such explanatory power; instead it is the entity that must reside in the brain in principle, just because Richard sings *Auld Lang Syne*.

If the meme is merely redundant with the behavior pattern it is supposed to cause, and therefore is unnecessary, does that do any harm? It does harm because it distracts researchers from studying the behavior itself. Dawkins imagines memes competing in the brain, in the way of "early replicating molecules, floating chaotically free in the primeval soup" (1989a, p. 196). As a result, he concludes prematurely that memes have no alleles. If instead we assume that what repeats is the singing of *Auld Lang Syne*, sometimes with errors, then other songs may compete with it in the behavior of an individual, as well as across individuals. Those competing songs are analogous to alleles, because they perform the same function (call it social bonding) as *Auld Lang Syne*. In general, two competing patterns of behavior may themselves—not some hypothetical entities in the brain—be considered alleles if they are both ways of achieving the same end or function—if there is more than one way to "skin the cat." If the term "meme" referred to a function (getting a job done) in our culture, such as social bonding or looking attractive, then it would usefully associate different ways of bonding (singing together, drinking together, etc.) or different ways of looking attractive (wearing stiletto heels with a mini-skirt, wearing a sarong with sandals, etc.) as competing variants or alleles.

A second objection may be called "mysterious workings" (see Baum, 1994b). Like any supposed representation, the meme as Dawkins conceives it is continuously present in the brain. Since the behavior it is supposed to cause occurs only occasionally, how does the meme cause the behavior? The situation is reminiscent of the notion that the rules of grammar somehow exist in the brain and cause utterances. The rules may determine some of the structure of the utterance, but they tell us nothing about the substance of the utterance or why it occurs at all. Similarly, the meme only represents the structure of a behavioral pattern; it tells us nothing about *when* and *why* the pattern occurs.

The notion of meme as an inner representation begs most of the questions that we need to answer if we are to explain the actual occurrence of cultural behavior. Why does this meme get stored and not some other? Why is it expressed in behavior only some of the time, and why on just the right occasion? These are really questions about the effects of past and present environment on behavior. The explanation of the behavior requires that we know its consequences, the context in which it is "appropriate" (i.e., the context in which it may be rewarded, for example), and the history by which that context came to control that behavior. Knowing the way behavior is "represented" in the brain might help, but it cannot substitute for an understanding of the way behavior is selected and facilitated by the environment.

A third problem is the falsity of the assumption that repetition requires representation. When behavior repeats from time to time, one may be tempted to assume that the behavior must be represented somewhere in the nervous system (e.g., Dawkins, 1989a). Yet other natural events repeat—sunrise, sunset, the seasons, elections, and taxes. Every year during September and October, tropical storms arise in the Caribbean, yet no one suggests that these events are represented somewhere in the water or the air. When the conditions recur, the events recur. Why should it be any different with behavior? Every morning I brush my teeth; when the conditions recur, the behavior recurs. If Richard is at a gathering on New Year's Eve, his singing of *Auld Lang Syne* recurs. Ultimately, of course, the full explanation of both behavioral occurrences requires us to attend to the history of my tooth brushing and Richard's singing of *Auld Lang Syne*—how it was taught and how environmental cues came to control it. The relevance of past events in no way necessitates memory as a stored representation (see Turvey, 1977; Watkins, 1990); behavioral patterns are probably more like storms than playback of recordings.

The influence of past events on present behavior doubtless explains much of the temptation to posit inner representations. People prefer to have their causes in the present, even if that means appealing to invisible causes. Memes as inner representations, however, create more problems than they solve, and raise pseudo-questions. If a representation is stored, then how is it retrieved? Who retrieves it, and why? The instruction, training, or experience that would explain the presence of the meme in the brain also constitutes an adequate explanation of the behavior, without the meme. The repeated circumstances that would explain the activation of the meme also constitute an adequate explanation of the repetition of the behavior, without the meme.

The compulsion to imagine inner representations also seems ironic. As genes are related to phenotype, so memes are supposed to be related to phenotype. Genes, however, unlike memes, cannot be said to represent phenotypic traits. Although DNA may be poetically referred to as the "blueprint of life," the most genes may be said to encode are the sequences of amino acids in various proteins (Dawkins, 1982). Beyond that, development is a complex sequence controlled by gene products and environment. Why should memes be different?

Which brings us to a fourth problem: the disanalogy between genes and memes. We understand that a gene in a parent is replicated in a zygote and then influences development of the offspring's phenotype. How would a meme be replicated? The environmental events—stimuli and consequences from other people—that would replicate the meme are the very same ones that would replicate the behavior. Understanding how the meme or the behavioral pattern is replicated requires an analysis of the same environmental influences. If Richard teaches a friend how to sing *Auld Lang Syne*, the song is now replicated in this friend, and Richard's cues and consequences explain the presence of the song in the friend's behavioral repertoire. More proximate stimuli and reinforcers will explain its occurrence thereafter on particular occasions (e.g., other people singing at a party and their approval in the form of friendly contact). Instead of a clear causal relationship like that between gene and phenotype, we have a sort of a muddle, in which the acquisition of the meme parallels the acquisition of the behavior and raises a chicken-and-egg problem.

These four problems also inhere in Pulliam's "culturgen" (1983) and in Boyd and Richerson's "information" (1985), with the exception that "culturgen" and "meme" could be redefined to refer to patterns of behavior, whereas "information" remains irremediably abstract. These criticisms, however, need to be offset by constructive proposals, focusing on concrete, observable occurrences such as behavior and its environmental context. The proposal outlined here relies on principles of behavior analysis.

Proposal

To develop a coherent account of cultural evolution, we need first to define the concept of evolutionary process in a way that is general enough to apply to both genetic evolution and cultural evolution, in the spirit of Dawkins's proposal in the *Selfish gene* (1989a). We may assume that any evolutionary process includes the elements of transmission, variation, and selection. The notion of replicator embodies the element of transmission—persistence through time as a result of copying. The notion of a pool of replicators embodies variation: within the pool, replicators that perform the same function, but in different ways. Two alleles, for example, both of which affect coat color, may differ in the color of coat they promote. Selection occurs when some of the variations in replicators produce varied consequences in the environment that in turn result in varied copying success. It might seem a bit odd to put better predator avoidance or better resource acquisition in the environment, but they belong there because they are effects in the environment, along with a better nest, better parental care, and better manipulation of a host (see Dawkins, 1982). These environmental consequences feed back to affect the copying success of the replicators that promote them. Better consequences mean better copying success (cf. Baum, 1973). Differential consequences mean differential copying success, which is selection.

Genetic evolution is well understood as an evolutionary process. To understand cultural evolution as an evolutionary process, we need to answer the same questions already answered for genetic evolution. What are its replicators? How are they transmitted? What is the pool of cultural replicators? How do the various replicators compete? And what is the principle of selection among variants?

In this chapter I will argue that the analogue to genotype in cultural evolution is behavior itself. This will take some getting used to by evolutionary biologists, who are accustomed always to think of behavior as phenotype. If, however, we ask what actually gets replicated, what actually passes from one person to another, arguably it is behavior itself. The units of behavior that get replicated or passed along, I shall call *practices*. A practice like vegetarianism, for example, satisfies all of Dawkins's (1982) requirements for an "active germ-line replicator;" it is copied and promotes its own copying through its consequences in health and social affiliation. Calling the units practices helps to distinguish them from idiosyncratic patterns acquired by individuals and unavailable for replication (what Dawkins, 1982, calls "dead-end replicators"). With Boyd and Richerson (1985), I include in culture all behavior patterns acquired as a result of membership in a group and exclude those acquired in other ways.

The pool of replicators, analogous to the gene pool, is the pool of practices that occur in the group and constitutes the group's culture. Another useful analogy might be to the genome, because practices, like genes, are interdependent. Selection may operate on clusters of genes or even the entire genome (Dawkins, 1982). The analog to the genome would be all the practices that occur in an individual's repertoire at a certain point in time (actually, a period long enough to be sampled, but short enough to be considered stable; see Baum, 1994b).

Other analogies may be more difficult to draw. If practices are analogous to genes, then what is the cultural analog to phenotype? Perhaps it would be the culture's artifacts: art, technology, costumes, books, music, and so on. These are associated with a culture, but cannot be considered part of it. Rather, they are produced by practices of the culture, as a phenotype is produced by a genotype. A certain type of bowl may be distinctive of a certain culture, but the practice of making such bowls is part of the culture itself. Manufacturing practices produce artifacts like television sets, which in turn facilitate selection of other practices, such as entertainment, just as strong legs facilitate selection of patterns of capturing prey that entail running.

The parallels end there, however, because the relationship between phenotype and Dawkins's "vehicle" may be absent in culture. It seems inaccurate to say the artifacts of a culture carry the practices as a vehicle carries replicators. Rather, it seems that the vehicle of cultural evolution would usually be the individual, just as in genetic evolution. (I say the individual rather than the group, because it would be an individual, for example, who would carry a practice from one group to another.)

In summary, in genetic evolution behavior is regarded as phenotypic, influenced by but distinct from the underlying genes—the replicators, the entities actually copied and transmitted. In cultural evolution, however, behavior itself is copied and transmitted; the replicators (the analogs to genes) are themselves behavioral units. When one person imitates another with good results, the behavior is replicated in the imitator, just as a parent's gene is replicated in a child.

Behavioral units

B. F. Skinner (1981) suggested that the unit of culture is a contingency of reinforcement (a dependency of reinforcement on behavior in a certain context), one that is characteristic of the group and is social, in the sense that one individual arranges consequences for the behavior of another. In contrast with the meme or culturgen, a contingency exists in the individual's environment. The context-behavior-consequence combination that constitutes a contingency is thus available for direct scientific study. As a unit, the social contingency has much in common with traditional views that define culture as a collection of transmitted values, because, Skinner (1971) argued, the behavior patterns that a group calls "good" consist of those that the group usually reinforces, and the patterns that a group calls "bad" consist of those that the group usually punishes (see Zuriff, 1987, and Garrett, 1987, for discussion of the ethical issues involved).

Skinner's concept of contingency contrasts also with meme and culturgen in that it points to the instructional practices of the group, rather than to the behavior instructed. It relates a religious practice, for example, to the reinforcement that strengthens and maintains it, to the punishment for alternative practices, and to the context, particularly accompanying verbal explanations, in which it is reinforced and not punished. In such a view, every cultural practice is tied to the instruction that produces it, because if the unit of culture consists in a social contingency, then defining any particular unit requires specifying a social context and socially mediated consequences. In Skinner's view, the analog to the genotype-phenotype distinction would be the distinction between the instructional practices of the group, which like genes might be said to "promote" the practices they shape, and the practices shaped by those instructional practices, which like phenotypic traits develop from the interaction between instruction and other environmental influences.

Which units should be included in a definition of culture depends on how broad the definition should be. Whereas some definitions, such as Skinner's (1981), aim primarily at elucidating human culture (Glenn, 1991; Harris, 1980), others potentially include cultural phenomena in non-human species (e.g, Boyd & Richerson, 1985). A definition broader than Skinner's would embrace not only instructional practices but also behavior acquired by imitation. For instance, an inclusive definition equates culture with learned behavior acquired as a result of group membership (Boyd & Richerson, 1985). That would include the example of sweet-potato washing spreading through a group of monkeys, which involves imitation but presumably no instruction.

Transmission

Although behavior may be transmitted also by imitation, Skinner's definition relies solely on instruction. Distinguishing between the two may be crucial. The instructor both creates the context and supplies reinforcement for a cultural practice, whereas the model imitated provides only the context, and the non-social environment, or at least someone other than the modeler, supplies reinforcement. Washing sweet potatoes removes dirt and perhaps adds taste; the monkey being imitated has no influence on these consequences. Imitation induces the potato washing, but reinforcement of the behavior in no way depends on behavior of the individual imitated. In contrast, an instructor may model behavior, or induce it in other ways, and having induced it, reinforces it. A parent might show a child how to wash an apple and tell the child to wash an apple before eating it, but the social interaction cannot stop there, because the parent must praise the child for compliance, or at least withhold the punishment that he or she would administer for non-compliance. Of course, washing apples has non-social consequences also, such as long-term health, but these are usually too deferred to establish apple-washing (see below). Thus a more immediate reinforcer is required. The instructor supplies it.

The inclusive definition of culture (Boyd & Richerson, 1985) has the virtue that it allows for culture and its rudiments in animals, but has the weakness of glossing over the distinction between instruction and imitation. Some cultures may indeed be "imitation-only," whereas others also include instruction (Baum, 1994b). Whether instructional practices occur in the cultures of species other than humans is an open empirical question. The inclusive definition, applied to human culture, reminds us that some behavior may be acquired by group membership without the need for socially mediated consequences, because the non-social environment supplies relatively immediate consequences. When one person eats what he sees another person eating, more often than not he has a good gustatory experience. If he eats the food and gets sick or discovers that it tastes horrible, he avoids eating it again. Even in this imitation-only acquisition, imitation cannot transmit behavior by itself, however; rather it induces behavior that persists only if it is reinforced. In instruction, too, the induced behavior persists only if it is reinforced, but instruction includes other means of inducing reinforceable behavior besides imitation, and the reinforcement is social.

In contrast with imitation-only acquisition, much behavior in humans is acquired as a result of contingencies in which both the context and the consequences are mediated socially. Imitation-only acquisition fails when the non-social consequences are too deferred to strengthen behavior in the short term. Learning to sing a traditional song along with a group of associates is reinforced immediately by signs of approval and acceptance (smiles, etc.), but the long-term advantages of affiliation, cashed out in the currency of reproductive success, are the key to understanding the existence of such reinforcement practices (that is, such short-term cultural contingencies).

Cultural contingencies

To understand exactly what is meant by a contingency and to introduce the technical terms involved, let us consider an example. Herrnstein and Loveland (1964) trained pigeons in a discrimination that required them to peck at a response key if a picture contained a human being. The pigeon faced a panel on which the response key, operated by pecks, was mounted next to a small screen on which slides were projected from behind. Below, an opening to an electrically operated hopper allowed the pigeon access to grain for a few seconds. Slides were obtained from *National Geographic* magazine, and showed scenes from all over the world (jungles, deserts, cities, mountains), only some of which contained people (of all ethnic groups, dressed or undressed, old or young, in groups or alone). A slide was shown to the pigeon for a brief period, and if the slide contained a person according to the judgment of the experimenters, pecks at the key occasionally operated the hopper, allowing the pigeon to feed. If the slide contained no person, pecks were ineffective. After a few sessions with forty positive and forty negative instances shown in random order, the pigeons were pecking more often in the presence of the slides with people than the slides without. When the pigeons were well trained on the fixed set of slides, Herrnstein and Loveland began showing slides that the birds had never encountered before. The pigeons continued to discriminate between the slides with people and those without, even though every slide was new. Herrnstein and Loveland called the resulting discrimination a "concept," but we need only to consider a person in the slide as the context in which pecking was reinforced. Technically, a person in the slide would be called a *discriminative stimulus* (S^D); here I am using the word "context" as synonymous with "discriminative stimulus."

The example of "person" illustrates an important point about the context or S^D: It is a class of events, rather than a unique event. Although we have no idea exactly what cues distinguish person from non-person, the important point is that all the instances of person are functionally equivalent. They all signal the possibility of reinforcement, and they all have the same effect on behavior. It is well-known that such equivalence can be trained, even in pigeons (Vaughan, 1988; Wasserman, DeVolder, & Coppage, 1992). Had the example been a discrimination between a green light and a red light, the generic nature of the stimulus might have been less obvious, but even "a green light" varies from one occasion to another—it is viewed from various angles, ambient light may change, and line voltage may change. The stimulus always is a class of functionally equivalent events (Skinner, 1935/1961).

All three components (context, behavior, and reinforcement) of the three-term contingency have this generic character. One cannot sing a song the same way twice; "singing *Auld Lang Syne*" names a class of events, not a unique event. Even a pigeon's key peck constitutes such a class, because the pigeon never pecks the key exactly the same way twice. Operant behavior patterns in general and cultural practices in particular are best viewed as classes defined by their environmental effects—by the "job" they "get done" (Guerin, 1997). Singing *Auld Lang Syne* may be a sub-category of the larger class "singing with a group," which in turn might be a sub-category of the class "affiliating with a group," which might often be the most useful level at which to characterize the behavior (Baum, 1995b, 1997; Rachlin, 1994).

As the stimulus and behavior are generic and hierarchical, so too are the consequences. Even food delivery constitutes a class of events—the same food cannot occur twice. Smiles, nods, and vocalizations may all belong to the category "approval." It is probably of little importance whether some of these are fixed-action patterns and others more obviously modified by experience.

All three terms have also the property of temporal extension. The ways members of a group make pots, build houses, or adorn their bodies constitute practices characteristic of the group because they are prompted and reinforced by the group, even though no one could say these practices occur at any particular moment. Making a pot, building a house, or adorning one's body takes time. Similarly, the context and consequences of such practices cannot be said to occur at particular moments, but rather extend through time. The context for building a house in a certain way stretches over years and includes many events of instructing, modeling, and requesting, as well as presence of materials, tools, and tenants. The praise and appreciation one may receive for adorning one's body occurs intermittently and repeatedly, possibly over days, weeks, or years (Baum, 1997; Rachlin, 1994).

Similarly, in situations of instruction, in which the instructor supplies both the context and the reinforcement for appropriate behavior on the part of the instructee, the instructions, behavior, and consequences all are generic and temporally extended. The instructor may set the context by saying (and showing), "This is the way to shape the base of the pot," or "Do it like this," or simply "Look," and these are all equivalent in their effects on the behavior of the instructee. The instructor may reinforce by supplying a pat on the back, a vocalization like "Well done," or by withholding punishment. The context is set and re-set, in a variety of ways, again and again through time, and the consequences are supplied and omitted in a variety of ways, again and again through time.

Although the concept of three-term contingency allows us to understand acquisition of culturally appropriate (i.e., reinforced) behavior as a result of instruction, we still need to understand the behavior of the instructor. At a global level, instruction itself is a practice of the culture and therefore may be instructed. In the short term, the appropriate behavior on the part of the instructee reinforces the behavior of the instructor. Viewed this way, instruction would coincide with verbal behavior in Skinner's (1957) sense, because the instructor would be seen as a speaker whose behavior is reinforced by the behavior of the listener (Skinner, 1957; Baum, 1994b). The common practice of third parties' reinforcing instruction with money and goods reminds us also that the instructor's behavior is often valuable to the whole group. Why instruction occurs in the first place, however, and how it is valuable to the instructor and the group, cannot be explained just in terms of relatively short-term considerations. Evolutionary theory demands an ultimate explanation tied to fitness (i.e., gene replication).

Genes and culture

For consistency with notions of selfish genes and extended phenotypes, we need to suppose that culture confers a benefit on genes that promote it. It performs a function that gives those genes an advantage over their alleles. But how could genes cause culture? And what function could culture perform that would feed back differentially to select genes for culture?

Genes may cause culture by promoting three phenotypic effects: sensory specializations, imitation, and social reinforcers. Sensory specialization is nothing new; it occurs throughout the animal kingdom to tune behavioral responses particularly to cues from conspecifics, prey, and other important features of the environment. Individuals typically respond only to mating calls and other courtship displays characteristic of their species, for example. No one should be surprised that humans' auditory systems are tuned particularly to speech sounds, that infants come into the world making phonetic distinctions, and that visual, auditory, and chemical cues from one person may cause physiological responses in another. Sensitivity to some cues generated by the behavior of other humans may be inherited from our pre-hominid past. Sensitivity to speech must have coincided with the development of verbal behavior.

Imitation may be viewed as an adaptation. Its occurrence extends beyond humans and primates, even to pigeons (Epstein, 1984). Boyd and Richerson (1985) argue that imitation confers an advantage as a shortcut to learning. In a species with relatively stable social groups, an infant or an immigrant who imitates will quickly learn to behave as the others do. If the behavior acquired by imitating group members is also adaptive within the particular environmental conditions, then genes that promote imitation will be selected over the alleles that fail to promote imitation. Correct foods will be eaten, predators and other dangers will be avoided, and so on. Imitation plus stable social groups makes for at least the rudiments of culture in other species besides humans.

The third phenotypic effect, social reinforcement, is the key to instruction. Without social reinforcers, instruction would be difficult, if not impossible. They have the great advantages over tangible reinforcers like food of being constantly available, immediately presentable, and slow to satiate. They allow rapid interchange between instructor and instructee, resulting in rapid shaping of the instructee's behavior. We may suppose that the primary social reinforcers, such as smiles and exclamations, being fixed-action patterns, have a genetic basis. More importantly, the susceptibility to reinforcement by the cues produced would also have a genetic basis. How do the genes that promote these cues and the susceptibility to them cause culture? Once such cues can feed back on behavior differentially, to shape it, their employment becomes almost inevitable, because they offer their "users" virtually endless opportunities to manipulate the behavior of others for selfish ends. More technically, altering the behavior of others often provides reinforcement for the behavior that achieves the altering. As a result, such "manipulation" should be strengthened and maintained in members of the group.

The susceptibility to reinforcement by social cues ("docility") should nonetheless be seen as adaptive (Simon, 1990). It is presumably selected because it benefits the genes of its possessor. Having your behavior manipulated by other group members is good (within limits). That is one way to understand the virtues of imitation; when A imitates B, A's behavior is manipulated by B's, but A benefits. Similarly, when B instructs A in some useful way, A benefits, even though B benefits too and might be said to manipulate A's behavior. The mother who teaches her child good hygiene manipulates the child's behavior, but the child's genes stand to benefit as much as hers do.

In sum, the answer to the first question— how genes cause culture— is that by providing three basic elements (sensory specialization, imitation, and social reinforcement), they ensure its development. The behavior of a child or an immigrant placed in a group in which those three elements exist will be shaped into the group's practices. Useful new practices that arise will spread, and will do so automatically. Psychology textbooks abound with examples of research in which a group of total strangers organizes itself and brings the behavior of its members into line in a matter of hours, if not minutes. The sensitivity of human behavior to cues from other humans is such that manipulation and organization are inevitable.

The second question—what biological function culture performs—is more complicated. A general answer is implicit in the notions of "shortcuts to learning" and "useful practices." A more specific answer must focus on what is learned and why it might be useful. To be beneficial, behavioral patterns must make contact with important long-term consequences that affect the copying of genes. But how is such long-term beneficial behavior acquired, when the consequences are so deferred?

The short answer is, "Rules" (Skinner, 1969a; Baum, 1995b). A discriminative stimulus produced by the behavior of an individual who may also supply consequences (up to now called the "instructor"), and whose behavior is reinforced by appropriate behavior on the part of a second individual (up to now called the "instructee"), is a *rule*. This term allows us to get away from talking about "manipulation" and "instruction," with a gain in precision. It reminds us that a culture is as much characterized by what group members say to one another as by non-verbal products like pots, televisions, and airplanes.

For example, in the United States the slogans that appear on license plates are often revealing. Maine's plates say on them "Vacationland," reflecting the economy's dependence on tourism, and Massachusetts's say, "The Spirit of America," probably advertising its prominent role in American history. New Hampshire's citizens, contrary to the surrounding states, choose to put personal advice on their plates: "Live Free or Die." This slogan gives a hint of the local culture, and sheds light on some of the state's peculiarities, such as the absence of any personal income tax or sales tax, and, with a population of less than a million, its possession of the third largest legislative body in the world.

Slogans, like other rules, are stimuli, produced by the behavior of a speaker and controlling behavior of a listener. They accomplish nothing by themselves, but the behavior they control, which I shall call "rule-following," achieves important ends—ends that impact reproductive success. The importance of those ends also lends a similar, if less direct, efficacy to the types of behavior that produce rule-following, which I shall call "rule-giving" and "rule-making" (Baum, 1995b). Rule-following, rule-giving, and rule-making are the key to understanding what the units of culture (practices) are, how they are transmitted, what function they perform, and how they are selected.

Rule-following

In all cultures, children are taught to obey adults (Simon, 1990). Rule-following is reinforced massively and repeatedly. The child who fails to learn to follow rules fails in all walks of life, and the parents who fail to teach rule-following are considered negligent. These commonplace observations point toward consequences of rule-following beyond the short term. If a child learns to wash fruit before eating it because the parents give the rule and reinforce the behavior, it is equally true that the parents provide instructions and reinforcement for the sake of the child's long-term health and reproductive potential.

This explanation of rule-following and rule-giving hinges on the ultimate consequence of enhanced gene copying.

Rule-giving arises because the long-term consequences of reproduction-enhancing behavior, such as washing fruit, are too deferred to reinforce the target behavior (Rachlin, 1995a). In the short run, washing fruit takes time and effort, costs that militate against its occurrence. The socially mediated consequences act to overcome these immediate costs. The situation is more challenging still when misbehavior—detrimental to reproductive potential in the long run—is immediately reinforcing, and socially mediated contingencies need to overcome the "temptation" to obtain the short-term reinforcement. However nutritious fresh fruits and vegetables are in the long run, junk food tastes better.

Figure 21.1 details the problem created by the effects of delay on reinforcement. Even though a consequence like good health might have tremendous effect or weight (V_L) when presented immediately (at time *c*), the more delayed this consequence is, and in general the less obvious it is, the less its effectiveness (Logue, 1988). At time *b* for example, when the consequence is off in the future (i.e., delayed), it is much less effective, and at time *a*, still less so. The curve relating value to delay is called a *discounting function*, and is well known to approximate a hyperbola (Logue, 1995). Because of value discounting, a relatively minor consequence like sweet taste (V_S) that follows behavior shortly (time *b*) is much more effective than a major but delayed consequence. At point *a*, when both consequences are discounted, the large, long-term reward (LL) is preferred, but as point *b* draws near, when the small, short-term reward (SS) will be available, preference switches (e.g., Ainslie & Herrnstein, 1981). This local switch causes a dilemma; one may take SS, even though one's best interest is to wait for LL. One may choose to spend money on a small item now instead of saving toward a large goal, or one may choose to avoid the dentist now instead of avoiding major dental work in the future.

Pigeons, for example, faced with a choice between immediate 2-s access to food versus 4-s access delayed by 4 s, almost invariably choose the immediate 2-s reinforcer (Rachlin & Green, 1972). When they are allowed to make their choice at an earlier time, however (point *a*), they choose the larger reinforcer. Humans, faced with similar choices, behave similarly (Logue, 1995). The behavior of choosing LL is termed *self-control*, whereas the behavior of

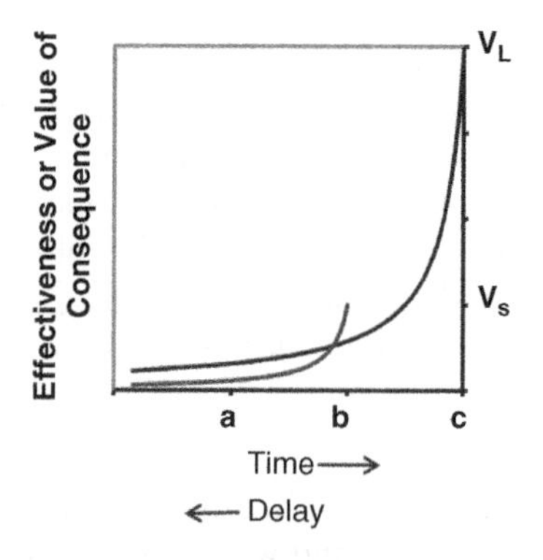

Figure 21.1 The effects of delay on the value of consequences
Note: A large reinforcer's immediate value (V_L) at time *c* is discounted at earlier times *a* and *b*, when it is delayed. Consequently, a small reinforcer's value (V_S) may exceed the value of the large reinforcer at time *b*, when the small reinforcer is immediate and the large reinforcer is delayed. This creates a problem in self-control. If a choice could be made at time *a*, when both reinforcers are discounted, no problem would arise, but such commitment is usually impossible. The discounting functions are known to approximate hyperbolas.

choosing SS is termed *impulsiveness. One of the principal functions of culture is to promote self-control in the face of choices that affect reproductive potential in the long run.*

Figure 21.2 diagrams the short-term and long-term contingencies in one situation calling for self-control. On the one hand, in the presence of junk food ($S^{D}{}_{I}$) we have the impulsive tendency to eat junk food, with the minor, short-term reinforcement of good taste (S^{R}, short arrow) and the long-term, major reproduction-reducing outcome of poor health (S^{P}, long arrow). On the other hand, in the presence of fruits and vegetables ($S^{D}{}_{S\text{-}C}$) we have a perhaps weaker tendency to eat healthy food (self-control), with the long-term major reproduction-enhancing outcome (S^{R}) of good health (Rachlin, 1995b).

Research on self-control demonstrates that for humans faced with the sort of choice shown in Figure 21.2, the frequency of eating junk food would be high and poor health would be likely, if no other intervention occurred. As the shaded boxes in Figure 21.2 illustrate, however, another contingency may be operative. Just as immediate good taste may function as a powerful reinforcer, so immediate social reinforcers like approval may have great power. A rule such as "Eat a good diet" enhances the tendency to eat fruits and vegetables (self-control) because in its presence the relevant behavior is followed by approval (social reinforcement S^{R}). The relatively minor but immediate social consequence may overcome the relatively minor but immediate taste consequence, leading to the major long-term consequence of good health. Fostering self-control in the face of the temptation to impulsiveness is one of the main functions of rules, their more general function being to foster adaptive behavior.

To understand rule selection, we must assume that rules compete. Suppose some group members some of the time give the rule "Eat what you want, but just avoid

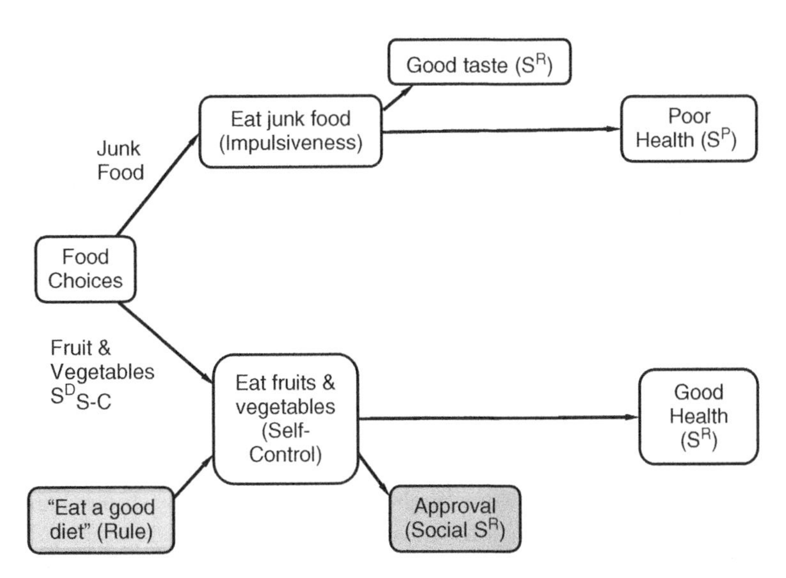

Figure 21.2 A typical problem in self-control offset by the rule an instructor gives; a cultural contingency

Note: The choice is between impulsiveness, which produces minor reinforcement immediately (short arrow) but major punishment in the long run (long arrow), and self-control, which produces major reinforcement in the long run. The instructor gives the rule, a verbal discriminative stimulus, which induces the behavior of self-control, and also gives immediate social reinforcement, which counteracts the minor immediate reinforcement gained by impulsiveness. The result is that self-control occurs even without any opportunity for prior commitment.

carbohydrates." This rule would compete with "Eat vegetables, and avoid excess fat and sugar," implicit in Figure 21.2. Following either rule may result in weight loss, a potentially health-enhancing outcome. Following the latter rule, however, might ensure ingestion of essential vitamins and minerals. If, over time and people, the "Eat vegetables" rule results in better health, following that rule will increase in frequency, and following the competing rule will decrease in frequency.

The tendencies to give the competing rules will change similarly. Giving the "Eat vegetables" rule will increase in frequency, while the frequency of giving the other rule will decrease, because the reinforcement of rule-giving derives from the instructee's rule-following and thus indirectly from its long-term consequences. The situation depicted in Figure 21.2 really contains two social contingencies: an explicit one, governing rule-following, and an implicit one, governing rule-giving. Since rule-giving is as much operant behavior as rule-following, rule-giving must be reinforced; reinforcement is provided by the instructee's rule-following (Skinner, 1957, 1969a).

By the same token, Figure 21.2 illustrates two different kinds of cultural practices. A first kind is the "garden-variety" practice (rule-following), the sort that might be spoken of as just "what we do." The second kind is an *instructional* practice, rule-giving, that results in transmission of the garden-variety practice—imitation being the other means.

Rule-giving

The contingencies governing rule-giving are harder to discern than those governing rule-following. We may begin with the relatively simple case of rule-giving among relatives. Why does a parent tell a child to wash his hands before eating? It seems clear that the parent is concerned with the child's welfare, and indeed we may observe further that parents too little concerned with their children's welfare would be selected against. Thus, the ultimate reason for rule-giving is the enhancement of the reproductive potential of both the rule-giver and the rule-follower. The proximate reason is that rule-giving (giving instructions, orders, requests, advice) is reinforced by the compliance of the rule-follower (cf. Alcock, 1998).

Recognizing the consequences of rule-giving tells only half the story; we need to discover also the relevant context or discriminative stimulus (S^D). Most people in the United States have little direct experience with relationships between illness and washing fruit before eating or between illness and washing hands. They rarely say, "Wash your hands so you will stay healthy." Instead, they talk about "dirt" and what is "nice" or "right." The environment is too clean for the connection to be obvious. It only becomes so when Americans travel to places like India, where one may see open sewers, fruit in the market covered with flies, and people in the street who clearly are ill. Since these sights are hidden from most Americans, their rule-giving has to be based more on what they hear from others than on their own non-social experience. We teach our children to wash their hands before eating because we were taught by our parents to wash our hands before eating. Technically, the context (S^D) for most rule-giving consists in other rules, verbal stimuli produced by other people—such as "Good parenting includes teaching a child good personal hygiene" (Guerin, 1992).

Rule-making

Rule-giving is part of culture (the particular rules given distinguishing one culture from another), and rule-giving is transmitted from person to person within a cultural group, but someone has to be the first to give the rule. The first-time utterance of the rule has to derive from the non-social environment. A more detailed analysis requires us to consider how a behavior-consequence relationship that is obscure, because it must be seen either over time or across people, may come to control formulation of a rule new at least to the speaker. This form of new-rule-giving may be called *rule-making*.

The basis of rule-making is recognition of repetition. Before we learn to make rules, we learn to categorize events. We come to name occurrences such as "upset stomach," "fever," "winning," "losing," and so on. The reinforcement for such naming resembles the reinforcement for other rule-governed behavior; it is socially mediated by the behavior of people who say, "This is a shoe," "Do you see the shoe?" "Please bring me the shoe," and later, "Is your stomach upset?" and "Did you win the game?" (Horne & Lowe, 1996). Later still, we begin to name conjunctions of events, such as "I ate too much, and I got an upset stomach" or "Billy worked really hard and got an A." These episodes form the basis for what are commonly called "causal inferences" and that I am calling rule-making.

Let us consider a simple example, the discovery of dental hygiene. It may be a particularly revealing example, because the relatively recent incorporation of tooth-brushing in American culture at the group level may parallel one's experience as an individual. My parents, coming from Old World families, never acquired the habit of tooth-brushing and also never demanded that I brush regularly, but they made sure that I visited the dentist once a year. Inevitably, I had cavities that required filling, some years only a few, some years many. Over the course of several years, I began to notice a pattern to this variation: The more carefully I brushed, the fewer the cavities. Figure 21.3 summarizes the experience in a chart. It includes another discovery: If I stopped eating candy and drinking soda, the annual number of cavities fell to an even lower level. The first comparison resulted in an increase in the frequency and care with which I brushed. The second comparison resulted in a large decrease in my consumption of sweets.

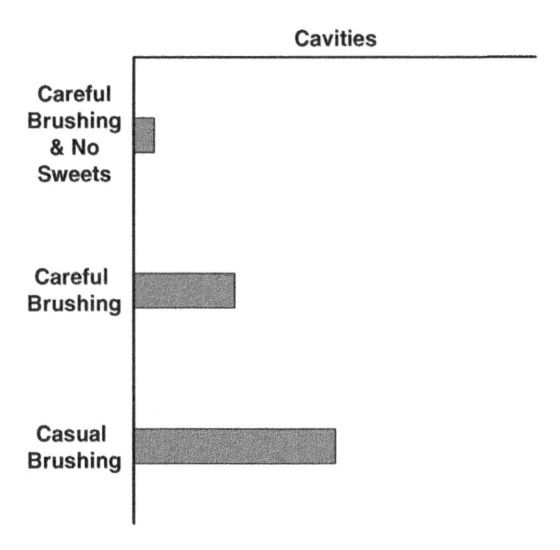

Figure 21.3 An example of a behavior-consequence relation that may act as a discriminative stimulus controlling rule-making
Note: Variation in behavior (tooth-brushing and eating sweets here) covaries with variation in consequences (cavities here). The whole aggregate of conjunctions functions as a discriminative stimulus as a result of a history, usually specifically trained, of examining such data and formulating conclusions (i.e., making rules).

Such changes in behavior as a result of behavior-environment covariation may be viewed as the shaping of behavior by environmental consequences. The way that tooth-brushing was reinforced by a reduction in cavities, and consumption of sweets was punished by continued cavities, mirrors what one might expect even in a non-verbal organism, although probably over a less extended time span. Only when the covariation affected my verbal behavior could we state unambiguously that I made causal inferences from my experiences. Only when I began to preach about the virtues of tooth-brushing and the evils of sweets, which I referred to as "tooth rot," might we incline to say that I engaged in rule-making.

Still, however, we should hold back on calling it rule-making, because another step is required: transmission—at least of the behavior, but also, following Skinner's reasoning, transmission of a social contingency; that is, transmission of the triad, rule plus behavior plus rule-giver-mediated consequences. My friends soon tired of hearing my sermons about brushing and sweets, and punished the behavior by calling me "Old Rot Tooth." My children, however, were in no position to respond this way. I taught them all to brush their teeth at the earliest possible age and exhorted them to be sure to brush regularly and well. As adults they all brush regularly, and they have had almost no cavities. (The absence of cavities may also be due to fluoridation of the water in the town where we lived, but this only shows that a rule may cease to be supported by someone else's experience in the future.)

The real proof of rule-making, however, occurs when it becomes rule-giving in the next generation. My oldest daughter is now teaching her four-year-old son to brush and is exhorting him as I did her. Since she lacks direct experience with cavities and her only experience with tooth-brushing is the result of my rule-making and consequence-supplying, we may conclude that her behavior now constitutes rule-giving.

How do we explain rule-making and its transmission as rule-giving with the concepts at hand? First, rule-making is trained; it is operant behavior. It builds on the naming of conjunctions already discussed, as when children are asked questions like, "Did you have fun in the park?" and "What did you like best there?" These in turn lead up to questions asking "why" ("Why do you like the zoo?") and answers using the word "because" ("Because I like the monkeys.") Although rule-making requires no formal education, such training constitutes one important function of schools. Children are trained to make causal inferences and to draw conclusions from data: "Why do you think the plant we kept in the sun is doing fine, whereas the plant we put in the dark is doing badly?" Correct inferences may be reinforced with praise and high grades. As a result, most citizens are adept at rule-making by the end of high school. Those who attend college get further training, and those who go into the sciences get still more.

Like all operant behavior, rule-making is under stimulus control. Skinner (1953, 1957) regarded the behavior of scientists as primarily the formation of discriminations. The relevant discriminative stimuli are called "data." They consist not of a single event, but of a succession of conjunctions like those represented in Figure 21.3. The entire series of visits to the dentist, in which the variations in tooth-brushing and sweets-consumption correlated with variations in number of cavities, increases the likelihood that eventually I will say, based on patterns shaped in school, that brushing teeth regularly and well leads to fewer cavities.

Figure 21.4 illustrates the relationship between rule-making and rule-giving. The top line depicts rule-making in general. The second line illustrates how the entire set of events in Figure 21.3 stands in the place of a discriminative stimulus (S^D). The latter

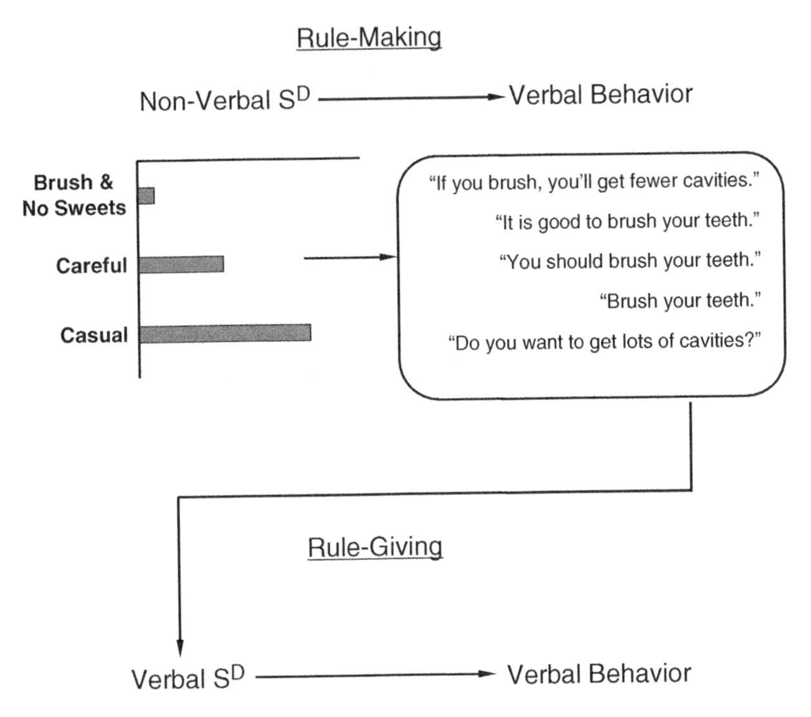

Figure 21.4 How rule-making leads to rule-giving
Note: Top line: In rule-making, a non-verbal discriminative stimulus (S^D), in the form of a behavior-consequence relation, induces verbal behavior, in the form of a class of functionally equivalent utterances. Second line: an example of a behavior-consequence relation and the class of utterances it induces. Third line: The utterances produce (vertical arrow) verbal discriminative stimuli functionally equivalent to one another in that they induce the same verbal behavior (an instance of giving a rule, such as in Figure 21.2).

controls a class of verbal behavior including declarative forms ("If you brush, then you'll get fewer cavities") and other functionally equivalent variants (some declarative value judgments, some exhortations, some commands, and even some questions). The third line shows how this class of verbal behavior produces a class of verbal discriminative stimuli, heard by the next generation, which control the same verbal behavior as in the original rule-making, but now released from its non-verbal origins (Guerin, 1992).

Although the discussion of rule-making and rule-giving so far might seem to suggest that the rules tend to be "true," rules need not be veridical. First, a rule may just be wrong. Based on incomplete data, one may draw an incorrect conclusion that is corrected later. The medical profession may first recommend blood-letting as a treatment and later repudiate it. Second, the function of a rule may differ from its surface reading. A rule like, "If you eat pork, God will punish you," needs to be examined in the light of the actual consequences of pork avoidance in this world. It probably represents a type of rule related to group cohesion and segregation, functionally equivalent to, "If you are to belong to this group and reap the benefits of membership, then you must behave so" (Guerin, 1998). Rules of that sort enable members to identify one another and to avoid squandering resources on people who are unlikely to reciprocate (Dawkins, 1989a).

Rule-giving and altruism

We can now address the question of why rule-giving occurs between non-relatives. Rule-giving is a form of altruism (Dawkins, 1989a; Simon, 1990; Wilson, 1975). It costs the speaker little and benefits the listener (rule-follower) greatly in the long run. If the two actors are unrelated, the rule-giver's reproductive potential may be enhanced eventually by reciprocal rule-giving from the listener or other members of the group. I may tell you where to get a good buy on automobile tires today; you may advise me about the best schools for my children tomorrow. In more everyday terms, group members share useful information with one another, to their mutual benefit in the long run.

The delays in reciprocal altruism open the door to cheating, usually failure to reciprocate. In rule-giving, however, another type of cheating is possible: The speaker may give the listener advice that will harm the listener in the long run. The speaker does so "knowingly" when reducing the listener's fitness in the long run reinforces the speaker's false rule-giving. Such behavior accords with the competitive aspects of evolution, because one way to increase the relative fitness of one's genes is to reduce the fitness of the genes of one's competitors. Accepting the minor cost of giving misinformation in return for a large reduction in the competitor's well-being would be called *spite* (Wilson, 1975). Individuals guard against spite by discriminating between immediate group members and strangers, between those whose rule-giving is likely to be helpful and those whose rule-giving is uncertain. As one interacts repeatedly with another person, this person's rule-giving eventually comes to exert strong stimulus control over one's rule-following or none at all. This fact of life is summarized in the saying, "Fool me once, shame on you; fool me twice, shame on me" (a rule about rule-following).

Units, transmission, and selection

In sum, the replicators in cultural evolution are the behavioral units that I have called "practices." A cultural practice is a unit of operant behavior, whose full definition includes not only a class of actions, but also the relevant context and reinforcement. As Figure 21.2 illustrates, the context of a practice is often a rule, and the reinforcement is socially mediated (e.g., approval). If the behavior occurs enough, the long-term contingency may take over. In such cases the relevant context for behavior becomes the non-social discriminative stimulus labeled $S^{D}_{S\text{-}C}$ in Figure 21.2 (fruits and vegetables). Although some psychologists would say that the rule has been "internalized," all that has occurred is a transfer of control from an obvious, proximate contingency to an extended or ultimate contingency. The person who is spontaneously kind to strangers has not "internalized altruism;" rather, his behavior has come under control of social contingencies in which altruism is ultimately praised and materially rewarded with status and its accompanying opportunities. Control remains in the environment (for further discussion and experimental evidence, see Logue, 1995).

Transmission of cultural practices occurs in two ways: imitation and instruction. Imitation is probably the less important of the two for human culture, because it relies on non-social consequences to strengthen behavior. Its scope is limited in comparison with instruction, which, because it relies on contrived contingencies and social reinforcers, opens the door to endless innovation. Instruction depends on a history of obedience training and transmits a practice by inducing it with a rule and reinforcing it when it occurs. Besides this rule-governed behavior itself, the practice of instructing the behavior (rule-giving) is also transmitted. Social reinforcement is often arranged for rule-giving.

Not only are people encouraged to use birth control, some of them are also instructed in how to encourage others to use birth control.

Finally, practices are subject to selection because some function better than others. Practices may be selected in clusters, just as other replicators are, because they are interdependent and perform several related functions. Extended families, for example, might foster arranged marriages: too many individuals' lives are affected by a marriage to leave the relevant decision to the bride and groom. Whether selected individually or as a cluster, however, practices are selected by their consequences. Whichever practices are more reinforced than their competitors (alternative ways of performing the same function) increase in frequency. They increase in the behavior of individuals, as when an artist gradually shifts toward non-representational painting or a scientists finds more and more occasions to speak of selfish genes, and they increase across individuals, as when new students come out of art schools painting non-representationally, or when new students come out of graduate programs already speaking of selfish genes.

Selection

To understand the selection that goes on in cultural evolution, it is essential to identify the reinforcers that do the selecting. If these reinforcers were arbitrary, culture could evolve in any direction, independent of the human genome and perhaps even antithetical to it. Dawkins (1989a) proposed such a view. Boyd and Richerson (1985), in contrast, argued that cultural evolution cannot but be ultimately limited by natural selection. It may wander widely, but it wanders within confines set by the phylogenetic history of our species, because the reinforcers that control human behavior are far from arbitrary.

The importance of reinforcement

Both fixed-action patterns and operant behavior depend on contingencies in the environment. The difference is that the contingencies to which fixed-action patterns are adapted were stable enough to make contact with genetic selection, with the result that fixed-action patterns are relatively immune to the vagaries of short-term environmental variance and are available at crucial times, such as in the presence of a predator or a potential mate. Fixed-action patterns depend only on context, in the form of releasers, and have no need of consequences. Operant behavior, in contrast, is shaped by consequences within a lifetime, and therefore is highly sensitive to short-term changes in contingencies. Operant behavior may nevertheless originate in fixed-action patterns. For example, a baby gull's pecking at its parent's beak may improve because it produces food more reliably, thereby turning into operant pecking (for a general discussion, see Teitelbaum, 1977).

The ultimate consequences of the cultural practices pointed to in Figure 21.2—health, resources, relationships, reproduction—which affect the replication of genes, are usually too deferred or extended to influence operant behavior within a person's lifetime. Feeling healthy, acquiring a mate, amassing resources, and establishing friendships are all results of continued effort over extended time periods. They often exist as alternatives to short-sighted, impulsive behavior and therefore call for self-control (Rachlin, 1995a).

The advantages of self-control provide a role for social reinforcers, which presumably result from our ancestors' history of living in groups. Boyd and Richerson (1985) suggest that culture originated initially in the advantage conferred by being able to learn from others in a variable environment. A child growing up in a group or an adult joining a group might do well to behave as those around behave. If such a tendency allowed rapid adaptation, it might also open up additional reproductive opportunities, and genes that favored it would spread. Those genes would promote manipulation of their vehicles' behavior by other group members (Dawkins, 1982).

So a possible first stage in the evolution of culture would be imitation reinforced by non-social reinforcers, as in non-human primates. Exactly how the transition from this first stage to manipulation by social reinforcers would have occurred is unclear, but it probably happened as a result of the advantages of opening behavior to manipulation, coupled with advantages to the manipulator (Simon, 1990). How good for a parent to be able to shape a child's behavior with reinforcement and punishment, to adapt it to whatever the environment requires! How difficult to do that without reinforcers and punishers that are convenient, low-cost, and immediate! If the result benefited one's children, thereby benefiting the parent's genes, the genes for susceptibility to parental cues as consequences could spread. Once established, anyone in the group might be able to benefit by manipulating the behavior of others with those same facial expressions, intonations, and so on. Those reinforcers and punishers would become the dominant means by which cultural practices were established and transmitted. Indeed, no one could live in a group and fail to have his behavior shaped toward group norms. Besides making culture possible, social reinforcers and punishers make culture inevitable.

Once we appreciate the importance of reinforcement and couple it with the idea of stimulus control (context), it becomes neither necessary nor likely that the recurrence of cultural practices depends on internal representations such as memes. Whatever the underlying neural basis for the recurrence of operant behavioral patterns might be, it is unlikely to resemble the mechanism for a fixed-action pattern. Sensitivity to consequences guarantees that cultural practices, like all operant behavior, involve more or less constant feedback in the form of interaction with the environment (Baum, 1973, 1989).

Unfortunately, some scholars writing about culture, failing to recognize the role of behavior-environment feedback, treat behavior as if the environment released it in an input-output fashion. They thus postulate not only representations, but also miniature machines, called "devices" and "modules," that somehow produce behavior as a result of environmental inputs. The results have been disastrously misleading.

Devices and modules

Let us consider a well-worn example: the so-called language acquisition device, which has become standard fare among cognitive psychologists (e.g., Pinker & Bloom, 1992). Each child is supposed to possess this device and to acquire language because of it. The device is supposed to take input from the environment (heard utterances of speakers) and to process it into the rules of grammar and a lexicon of meanings. With their help, a person becomes able to decode utterances heard and to encode utterances to be generated.

Pinker (1994) and others have made good use of the idea of a language acquisition device to argue that language is an adaptation, a product of natural selection (Pinker & Bloom, 1992). This is an important point, because it allows a richer understanding of the origins and functions of language. The point could have been made without the language acquisition device, but insofar as the notion helped, it was useful. That, however, is as far as its usefulness goes.

The criticisms of the meme made earlier apply to the language acquisition device in full force. Even if it really did extract grammar and meaning from the heard utterances, it would leave unanswered the questions that most need answering: why language is useful and why particular utterances occur. The language acquisition device tells us nothing of the behavior of speaking and its consequences.

The notion of the language acquisition device is not only unhelpful, however; it is positively misleading. Its pernicious effect lies in supposing that it is a unitary mechanism that might itself perform a function. Tooby and Cosmides (1992), in their lengthy critique of the standard social science model, emphasize that natural selection cannot produce organs serving general purposes. As a product of natural selection, the brain ought to consist of a collection of many special-purpose adaptations, for selection is opportunistic and must make do with whatever variation at hand. It can select only what is there and immediately useful. In other words, natural selection has no foresight and cannot plan ahead for contingencies other than the ones actually occurring in the environment. Hence its adaptations tend to be cobbled together, rather than elegantly designed.

This argument applies not only to the brain, but also to the language acquisition device. Rather than producing a unitary mechanism, natural selection would produce a collection of more specific adaptations—modifications to sensory systems, to effectors, and possibly portions of the brain—all of which aided the acquisition of language. The result would be less a unitary mechanism than a motley crew. Pinker and Bloom (1992) and Tooby and Cosmides (1992) nod toward this point, but treat it as trivial. They imply that the only issue is whether the device is localized or distributed. They miss the larger implication that a motley crew cannot perform a function; rather each member arises to aid and abet some function that is being performed. The usefulness of that function provides the basis for selecting its aides and abettors.

What function is being aided and abetted? Actually two partially independent functions (often lumped into "language acquisition") are aided and abetted: the acquisition of spoken language and the acquisition of receptive language, which correspond to verbal behavior and stimulus control by verbal stimuli. Language or verbal behavior is undoubtedly useful; it allows a speaker to manipulate the behavior of a listener for the good of both their genes. It is essential, however, to remember the difference between verbal behavior and other communication schemes. Courtship in birds and pheromones in insects offer instances in which creatures communicate and may be said to manipulate one another, but the consequences of the manipulation operate only in the course of natural selection across generations; they cannot alter the fixed-action patterns involved during the lifetimes of the individuals.

Verbal behavior, in contrast, is operant behavior; it is controlled by its consequences. It depends on reinforcement and punishment (Moerk, 1996). A person who speaks Spanish and English speaks each language only with those who understand it. Why? A person in need of the salt requests it of the person across the table. Why? The answers to such questions cannot be found in the language acquisition device or the motley crew of aides and abettors; they are to be found in the history of consequences in context that led up to the utterances. If the language acquisition device could perform the function of language acquisition the way it is said to, it would be possible in theory to put a child in front of a television set for two years and at the end have it speak the language heard. One need not do the experiment to know it would fail. When I was a child, my parents often spoke Yiddish in my presence. They did so because I couldn't understand; Yiddish was their secret language. Even though I heard it often and was motivated to learn it, I never acquired Yiddish. Hearing it had an effect; I was able to make the non-English sounds better than my classmates when I studied German and Russian in college. But I never learned Yiddish, because my parents only spoke it to each other, never *to me.* Verbal behavior depends on and consists of interaction with the environment. The adaptations produced by natural selection only make it certain that verbal behavior and verbal stimulus control will be acquired. They take care of the selective attention, raw material, variation, and social consequences required.

Similar criticisms apply to the "modules" that Tooby and Cosmides (1992) assume to be produced by natural selection. They list a variety of modules that are supposed to perform all sorts of functions, from mate selection to detection of cheating. Perhaps the most extreme example of a misleading imaginary module is what they call the "theory of mind" module. As with the language acquisition device, this module is supposed to take input from the environment and convert it into a theory, not of grammar, but of others' minds. Translated into behavior, this notion means that people behave (mostly talk) as if other people possessed minds—meaning inner thoughts, intentions, and beliefs. It comes down to observations such as the following: (a) Tom's seeing Jane pause and then initiate action, followed by his saying that Jane was "thinking;" (b) Tom's seeing Jane persist in behavior that customarily has produced particular results, followed by his saying that Jane is "trying to achieve" these results; (c) Tom's hearing Jane talk about God and seeing her go to confession, followed by his saying that she is religious. Other people's behavior provides cues that guide our behavior, notably our verbal behavior. Such social cues could be important enough that special sensitivity to them was selected, because effective social behavior could avoid rejection and open up opportunities for reproduction. But to attribute the sensitivity and the talk about mind to a "theory of mind" module seems gratuitous. Worse, it substitutes for the study of various behavioral facts a line of fruitless speculation about what the "theory of mind" is like (Baum, 1998; Heyes, 1998).

There is no substitute for an understanding of the interaction of behavior with the environment over time. Imaginary inner devices and modules cannot take the place of understanding the effects of reinforcement and punishment in shaping people's behavior in general, and shaping cultural practices in particular.

Conclusion

The notion that the units of culture and cultural transmission are stored somewhere in the brain is both inaccurate and misleading. The units passed from one group member to another consist of behavior—more accurately, operant behavior in context. Like any operant unit, a cultural practice thus contains three terms: a context or discriminative stimulus (S^D; a model or a rule), the effective behavior (defined as the class of variants that accomplish the particular consequences), and its consequences (the class of outcomes produced by the behavior in that context). Since copying transmits cultural behavior itself, an analog to genotype in cultural evolution must consist of behavioral units. The analog to phenotype might consist in the products of cultural practices. Contrary to Dawkins's account, this behavioral view contains an analog to alleles in the various competing ways of achieving a common result.

Practices may be distinguished on the basis of their mechanism of transmission (whether by imitation or by instruction; i.e., rule-following) and on the basis of their type of reinforcement (socially mediated or not). Imitation-based transmission occurs when a model provided by one group member induces similar behavior in another, and the imitator's behavior is reinforced automatically by environmental consequences, without intervention from the modeler. Instruction-based, or rule-based, transmission occurs when an "instructor" gives a rule (a verbal S^D) and also supplies immediate reinforcement of the behavior appropriate to this rule. Although practices transmitted by compliance or rule-following depend at least for a while on socially mediated reinforcement, they may eventually come under the control of long-term, environmentally based reinforcers more directly related to reproductive success. Both aspects of the instructor's behavior, rule-giving and consequence-supplying, constitute verbal behavior in Skinner's (1957) sense, because they are reinforced by the instructee's (the listener's) appropriate behavior or rule-following. The behaviors of rule-giving and rule-following may depend on socially mediated reinforcement in the short term, but must ultimately enhance reproductive success. Particular rules dominate in a culture if in the long run they enhance reproductive potential more than their competitors do.

Although novel practices may enter a culture by immigration or by mutation due to copying errors, a more reliable source is the behavior of rule-making, in which sequences of events or relations in the non-social environment play the role of discriminative stimuli for the verbal behavior of the rule-maker. If a subset of the group engages in rule-making, new practices will enter the culture on a regular basis, and then persist if they produce reinforcers ultimately tied to reproductive success. Their spread among non-relatives, however, must depend on rule-giving that is a form of reciprocal altruism. The benefits of "sharing information" (i.e., reciprocal rule-giving) tend to accrue to all group members, whether genetically related or not.

Understanding all cultural phenomena thus depends on an understanding of the role of reinforcement and punishment. Cultural practices are tied to natural selection because the reinforcers and punishers that shape them were selected in the course of our species' phylogeny. Low-cost social reinforcers and punishers not only made culture possible; they made it inevitable. Many other adaptations have been selected to aid and abet the process of shaping by consequences; these in no way take the place of that interactive process. Casting them as "devices" or "modules" only detracts from an understanding of the development and transmission of cultural practices.

22

Behavior Analysis, Darwinian Evolutionary Processes, and the Diversity of Human Behavior

Foreword

This paper aimed to lay out a general description of a Darwinian evolutionary process. I tried to show how genetic evolution, cultural evolution, and behavioral evolution could all be understood in the same terms. Also, I aimed to show how the three are interrelated. The accompanying diagrams sketched the ways the three processes interact. They are necessarily simplified, particularly for two reasons. First, the interactions occur in two directions some of the time. The diagrams only show how genetic evolution constrains cultural evolution, but culture sometimes affects selection on genes—e.g. if a cultural practice like birth control reduces the rate of reproduction differentially within a society. Also, the diagrams only show how culture generally constrains individual behavior, but do not show how individual innovations contribute to culture also.

Secondly, the paper does not include much information about recent advances in understanding the role of environment in ontogeny (e.g. Sultan, 2017, 2019). This omission was probably inevitable, because the paper was already long. Nothing in the paper should be read as contradicting the growing understanding of the mechanisms by which the genome interacts with environmental factors to produce the phenotype.

I included an explanation of the Price equation because this equation describes evolutionary selection in extremely general terms. A longer treatment of the Price equation as it applies to behavior appeared around the same time as this book (Baum, 2017c).

Abstract

The resemblance among behavioral evolution (i.e. "shaping"), cultural evolution, and genetic evolution emerges in the light of a general concept of evolutionary process. Every evolutionary process consists of three elements: variation, recurrence, and selection. Evolutionarily significant variation occurs among substitutable variants within a pool. These variants are defined by differences in environmental effects. Although the metaphor of copying characterizes recurrence in genetic evolution, replication is only one type of recurrence. Selection occurs when recurrence is differential. Differences in environmental effects produce differences in recurrence, and those differences feed back to affect the composition of the pool of variants. Genetic, cultural, and behavioral evolution all admit of the distinction between proximate and ultimate explanations. The three evolutionary processes may be seen as nested: cultural evolution within genetic evolution,

Science and Philosophy of Behavior: Selected Papers, First Edition. William M. Baum.

and behavioral evolution within cultural evolution. A complete understanding of human behavior requires six types of explanation: proximate and ultimate explanations in all three processes.

Keywords: evolution, general evolutionary process, cultural evolution, behavioral evolution, operant behavior, substitutable variant, proximate explanation, ultimate explanation, nested processes

Source: I thank M. Ghiselin, S. Glenn, K. Panchnathan, P. Richerson, and H. Rachlin for helpful comments on an earlier draft. Originally published in M. Tibayrenc and F.J. Ayala (eds.), *On human nature: Psychology, ethics, politics, and religion*, pp. 397–415. New York: Academic Press, 2017. Reproduced with permission of Elsevier.

The diversity of human behavior stems from three Darwinian evolutionary processes: genetic evolution, cultural evolution, and behavioral evolution. Genetic evolution by natural selection is the paradigmatic example of descent with modification or selection by consequences (Skinner, 1981). Cultural evolution accounts for the diversity of practices across groups and through time (Boyd & Richerson, 1985). Behavioral evolution, the third example of selection by consequences, also called "shaping" of behavior, occurs within the lifetime of an individual, and is a primary focus of behavior analysis.

The science of behavior, or behavior analysis, is properly part of evolutionary biology. It is intimately tied to evolutionary theory in two fundamental ways. First, evolution is the explanation of why behavior and behavioral processes exist at all. It is the only way to understand the peculiarities and constraints that characterize matters like classical and operant conditioning (Baum, 2013). Second, behavioral explanations follow the same mode of explanation as do evolutionary explanations. Skinner (1981) called it "selection by consequences." It may also be called "historical explanation" (Baum & Heath, 1992).

Several writers have suggested that operant conditioning, or behavioral shaping, may be seen as parallel to natural selection. Donahoe (1999) suggests that even Thorndike may have recognized the resemblance between the law of effect and natural selection at the end of the nineteenth century. Skinner (1953; 1981) stated it overtly and made it the centerpiece to his view of behavior (Ringen, 1999). Gilbert (1970) drew out the parallel at length, and Staddon & Simmelhag (1971; Staddon, 1973) enlarged on it by distinguishing between "principles of variation"—i.e., those processes that give rise to behavioral variants—and "principles of selection"—those processes that cause selection among behavioral variants. More recently, Hull, Langman, and Glenn (2001), used the term "interactor" to apply both to organisms in genetic evolution and to behavioral units in behavioral evolution, and again compared behavioral evolution to genetic evolution. The parallel is of more than casual interest; it represents a revolution in thought.

Population Thinking

The importance of selection and history has been under-appreciated in the study of behavior within the traditions of psychology. One reason is that psychology, like other sciences, was influenced by philosophical views in which the world is thought to be composed of ideal types into which all particulars may be classified. In biology, typological thinking affected the concept of species. In psychology, typological thinking affected the concept of response, the unit of behavior. In biology, typological thinking eventually gave way to population thinking, which is the cornerstone of evolutionary theory and all selectionist theories. The difference between the two ways of thinking may be seen in their views on variation. In typological thinking, variation is "error" and dealt with as a nuisance to be eliminated by averaging. In population thinking, variation is central, and averaging is only an analytic convenience. As Ernst Mayr (1959) wrote, "For the typologist, the type (*eidos*) is real and the variation an illusion, while for the populationist the type (average) is an abstraction and only the variation is real" (p. 2). Arguably, Skinner's greatest contribution to the study of behavior was his rejection of the typological view of behavioral units (responses). Mayr (1970) observed, "The replacement of typological thinking by population thinking is perhaps the greatest conceptual revolution that has taken place in biology" (p. 5). The same may be said of the (ongoing) revolution in the study of behavior.

From an ontological point of view, a population is an individual—that is, an integral, functioning entity that can change and still retain its identity (Ghiselin, 1997). An individual is a whole with parts—as opposed to a class or category with criteria and instances. An individual changes when its parts change. The parts of a population are its members, and the population changes when its parts change. The population of mice in a particular meadow is an individual and is composed of the mice there. If the proportion of dark-colored mice increases relative to the proportion of light-colored mice, the population has changed or evolved, but it is still the population of mice in that meadow.

Population thinking means the appreciation of populations as units of evolutionary change. Evolutionary theory would be impossible without it. Darwin explained the origin of species by conceiving of species as comprised of populations that change across generations. The idea of change in the composition of a population allowed evolution to be explained as descent with modification, the result of variation with consequences for reproductive success.

The explanation of behavior as an outcome of selection by consequences similarly requires thinking about behavior as comprised of individuals with parts that may change (Baum, 2002, 2004, 2013; Glenn et al., 1992). Only if behavioral change is seen as change in the composition of a behavioral individual may it be explained as descent with modification. The behavioral analog to a population of organisms is an activity, because an activity is an individual with parts that may change when the parts change (Baum, 2002, 2004). A group's pottery manufacturing is an activity with parts like obtaining clay, processing the clay, shaping the clay, decorating the pot, and firing. If, say, the decorations or type of clay changed, the activity of manufacturing pottery will have changed. My tennis playing is an activity with parts like buying equipment, serving, positioning, and returning the ball. My serving may change if I toss the ball in the air differently or swing my racquet differently, and if my serving changes, my tennis playing changes. Behavior analysts think of changes in activities as the result of contingencies between activity parts and phylogenetically important events, and the effect of contingencies is often labeled "selection by consequences" (Baum, 2012a; Skinner, 1981).

Taking the unit of change as an individual—population or activity—allows one to think of evolutionary change as descent with modification. To have explanatory power, however, descent with modification or selection by consequences requires identification of an evolutionary process. Plausibility requires construction of specific explanations, which require specific mechanisms (Hull, 1988).

Darwinian Evolutionary Process

A Darwinian evolutionary process includes three basic elements: variation, recurrence, and selection. Each is necessary, and together they suffice to ensure evolutionary change (i.e., descent with modification).

Variation

Variation occurs within an individual thought of as a population or pool. A population of mice may be conceived of as a pool of genes, each mouse containing a set of genes, but because of recombination, a pool sufficiently fluid that the individual mice may often be ignored for purposes of explaining change within the population. Within the pool, different genes have different effects. Some influence coat color, some lung capacity, others the structure of the brain. The key variation within any such pool is the variation among units that affect the same trait but produce different phenotypic effects—for example, that influence coat color but cause coat color to be dark or light. Evolutionary change depends on the existence of *substitutable variants*. In genetic evolution, such substitutable variants are referred to as alleles. They are substitutable in the sense that one allele may replace another, and they are mutually exclusive in the sense that such replacement is all or none. Evolutionary change consists of change in the relative frequencies of substitutable variants within the pool or population.

The substitutable variants in behavior are the smaller parts of more extended activities, and the total activity over a span of time constitutes the pool. A group's pottery making over the course of a decade may include a variety of styles of decoration, and as time goes by, one of those styles may replace the others, changing the group's pottery making. My tennis serving over the course of a year may include variation in how high I toss the ball, and in time one toss height may prevail, changing both my serving and my tennis playing.

Functional definition

The units within the pool are defined, not by their structure, but by their function—that is, by their effects. In genetic evolution, the question arises as to how to break the genome into the constituent units that influence phenotype (Dawkins, 1989b). Although DNA has structure, one cannot tell which pieces should be called genes just by examining the structure. According to Dawkins (1989a), a gene must have three properties: fidelity (faithful copying), longevity (long-enough lifetime to be copied), and fecundity (frequent copying). The units that possess these properties, however, may be small or large pieces of DNA, may be contiguous in a chromosome or not, may even be in different chromosomes, and may even be in different organisms [as in Dawkins's example of parasites' affecting phenotypic traits in hosts (Dawkins, 1989a and b)]. Whatever pieces of DNA act in concert to produce the phenotypic effect and may be said to be faithfully copied, to endure well enough to be copied, and to be frequently copied constitute the gene. The alleles compete because they differ in fidelity, longevity, or fecundity. A degree of copying fidelity may be assumed, because it is necessary for transmission. Differences in longevity and fecundity, however, result from differences in the alleles' phenotypic effects. A darker coat color may increase its possessor's likelihood of surviving long enough to reproduce.

Phenotypic effects, though often thought of as effects on morphology, are better seen as effects on the environment (Dawkins, 1989b). A darker coat color affects the ability of predators to detect a mouse against the ground. All phenotypic effects are environmental effects, because they facilitate exploitation of resources, survival, or reproduction. The point becomes clearer if we focus on genetic effects on behavior. Building a better nest alters the environment of the nestlings. Dawkins uses the example of the beaver's dam construction, which creates the beaver pond, which reduces risk of predation and has a host of other good effects on the beaver's environment. Human beings change their environment in myriad ways, creating shelters, places for plants to grow, instructional institutions, and so on. All may be seen as effects on the environment that function (usually) to enhance exploitation of resources, survival, and reproduction.

This focus on function, in the form of environmental effects, solves what would otherwise be an intractable problem: how to define genes. In a broader view of selection by consequences, it solves the problem of defining the substitutable variants within the pool of variants.

The same problem of defining units arises in cultural evolution and in the evolution of individual behavior, both of which constitute Darwinian evolutionary processes. As a focus on the structure of DNA offers little guidance about the definition of genes, so the structure of cultural practices and individual operant activities tells little about how they should be defined. Not that structure is totally irrelevant; in all three processes, structure constrains the definitions. DNA sequences specify the ordering of strings of amino acids, different configurations coding for different amino acids. Cultural practices and individual activities divide along what Skinner (1938) called "natural lines of fracture" (p. 33), constrained by anatomy and arrangements in the nervous system. As the codons of DNA represent "natural lines of fracture" and specify minimal units that may be aggregated into genes, so fixed-action patterns and the structure of bones and muscles constrain what may be aggregated into the substitutable variants of culture and individual operant behavior (Skinner, 1969c). The making of a pot comprises motions of the fingers and hands, but only those motions that the structure of the fingers and hands allows. The physical motions, however, are little help when it comes to defining the evolutionary unit. That will be the making of a certain kind of pot, and the substitutable variants will be the making of other kinds of pot, some of which may leak less, may be easier to handle, or may serve a social function by virtue of design. These variants compete within the culture pool, just as alleles compete within the gene pool.

All three units—genes, cultural practices, and individual operant activities—are defined in terms of environmental effects. All three are defined by what they accomplish in the world with which they make contact. Dawkins (1989b) made this clear about genes in his discussion of "extended phenotype." Guerin (1997) clarified the point for cultural practices when he argued that the functional unit of culture is "getting a job done." Every culture comprises "jobs" that must get done. At the most general level, the jobs might be reproduction, obtaining resources, protection from weather and enemies, and maintaining group cohesion. These general jobs subsume more specific jobs, such as child rearing, transport, and ownership. Whatever the level of generality that suits the analysis of culture, the substitutable variants will be different ways of getting the same job done. Different ways of raising children, of transporting oneself and goods, or of demonstrating group membership may compete with one another and may differ in their outcomes. Skinner (1938; 1953; 1957) explicitly defined operant "responses" (activities here) according to their environmental effects. One might say, following Guerin (1997), that a rat's lever pressing gets a job done. Although depression of a lever constitutes a discrete unit of behavior, more extended patterns produce more extended results. In the laboratory, several presses may be required for a bit of food, constituting a larger unit. In the everyday world, extended accomplishments always entail more specific accomplishments.

Helping an unhappy customer entails listening to the complaint, making suggestions, talking to suppliers, and so on. Making a living might entail finding a job, going to work every day, specifying one's duties, and so on. Giving directions entails various utterances—statements about location, queries, and descriptions of action. Different ways of helping the customer, of making a living, or of giving directions constitute substitutable variants that may compete and may differ in their results. A customer may come away more satisfied, one may make a better living, directions may be given more clearly.

Pooling

All three processes, genetic, cultural, and behavioral evolution, require a pool of variation that includes substitutable variants. In general, the pool generates a frequency distribution—a profile of the frequencies of various types—that may change with time. For talking about evolutionary change, the important relative frequencies are those of the substitutable variants. Change results from competition. Figure 22.1 illustrates the general idea. The top diagram represents a hypothetical pool, within which are substitutable variants A, B, C, and D, which occur at different frequencies. They could represent four different alleles for four different structures of an enzyme, or four different ways to catch termites (e.g., with fingers, with a leaf, with a stick, or with a leaf stem). The relative frequencies are shown in the middle graph (open bars; the shaded bars will be discussed below). Allele (variant) B occurs at the highest frequency, then C, then D, and the least frequent is allele A. This pattern of relative frequencies may remain stable with time or may change, depending on selection.

Whereas the middle panel of Figure 22.1 assumes discontinuous variation, the open bars in the bottom panel illustrate a hypothetical frequency distribution for variation fine enough to be considered continuous. It could represent variation in the genes affecting height or variation in squeezing clay that affects the thickness of the walls of pots. Although the variants cannot be grouped into discrete categories, they still exhibit a pattern of frequencies that may remain stable or change, depending on selection.

Figure 22.1 illustrates the general idea of a pool characterized by a pattern of frequencies, which constitutes the essential element of variation in any Darwinian evolutionary process. Genetic evolution, cultural evolution, and behavioral evolution (shaping) all assume such a pattern of variation, although they differ in details, such as whether the units are localized or extended and whether variation is continuous or discontinuous. Genes are usually thought of as particular locations on chromosomes (i.e., localized units) and alleles as differing in physical structure (i.e., varying discontinuously), but when enough genes act in concert and vary, the substitutable variation becomes (approximately) continuous. Under some circumstances, such as parthenogenesis, the unit of variation may even be the entire genome (Dawkins, 1989b). Then the substitutable variation is certain to seem continuous. Although Dawkins (1989a; 1989b) suggested a discrete unit of cultural variation analogous to the gene, the meme, nothing requires that the units of culture be localized or that variation in culture be discontinuous, any more than in genetic evolution. Boyd and Richerson (1985), for example, describe mathematical models of cultural evolution that assume variation to be continuous. In her classic book, *Patterns of Culture*, the anthropologist Ruth Benedict (1934) described the patterns of behavior in culture with the word "custom." A custom or a practice cannot be localized to particular moments of time; it is an extended pattern of behavior that can only be observed over a substantial period.

With individual operant behavior too, we are under no compulsion to assume discontinuous variation or localized units. That variation may be continuous is recognized in

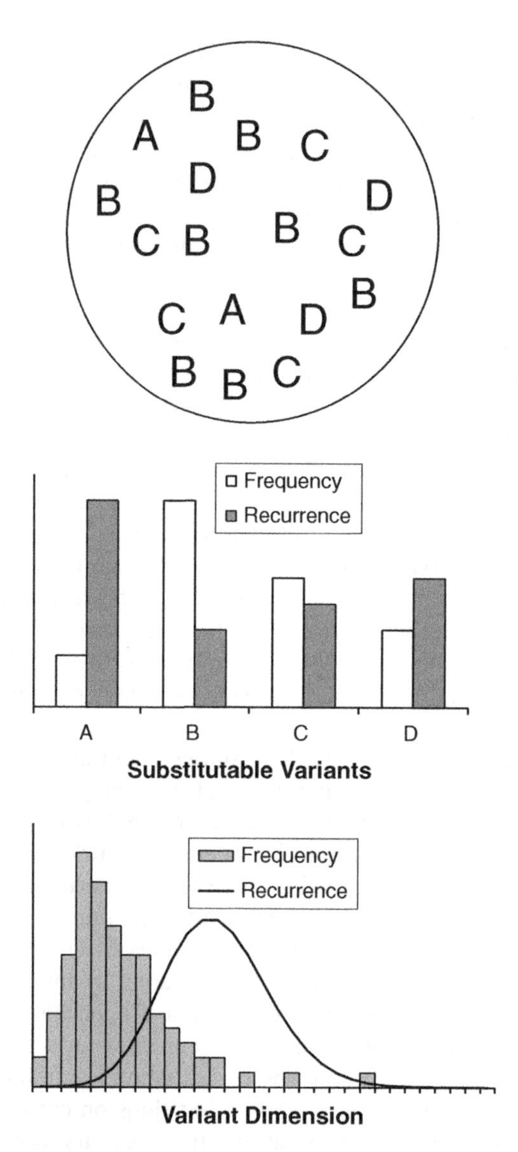

Figure 22.1 Population, frequency, and recurrence
Note: Top: a population with 4 substitutable variants, A, B, C, and D. Middle: Frequencies and recurrences of the variants. Bottom: Frequency and recurrence with continuous variation.

the study of response dimensions such as force, duration, and rate. The assumption that behavior must be defined in terms of momentary response units, however, has persisted for over a hundred years. Nineteenth-century connectionism, in the forms of associationism and reflexology, promoted a view of behavior as composed of localized units such as ideas, sensations, and responses. That view was an accident of history, however, and may be overcome. Baum and Rachlin (1969), for example, suggested that behavior be divided into periods filled with various activities. Favorite responses of the laboratory, such as the lever press and the key peck may be recast as activities, lever-pressing and key-pecking, which extend in time (Baum, 1976). Rachlin (1994) suggested that behavior generally be thought of as organized into patterns extended in time (Baum, 1995a, 1997,

2002, 2004, 2013). As with the customs of culture, an individual's daily behavior may be divided into activities like working, attending to family, and entertainment, none of which may be localized to moments in time.

Frequency distributions like those in Figure 22.1 presuppose that one may characterize the variation in a gene pool or behavior pool by taking a sort of snapshot of the frequencies at a moment in time. This might seem necessary for analytical purposes, because the frequencies may be changing. In practice, however, the snapshot is an abstraction, just as instantaneous velocity in physics is an abstraction, because any real sample is extended in time. In a gene pool, birth and death constantly alter the picture. For a behavioral pool, temporal extension is unavoidable. The customs of a culture or the habits of an individual cannot occur at a moment in time. The problem of taking a "snapshot" of frequencies is solved by choosing a time period long enough to provide an adequate sample but short enough that the pool may be thought of as unchanging for that duration. Depending on rate of change, one might gather data over a period of weeks, months, or years. As long as change in the pattern of frequencies during a period of sampling is negligible in comparison with change between samples, the course of change in the pool may be studied. That population thinking is central to genetic evolution has long been recognized. Population thinking, however, is equally central to understanding evolution of cultural behavior and individual operant behavior (Glenn et al., 1992; Hull et al., 2001).

Cultural practices and individual operant behavior

The line between cultural practices and operant behavior patterns may be fine. Indeed, Skinner (1981) argued that cultural patterns are operant patterns. Since cultural practices presumably are maintained by their consequences, they might be considered operant patterns by definition. They may be distinguished, however, for the purpose of discussing cultural evolution by two characteristics. First, cultural practices are the possession of a group. To be called a cultural practice, a behavior pattern must occur in the members of a group—possibly a subgroup within a larger group, but definitely in more than one individual. Second, cultural practices are transmitted from member to member in the group. Unlike individual operant patterns, which one may say are transmitted to the same individual at different times, cultural practices are transmitted from one individual to another and may occur in different individuals at the same time.

Recurrence

Recurrence means "occurring again" or "coming up again." It is a general term for the tendency of a type of unit to reappear, with variation, time after time in the population—a more general term than "replication," "transmission," or "retention" (Baum, 2001; Campbell, 1965).

The "copying" metaphor

Genetic evolution is usually taken as the paradigmatic example of a Darwinian evolutionary process. As a prototype, genetic evolution has both advantages and disadvantages. An advantage is that it has received enough attention to be both familiar and relatively well understood. A disadvantage is that, being comparatively well understood, some of its peculiarities are too easily mistaken for properties of evolutionary processes in general. This is particularly true of its mode of recurrence, which is often described with the metaphor of "copying."

Evolutionary recurrence need little resemble copying a page in a photocopier. The word "replication" may be a synonym for copying, but may also be equated to reconstruction, which, for DNA, would be closer to the mark. We know about the uncoiling of the strands and the assembling of amino acids into replicate strands, with some recombination and occasional errors. In meiosis, the entire genome is replicated and divided into haploid components at once. The entire haploid genome is passed as a whole to the next generation. We need to ask, however, which properties of genetic transmission are essential for evolution. Darwin knew nothing of what we know about genetics today. Yet he was the author of the theory of evolution. His ignorance of genetics in no way prevented him from seeing the role of recurrence.

Darwin understood what was necessary: that the traits of parents tend to recur in offspring. More generally, the traits of one generation are passed on to the next generation. Even the words "parent," "offspring," and "generation," impede generalization. The essential part is recurrence through time. Traits tend to "breed true." In the gene pool of a population of mice, the genes that make for a dark coat tend to recur through time, with the result that dark coats tend to recur (and possibly lower predation rate). In the beavers' gene pool, the genes that make for dam constructing tend to recur, and if substitutable variants exist, each variant tends to recur, although selection may favor one variant over others. True, the recurrence of genes is the result of the details of reproduction, and those details are essential to understanding many aspects of genetic evolution. For a general definition of evolutionary process, however, what counts is the tendency of types in a population to recur in time.

Today we forgive Darwin for having believed in the inheritance of acquired characteristics, but writers about culture point out that such inheritance does occur in cultural evolution (e.g., Boyd & Richerson, 1985). Again, however, the mechanism of recurrence is only important for understanding the details of cultural evolution. That it entails the inheritance of acquired characteristics helps us to understand the ways in which cultural evolution differs from genetic evolution. It is faster, for example, because transmission may occur between any individuals, not just parents and offspring, and may occur throughout the lifetime of an individual. It occurs as a result of imitation and instruction, although we might argue about exactly what those processes entail (Baum, 2000; 2005). The key, however, is that if a potter makes a certain kind of pot, his students will make the same kind of pot, and their students also, and so on. The practice recurs through time in the culture's pool of practices.

Suppose, instead of genetic evolution, we took behavioral evolution (i.e., shaping) as the paradigmatic example of an evolutionary process. We would see immediately that copying or replication is just one means of recurrence. A pigeon in an experiment pecks at a key time after time. A person goes to work day after day. Too little is known about the nervous system to say what the mechanism is by which behavior recurs in the same organism from time to time. Luckily, we may proceed with studying behavior without having any idea how recurrence is accomplished. The situation resembles that in biology prior to the synthesis of genetics with evolution in the early part of the twentieth century.

Imperfect recurrence

Part of the reason for variation is that recurrence is often imperfect. Hull (1988) defines a replicator as "an entity that passes on its structure largely intact in successive replications" (p. 408). A paraphrase for the sake of generality might substitute "recurrences" for "replications." Emphasis, however, should fall on the word "largely." Accidents happen. In genetic evolution, reconstruction of DNA sometimes goes awry. In cultural evolution, imitation and instruction may be inexact. In behavioral evolution, too, variation is intrinsic to recurrence, because context never exactly repeats, reconstruction even of a stereotyped activity like key-pecking varies, and accidents happen. Recurrence need only have high fidelity; it need not be perfect. Indeed, one might argue that it must be imperfect, if the pool is to include sufficient variation to produce novelty. Without novel types, a pool's response to selection must eventually cease.

Recurrence in behavior

In genetic and cultural evolution, the units (genes or practices) may outlive their possessors [their "vehicles" in Dawkins's (1989a) parlance; their "interactors" in Hull's (1988)]. Individual operant behavior, in contrast, must die with its possessor—by definition, because if it lived on in others, it would be considered cultural. In no way, however, does this disqualify behavioral shaping as an evolutionary process. The key elements remain: a pool of variation, substitutable variants, and recurrence of variants through time. The frequency distributions in Figure 22.1 might describe genetic variation, cultural variation, or behavioral variation. Whether they remain stable or change with time depends on their relation to selection.

Selection

The key to selection is differential recurrence. If substitutable variant A1 tends, on the average and over time, to recur more frequently than A2, the relative frequency of A1 increases at the expense of the relative frequency of A2. This happens if the size of the pool of variants is fixed or increases more slowly than the variants' rate of recurrence. Usually one assumes the pool size to be fixed, because the size of a population of mice, for example, is limited by the environment's carrying capacity—that is, the resources available to support the population. A behavioral pool, of cultural variants or individual behavioral variants, is fixed by limits on time, because only so much behavior can occur in a 24-hour day. Consequently, when one substitutable variant increases in frequency, another must decrease. If one variant is more successful, it tends to replace the other.

One substitutable variant succeeds over another by virtue of superior environmental effects. For genetic evolution, this is where longevity and fecundity come in. Genes endure if their vehicles (interactors) endure, and recur if their vehicles (interactors) reproduce. The specific reasons for enhanced recurrence may be extremely varied—better defending or capturing of resources, better avoidance of predators, better defense against parasites, better mate selection, more mates, more offspring, better care of offspring, and so on. All of these superior environmental effects ultimately increase the recurrence of the alleles that produce them and thus decrease the recurrence of the less successful alleles. For cultural and behavioral evolution, the same sorts of environmental effects act on the pool, but less directly and, therefore, with some slippage.

Selection in culture

The consequences of cultural practices that vary, making one variant more successful than another, may be thought of as reinforcers and punishers. Practices are maintained proximally by social reinforcement and punishment (Baum, 1995b, 2000, 2005; Skinner, 1981). "Our group shuns the eating of pork" is backed up with punishment (e.g., disapproval, ostracism) for eating pork. "This mode of adornment is correct" is backed up with reinforcement (e.g., status, mating opportunities) for adorning oneself so. Ultimately, the maintaining and differentiating social reinforcement and punishment are cashed out in terms of reproductive success of the vehicles (interactors) that carry the genes producing susceptibility to social consequences ("docility;" Simon, 1990). If group membership is beneficial to reproductive success, then group cohesion and cooperation are beneficial, and practices that make for group cohesion and cooperation are selected as a result of the members' docility—their susceptibility to social consequences. Indeed, Boyd and Richerson (1992) demonstrated theoretically that social punishment allows selection of cooperation and just about any other behavior salutary to the group. [This summary omits other adaptations, such as sensory specializations and imitation (Baum, 2000; 2005).]

Dawkins (1989a) argued that the slippage between proximal social consequences and ultimate reproductive consequences allows cultural evolution to proceed independently of genetic evolution. The logic of genetic evolution goes against such a view (El Mouden et al., 2014). If some alleles make for more docility than others, the most successful alleles lie somewhere in the middle of the range of possible levels of docility, somewhere in-between none and "anything goes." Too much docility means too little certainty of reproduction, because too much docility allows the spread and maintenance of customs that reduce reproduction. Thus, ultimately, alleles producing too much docility are selected against. [Boyd and Richerson (1985) make a similar point in their discussion of the tension between imitation and individual learning.]

One factor that curbs docility is primary reinforcement and punishment. As Dawkins (1989a; p. 57) suggests, consequences like "sweet taste in the mouth, orgasm, mild temperature, smiling child" or "various sorts of pain, nausea, empty stomach, screaming child" may have a genetic base because their presence generally affects the likelihood of successful reproduction. Such stimuli constitute proximate tokens of ultimate reproductive success or failure, called elsewhere *phylogenetically important events* (PIE; Baum, 2005, 2012a). If an interactor (vehicle) could be put together that would increase any behavior that produced the good environmental effects (circumstances that enhance reproductive success) and avoided the bad environmental effects (circumstances that depress reproductive success), such an interactor (vehicle) would prosper, and genes it carried would be more likely to recur in subsequent generations. As a result, genes that make for susceptibility to reinforcement by fitness-enhancing PIEs are selected. As a result, neither individual behavior nor cultural customs will stray too far from patterns that maintain the frequency of those PIEs. They will stray, however, for two reasons. First, the PIEs' bearing on reproductive success is far from certain. Sweet taste is a reinforcer, and some consumption causes no problems, but over-indulgence in sweets undermines health. Second, PIEs may conflict with one another. The same behavior that produces a smiling child may also produce an empty stomach. Which should win out may be far from clear; only the calculus of long-term reproduction can tell. When short-term and long-term consequences conflict this way, dysfunctional behavior may arise (Baum, 2016a; Rachlin, 1995a). Whether it persists or not depends on whether patterns that enhance fitness in the long run, such as eating fruits and vegetables, replace the dysfunctional ones, such as eating junk food (Baum, 2000, 2005, 2016a; Rachlin, 1995b). Although a person may die before dysfunctional behavior is replaced, correction in culture lies under no such limitation. It may take a few generations.

Primacy of reproductive success

Maladaptive customs tend to drop out of the culture pool for two reasons. First, failure to correct results in natural selection. Those who over-indulge in sweets tend to become sick with diabetes and leave fewer offspring as a result (Diamond, 1992). Maladaptive customs decrease in frequency when their vehicles (interactors) leave fewer descendants to continue the customs. Groups that follow dangerous or abstinent practices, such as Quantrill's Raiders or the Shakers, tend to disappear. Second, but probably more important, is that maladaptive customs contrast with their competitors (substitutable variants) in the calculus of reinforcement and punishment—i.e., in the tokens of fitness. Prohibition in the U. S. was a response to the ruinous effects of over-indulgence in alcohol; today educational campaigns encourage patterns of moderation or abstinence, based on better quality of life. When a maladaptive custom like smoking tobacco spreads through a culture because of short-term reinforcement, its long-term punishing effects eventually come into focus in the form of social contingencies that punish it more immediately (or reinforce alternatives like nicotine patches more immediately).

The power of reinforcers and punishers as tokens of long-term reproductive success makes persistence of maladaptive customs unlikely. The genetic underpinnings of reinforcement and punishment argue against Dawkins's (1989a; 1989b) conjecture that cultural evolution proceeds independently of genetic evolution. The genes that underlie learning and culture open the door, so to speak, to environmental influence, but they do not fling it wide.

Selection in all three processes—genetic, cultural, and behavioral evolution—may be traced back to reproductive success of genetic variants. In genetic evolution, the relation is direct; in cultural and behavioral evolution, the relation is less direct, but still present. When genes that make behavior sensitive to its consequences are selected, that must mean that the advantages of behavioral flexibility outweigh the costs it imposes in terms of energy and risk of error. If not, competitor genes that allow less flexibility would prevail. So, tokens of reproductive success (PIEs) attain the status of reinforcers, because variation in the environment rewards flexibility in the means to them. Signs of health, resources, and social relationships became powerful reinforcers, because the interactors (vehicles) for which they were reinforcers reproduced more often. A parallel argument applies to punishers, such as nausea, pain, snakes, and frowns. The same mechanisms of reinforcement and punishment that select among substitutable variants in behavioral evolution select also in cultural evolution. The differences are that cultural evolution entails transmission from individual to individual—imitation and instruction—and therefore depends heavily on social stimuli for provenance and social reinforcement and punishment for selection.

In all three of the processes we are discussing, a question arises about the extent of the vehicles or interactors. Hull (1988) defines an interactor as "an entity that interacts as a cohesive whole with its environment in such a way that this interaction *causes* replication [i.e., recurrence] to be differential" (p. 408; italics in the original; bracketed material added). But how large a unit can interact as a "cohesive whole"? Group selection has been treated with skepticism among evolutionary biologists, because it should normally be too weak to have much effect on gene frequencies. To work, it would require that gene flow into the group be negligible. That might be true of colonies of eusocial insects, for example (Seeley, 1989), but it would be unlikely for most groups, because of immigration of new members into the group. In cultural evolution, group cohesion is more likely, because imitation and instruction maintain a high degree of conformity within the group, even if new members join (Richerson & Boyd, 1992). In cultural group selection, groups with certain practices (e.g., a type of food cultivation or cooperation in obtaining resources) may out-compete other groups. The advantaged group may increase in frequency if it reproduces by fission and if competing groups tend to dissolve.

In evolution of individual operant behavior, where the distinction between replicators and interactors disappears, the analog to group selection is the temporally extended contingency. Extended contingencies are often discussed in terms of delays of reinforcement. The more temporally extended the behavioral pattern, the more delayed the consequences. Alternatively, one may consider delay to be incidental and the crucial aspect of extended patterns to be their cohesiveness. Extended behavioral patterns, such as eating a good diet, are notoriously difficult to maintain. Rachlin (1995b) argues that good extended patterns (often called "self-control") have a greater long-term payoff than short-term patterns (called "impulsiveness" or "defections" from the larger pattern). A defection from eating a good diet—eating an ice cream sundae—has an immediate payoff, but poorer consequences in the long run than eating a good diet (Baum, 2016a). Selection on extended patterns is weak for the same reason that group selection on genes is weak. The boundaries of the group or of the extended pattern tend to be permeable—to immigration or to defection—the greater the permeability, the weaker is selection. This may be overcome in behavioral evolution by introducing relatively short-term contingencies that maintain the extended pattern intact (e.g., reminders about one's diet).

Why not phenotypic plasticity?

To explain the variation of culture or behavior from one environment to another, the alternative to selection is phenotypic plasticity. Cosmides and Tooby (1992; Tooby & Cosmides, 1992), for example, put forward the idea that the human brain contains a large number of "modules," "algorithms," or "mechanisms" that produce behavior depending on environmental conditions. They contend, "If human thought falls into recurrent patterns from place to place and from time to time, this is because it is the expression of, and anchored in, universal psychological mechanisms (p. 216)." They refer to such patterns as "evoked culture." The conception has been compared to a jukebox containing recordings that can be played whenever called upon (Wilson, 1999). When food availability is highly variable, the "social contract algorithm" plays out food sharing; when food availability is stable, the jukebox plays keeping food within the family. Such explanations fail on two grounds. First, they are implausible. They exaggerate what is probably true, that genes constrain evolution of cultural practices and of individual behavior, to make a claim that would skip over the obvious effects of consequences on behavior. Instead of sensitivity to feedback, they substitute input-output rules; given a certain environmental input, the mechanism produces a certain output. Such a view fails to explain an obvious fact: that culture evolves even if the environment remains constant. Second, alternative explanations are more plausible. If human behavior "falls into recurrent patterns," that may be because of convergent cultural or behavioral evolution. Similar contingencies select similar behavior.

Price's Equation

George Price (1970, 1972) derived an equation, using straightforward algebra, that expresses precisely the meaning of natural selection. Imagine a population or group of N members at two time periods or "generations." A trait or property of these members x varies among them. The trait could be the presence (x=1) or absence (x=0) of a particular allele (x= 0, 0.5, or 1.0 for diploidy), or a quantitative trait like body size or coloration. At Time 1, the mean of x across individuals is $\bar{x}$. Each individual i contributes w_i surviving offspring to the population at Time 2—i's fitness or recurrence. Define a variable v_i that equals $w_i/\bar{w}$, the relative fitness of i. Each v_i is a weight; their sum equals N and their mean equals 1.0. The mean of x at Time 2 $\bar{x}'$ equals:

$$\Sigma_i v_i x_i / N + \Sigma_i v_i \Delta x_i / N,$$

where Δx_i is the change in x_i for member i from Time 1 to Time 2 due to imperfect fidelity of recurrence. Using the definition of covariance ($\Sigma_i v_i x_i / N - \bar{v}\bar{x}$), and taking the difference $\Delta\bar{x} = \bar{x}' - \bar{x}$ leads to the simplest form of the Price equation [see Price (1970) and McElreath and Boyd (2007; chapter 6) for more detail]:

$$\Delta\bar{x} = cov(v_i, x_i) + E_v(\Delta x_i). \tag{22.1}$$

Equation 22.1 says that the mean change in x equals the dependence of relative fitness on x_i plus an expected value that gives the mean change in x apart from the covariance. The covariance represents selection; if relative fitness varies positively with x_i, $\Delta\bar{x}$ is positive, and the trait increases in the population. If the covariance is negative, the trait decreases, and if the covariance is zero, the trait remains stable. The second term on the right, which expresses the tendency of x to change as it recurs from Time 1 to Time 2 due to factors usually rare or negligible (e.g., mutation or meiotic drive), may be considered close to zero for genetic evolution.

Put in more general terms, Equation 22.1 says that if recurrence is differential with respect to various levels of a trait and fidelity of recurrence is high, then the trait recurs increasingly (positive covariance) or recurs decreasingly (negative covariance) in the population. Since the equation is an identity, it describes rather than predicts, but it is a valuable analytical tool for thinking about selection, for example, group selection.

Suppose that the group under consideration is one group among many, because the population is structured into groups. Since Equation 22.1 does not depend on the sort of individuals, Equation 22.1 would also apply to selection among groups. For the sake of clarity, we index groups with the subscript g and apply Equation 22.1 to get an expression defining $\Delta\bar{x}_g$ the change in average fitness of the group. The equation says that if the covariance between levels of x_g and relative fitness v_g is positive, a group with a higher mean of x increases in x and grows in size relative to other groups. For example, if darker coloration is advantageous in avoiding predators, a group with darker coloration will grow and become darker.

In applying Equation 22.1 to group selection, however, the second term on the right, $E_v(\Delta x_g)$, is no longer negligible, because the change in x_g apart from group selection includes changes in individual members due to selection at the level of individual members. Recognizing that x_g is a mean across members in the group that may change from Time 1 to Time 2, we treat Δx_g the same way as we treated Δx and arrive at:

$$\Delta\bar{x} = cov(v_g, x_g) + E_g[cov(v_{ig}, x_{ig}) + E_v(\Delta x_{ig})] \tag{22.2}$$

where the subscript ig indicates member i in group g [see Henrich (2004) and McElreath and Boyd (2007; chapter 6) for more detail].

The right-hand side of Equation 22.2 has two components: one for selection between groups and one for selection between members within groups, the expected value term. For genetic evolution, we may ignore the expected value of Δx_{ig} in Equation 22.2 as we did the expected value of Δx_i in Equation 22.1, because the change would be due to rare or negligible factors such as meiotic drive and mutation. Equation 22.2 then simplifies to the sum of a covariance for between-group selection and expected value of covariance for within-group selection.

An alternative form of the Price equation that uses the algebraic relation between covariance and the regression coefficient β may further clarify understanding of group selection [see Henrich (2004) and McElreath and Boyd (2007; chapter 6) for more detail]:

$$\Delta\bar{x} = \beta(v_g, x_g)var(x_g) + E[\beta(v_{ig}, x_{ig})var(x_{ig})] \quad (22.3)$$

Equation 22.3 makes clear that the dependence (β) between v and x and the variance in x must both be greater than zero for selection to occur. Each term on the right implies a role for each of the three ingredients of evolution: variation ($var(x)$), selection (β), and recurrence (v).

When applied to genetic evolution, Equation 22.3 clarifies why genes for altruistic and cooperative behavior are unlikely to be selected by genetic evolution on its own. Some traits may have both between-group and within-group components that act in concert—both have positive β—such as body size (helpful in inter-group combat) or tendency to stay close to the rest of the group (avoiding predators). Genes for altruism or cooperation, however, though helpful for the group, tend to reduce the fitness of individuals that behave so. Thus, β is positive for between-group selection, but negative for within-group selection. Since variance across groups $var(x_g)$ in particular is lowered by migration between groups, the groups have to be practically completely isolated for group selection to overcome negative individual selection. Since negative β is likely to be high for any individual incurring an immediate cost for helping or cooperating with strangers, and within-group variance $var(x_{ig})$ is unlikely to be zero, even a small amount of migration between groups will ensure that between-group selection is smaller than the within-group selection, making $\Delta\bar{x}$ negative.

Equations 22.1, 22.2, and 22.3 may be applied to cultural evolution, too, with suitable changes in definition. A cultural practice that might be considered all-or-none (e.g., allowing marriage between first cousins or not, eating pork or not, and primogeniture or not) would take the place of an allele that is present or not. As with genetic evolution, however, quantitative variation in a practice (e.g., a ritual performed daily or less often, for longer or shorter duration, with more or less of some ingredient, etc.) would do just as well for the variable x.

Recurrence of a practice from Time 1 to Time 2 goes, not just vertically as in genetic evolution, but obliquely and horizontally also (Boyd and Richerson, 1985; Richerson et al., 2015). Teachers, coaches, ministers, and peers both model practices and instruct practices that have proven successful in gaining the proximate tokens of ultimate fitness (reinforcers and avoidance of punishers). Particularly if these influential people are prestigious and display the trappings of success, this modeling and instructing spur the recurrence of the practice at Time 2, denoted, as before, w_i. As before, too, v_i is relative recurrence, $w_i/\bar{w}$, and if covariance between v_i and x_i is positive, the practice spreads. The second term on the right might be considered non-negligible if one wished to incorporate errors on the part of recipients, but usually may be considered negligible because errors would likely be insignificant compared with selection [see El Mouden et al. (2014; supplemental Document S1) for a detailed derivation of Equation 22.1 for culture and further discussion)].

Equation 22.1 would be about the spread of a practice through a cultural group. If we think of cultural groups as parts of a larger population, they may compete with one another for resources and members. Equations 22.2 and 22.3 will apply, with the same changes in meaning of the variables. Equation 22.2 contains the same two components of inter-group selection and within-group selection (El Mouden et al., 2014; Henrich, 2004). If a cultural practice benefits the group, and the more members practice it, or the more members' increased x_g increases the prevalence of x_g in the group (v_g), the more positive is the first term in Equation 22.2. Whether the covariance in the second term is positive or negative depends on whether it is beneficial or harmful to the individual who engages in the practice.

To understand how an altruistic or cooperative practice can spread in a population, Equation 22.3 shows what happens if the within-group covariance for members is negative. As with genetic group selection, the variances $var(x_g)$ and $var(x_{ig})$ are crucial. In cultural group selection, biases in imitation and instruction ("transmission biases") that promote conformity lower $var(x_{ig})$ to near zero, even in the face or considerable immigration between groups. For example, if immigrants into a group tend to adopt the practice that is most frequent or above average in the group or tend to adopt the practice of prestigious or successful group members, their conformity will keep $var(x_{ig})$ low (Henrich, 2004). Thus, the cost to the individual represented by the negative covariance of the second term may be offset by the positive covariance for the group, and altruistic and cooperative practices can spread in the population. The group in which members sacrifice for the group will gain more resources than other groups and will increase in numbers at the expense of other groups (Henrich, 2004).

Equation 22.2 represents "multi-level" selection, but just at two levels—groups and individuals (McElreath and Boyd, 2007; chapter 6). A practice may have many parts, and some of these parts may change without the practice taking on a new identity. For example, a manufacturing process or a ritual may drop or add elements; the American Pledge of Allegiance after many years changed to include the words "under God," but it is still the Pledge of Allegiance, and process improvement constantly upgrades manufacture of automobiles, but the practice remains the manufacture of automobiles (Baum, 2002). When practices change in their parts, one might consider a third level of selection: within-practice selection. In Equation 22.2, Δx_{ig} may be treated as change due to competition among alternative parts within the practice and expanded as before to result in a three-term Price equation (McElreath and Boyd, 2007; chapter 6).

As far as I know, no one has applied Price's equation to behavioral evolution. The entities that may vary would no longer be individual organisms, because we are dealing with the behavior of a single organism. The varying entities are time samples within a larger time interval that constitutes the population—seconds within a minute, minutes within an hour, hours within a day, weeks within a year, and so on (Baum, 1973, 2012a). The variable x_i that differs across time samples may be occurrence or not of an activity, if the time samples are short, or rate or time taken up by the activity, if the time samples are longer, or any other quantitative aspect of an activity. The mechanism of recurrence is induction, which for operant activities is based on positive covariance with reinforcers and negative covariance with punishers (Baum, 2012a). Equation 22.1 applied to behavioral evolution means that if time samples (indexed by i) including more of an activity in Time Interval 1 correspond to relatively more such time samples in Time Interval 2 (i.e., positive covariance with v_i in Equation 22.1), then the activity ($\bar{x}$) increases from Time Interval 1 (Population 1) to Time Interval 2 (Population 2)—$\Delta \bar{x}$ is positive. In a laboratory example, if a rat's rate of lever pressing is measured in 2-minute intervals, the rate varies and may increase from one hour to the next (Baum, 2012a; Figure 17). If a person's life changes by having children, spending more time with family daily in a period (population) of a month may vary positively with relative recurrence v_i in the next month, and the amount of time spent with family increases from month to month.

As with cultural group selection, Equations 22.2 and 22.3 apply to operant activities at a level or time scale of selection that may include selection at a lower level or shorter time scale. Cultural practices are, after all, operant activities. In an ontological perspective, operant activities and cultural practices are individuals with parts that work together to serve a function (Baum, 2002; Ghiselin, 1997). The analog to group selection is selection of whole activities extended in time, like playing tennis well or poorly and relating to one's spouse well or poorly. The analog of within-group selection is the selection of parts referred to above, because every activity is composed of parts that are themselves activities on a smaller time scale (Baum, 2002, 2012a; Baum & Davison, 2004). Thus, the second term in Equation 22.2 for behavioral evolution represents selection among parts of an activity. A tennis player's serve may improve when the ball is thrown up in a new way, and that variant may be selected. Active listening may improve one's relationship with one's spouse and be selected to replace a passive stance.

Equation 22.3 for behavioral evolution illuminates problems in self-control, including altruism and cooperation. The conflict between impulsivity and self-control translates into a positive first term on the right (self-control) and a negative second term (impulsivity). Avoiding a bad habit or cultivating a good habit (e.g., refusing a drink or a piece of cake or visiting the dentist) entails the cost of forgoing immediate enjoyment or of immediate discomfort, whereas the extended pattern (x_g; sobriety, dieting, or good health) has positive β, but the positive β must suffice to offset the negative β of the second term (for x_{ig}). This offsetting may occur if $var(x_{ig})$ is reduced to zero by following a rule that enforces good behavior (refusal of the drink or cake; doing the right thing) on particular occasions. Seen this way, altruism and cooperation are examples of good habits with immediate cost and long-term benefit (Baum, 2016a; Rachlin, 1995b, 2002).

Price's equation is incomplete or limited in two ways. First, it relies on linear regression (Price, 1970). The full-range relation between x_i and v_i may be nonlinear; often it will approach an asymptote, because a stable level of x in a population, due to dominance or frequency-dependent selection, will go along with reduced variance in v_i. When the population reaches equilibrium, the covariance between v_i and x_i disappears. Thus, Price's equation only has meaning when a population is in the process of evolving and we consider a limited range of x_i over which the relation between v_i and x_i is approximately linear. It doesn't apply to the whole process of change toward equilibrium. Second, by focusing on one particular allele, trait, or variable, the equation ignores the fate of other, competing, variants. In all three evolutionary processes, if one variant increases, usually others must decrease. Competition implies a limit—carrying capacity for genetic evolution, fixed or slowly growing population for cultural evolution, and fixed time interval for behavioral evolution. A complete picture requires considering the fates of all the substitutable, competing, variants.

Evolutionary Explanations of Behavior

Mayr (1961) distinguished between proximate and ultimate explanations of behavior. Alcock (1993) incorporated the distinction into a textbook on animal behavior. According to Alcock, proximate explanations explain "how mechanisms *within* an animal operate, enabling the creature to behave in a certain way" (p. 2; emphasis in the original). Ultimate explanations explain how those mechanisms evolved as a result of selection. Proximate explanations refer to physiology and development. Ultimate explanations refer to history and reproductive success.

Genetic proximate and ultimate explanations

Proximate explanations are about individual organisms, whereas ultimate explanations are about populations. The question, "Why do beavers build dams?" may be answered in two different ways. In one interpretation, the account would refer to stimuli from water and trees, the structure of the nervous system, hormones, and so on. It would explain how the cluster of genes that make for dam building express themselves in physiology and interact with the environment to ensure that Beaver X, alive today, builds a dam. That would be a proximate explanation. The ultimate explanation, in contrast, would make no reference to Beaver X, but would be about beavers as a species or about a population of beavers or about the gene pool of a population of beavers, to which Beaver X or its genes might belong. It would refer to the adaptive consequences of building dams, variation within populations long ago, and increase in the frequency of alleles promoting dam building. It would explain how those alleles became common in the gene pool. Proximate explanations are silent about where the mechanisms for dam building came from. Ultimate explanations are silent about Beaver X, except to say that Beaver X builds a dam because it is the nature of beavers to build dams, and then to explain where that nature came from. So to speak, ultimate explanations are about the forest, whereas proximate explanations are about the trees.

Proximate explanations explain the behavior of individual organisms in terms of present mechanisms, whereas ultimate explanations explain patterns of frequency within a pool, such as those shown in Figure 22.1, in terms of a process of differential success. Proximate explanations refer to causes in the present, whereas ultimate explanations rely on processes that may be called historical, because they extend in time (Baum & Heath, 1992). Evolutionary change results from continual operation of the process of selection over a period of time. That it takes time may be incidental, because it is a process of adjustment with a beginning in a disturbance (in the environment or in the population) and an end in a stabilized population. It is understood as a whole; at any point in-between, it is incomplete. Suppose the shaded bars in the middle panel of Figure 22.1 represent the relative rates of success (recurrence) from one time period (e.g., generation) to the next. Although variant A is lowest in frequency, it has the highest rate of recurrence. Given this pattern, we expect the frequency of variant A to increase, and the frequencies of B, C, and D to decrease. If relative recurrence is independent of frequency, the process will be incomplete until A dominates. (If relative recurrence is frequency dependent, a different equilibrium results, but that in no way affects this discussion.) Suppose the curve in the lower panel of Figure 22.1 shows relative recurrence for the continuous case. The situation is essentially the same; we expect the frequency distribution to shift to the right. Eventually, we expect its maximum to coincide with the maximum of the curve; then the distribution stabilizes, and selection acts to keep it stable.

Cultural proximate and ultimate explanations

Since every evolutionary process includes the distinction between the mechanism that produces the advantageous behavior and the history of advantage for that behavior, cultural evolution and behavioral evolution also admit of the distinction between proximate and ultimate explanations (Alessi, 1992). In cultural evolution, the mechanism by which a custom is transmitted explains why it persists (i.e., why it recurs), whereas the prevalence of the custom ultimately lies in a history of competition and selection. As with the question, "Why do beavers build dams?" the question, "Why does this tribe adorn themselves with tattoos?" may be interpreted and answered in two different ways. In the proximate interpretation, the question might be reworded as, "Why do the members of the Hell's Angels motorcycle club wear the club's tattoos?" Alternatively, "Why does Tom, who belongs to Hell's Angels, wear the tattoos?" Three or more mechanisms of transmission might be involved. Tom might imitate other members of the tribe. Other members might instruct Tom: "If you want to be accepted, you should get the tattoos." Once Tom began getting tattoos, reinforcement from tribe members might lead to getting more tattoos. In the ultimate interpretation, the question might be reworded as, "How did it come about that the Hell's Angels wear those tattoos?" The answer would refer to the history of selection in the culture pool. Tattooing may have competed with other forms of adornment, such as wearing black jackets, wearing hair in a ponytail, or speaking in a certain dialect. One or two members may have gotten tattoos. Other members who saw them frequently may have imitated them as a result. Boyd and Richerson (1985) call this *frequency-dependent bias*. The first members with tattoos may have been imitated by the other members or may have been able to instruct the other members because they held high status in the club. Boyd and Richerson (1985) call this tendency to imitate success *indirect bias*. Tattooing may have worked better to identify members because other groups might wear black jackets or ponytails and because dialects vary from region to region; tattoos are permanent and unambiguous. Boyd and Richerson (1985) would call reinforcement of a custom by such an environmental effect *direct bias*. Any combination of indirect bias, frequency-dependent bias, and direct bias would result in an increase in the frequency of wearing tattoos among the tribe, until the wearing of tattoos became virtually universal.

Nesting of evolutionary processes

Genetic evolution may be thought of as an overlay on cultural evolution. Ultimate and proximate genetic explanations may be constructed for the question, "Why do the Hell's Angels wear tattoos?" They would be analogous to the explanations of dam building in beavers. The proximate rewording of the question would be, "Why did Tom get himself tattooed?" As with Beaver X, the explanation refers to stimuli (e.g., from the behavior of other members), reinforcers (e.g., from the other members), development (e.g., early exposure to Hell's Angels), and physiology (e.g., brain mechanisms)—all of which might be traced to gene expression. The genetic ultimate explanation rewords the question, "How is the wearing of tattoos beneficial to the members of the tribe?" The answer might be that it constitutes symbolic marking of the group, distinguishing it and promoting cohesion within it, which in turn promotes group selection for various forms of cooperation, which enhance the success of the group and thus boost the members' reproductive success. It would include also the advantages to alleles that make for frequency-dependent bias, indirect bias, and direct bias in competition with alleles that do not (Boyd & Richerson, 1985; Richerson & Boyd, 2000). Other stories might be told, but they would all end with the same reference to reproductive success, because in a genetic ultimate explanation genes must be selected. The biasing effects of genes ensure that any custom that decreases reproductive success, even if common for a time, is likely to disappear eventually.

The general evolutionary process

Figure 22.2A diagrams an evolutionary process in general terms and shows the different focuses of proximate and ultimate explanations. Each of the variables, *V, F, E*, or *R*, may be thought of either as a pool of variants, as a frequency distribution across variants, or as a frequency of a particular variant relative to all others. In the parlance of cybernetics, they are operands and transforms, whereas the rectangles represent processes (transformations) that produce transforms from operands (Ashby, 1956). A pool of variants (process) results in *V*, a distribution of substitutable variants or relative frequency of a particular variant relative to all its competitors. A process of expression transforms *V* into *F*, a distribution of interactors (or vehicles) or frequency of one type of interactor relative to all others. The expression process receives input from the environment, omitted from Figure 22.2 in the interest of simplicity. The input is represented by levels of variable N ($N_1 - N_6$) in Figure 22.3, which illustrates the various transformations in Figure 22.2A. The top line of Figure 22.3 shows V_i composed of three variants, A, B, and C, having equal frequencies. The different levels of the environmental variable N affect

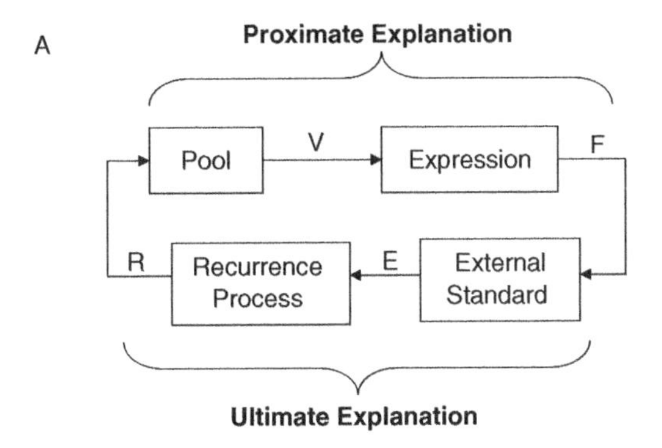

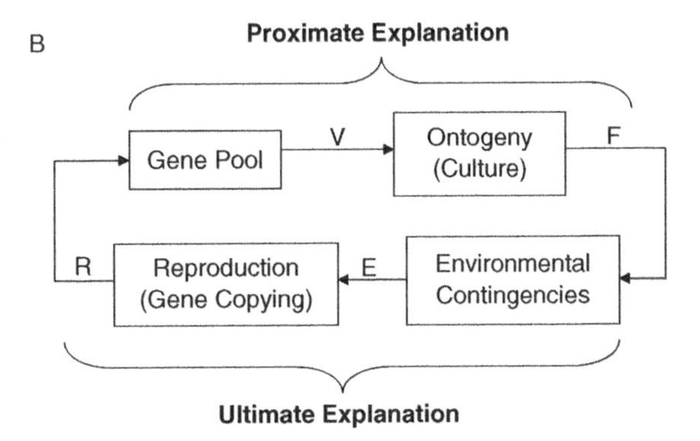

Figure 22.2 The general evolutionary process
Note: A: The general evolutionary process as a flow chart. A distribution of substitutable variants, *V*, is transformed by expression into a distribution of interactors, *F*, which an external standard transforms into a distribution of effects, *E*, which recurrence transforms into a distribution of recurrences, *R*, which enters or replaces the pool to produce a new transform *V*. B: The general evolutionary process chart applied to genetic evolution. A distribution of alleles (*V*) is transformed by ontogeny (e.g., culture) into a distribution of phenotypes (*F*), which environmental contingencies transform into a distribution of environmental effects (*E*), which gene copying transforms into a distribution of recurrences (*R*), which up-dates the gene pool and produces a new distribution of alleles.

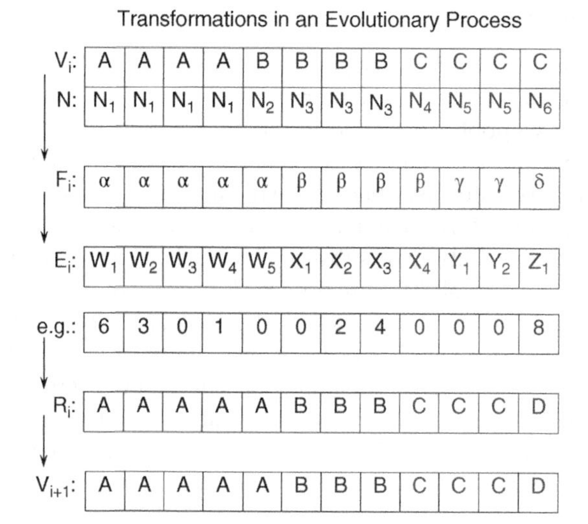

Figure 22.3 Illustration of the transformations of Figure 22.2
Note: In generation *i*, distribution *V* contains 3 substitutable variants, A, B, and C, with equal frequencies. The input N has 6 levels, $N_1 - N_6$, which enter the transformation to *F*, a distribution of interactors that contains 4 different interactors, α, β, γ, and δ, in unequal frequencies. This is transformed to a distribution of effects, labeled W, X, Y, and Z to correspond with α, β, γ, and δ. In the example, the effects for A, $W_1 - W_4$, sum to 10, the effects for B, $W_5 - X_3$, sum to 6, and the effects for C, $X_4 - Z_1$, sum to 8. These sums determine the frequencies in the distribution of recurrences *R*. A new variant (accident), D, also arises in *R*. In this example, the recurrence distribution for generation *i* becomes the distribution of substitutable variants for generation *i+1*.

the outcomes in F_i; so that B coupled with N_2 goes to α, the same outcome as A coupled with N_1, whereas B coupled with N_3 goes to β, and C transforms to β, γ, or δ, depending on *N*. In this illustration, the variation in *N* results in a distribution F_i that has more variety than V_i.

A set of external standards or contingencies, which might be characterized as a set of "if-then" rules, transforms *F* into *E*, a distribution of external effects. These are shown in Figure 22.3 as levels of *W*, *X*, *Y*, and *Z*, just to emphasize that they are transforms of the elements of F_i: *W* goes with α, *X* with β, *Y* with γ, and *Z* with δ. They may be thought of as levels of success—e.g., reinforcer rates or numbers of matings. *E* feeds back to a recurrence process, which results in R_i, a distribution of recurrence rates or a relative recurrence rate. Following the example in Figure 22.3, if the elements of *E* associated with *A* ($W_1 - W_4$) have a collective weight of 10 (6 + 3 + 1), those associated with B have a collective weight of 6, and those associated with C have a collective weight of 8 (entirely due to the fortunate occurrence of N_6, which resulted in transform δ in F_i), then the frequencies in R_i reflect these collective success rates. The appearance of variant D, however, has nothing to do with these success rates, but rather represents some kind of accident: a mutation, copying error, or external force. R_i closes the loop by entering the pooling process to result in the new distribution of variants V_{i+1}.

If this system is disturbed by a change in the pool (e.g., mutation or immigration) or in the external standards (e.g., a change in climate), resulting in disequilibrium like that depicted in Figure 22.1, it will tend to move back toward equilibrium, because iterations of the feedback loop cause the composition of the pool to change until it comes into accord with the external standards—i.e., to an optimal fit.

Figure 22.2A also illuminates the roles of proximate and ultimate explanations (braces). The links from the pool to the external standards—i.e., from *V* to *F*—constitute the focus of proximate explanations. They pose the question, "By what mechanisms does a distribution of variants result in a distribution of interactors?" The links from the external standards to the pool constitute the focus of ultimate explanations. They pose the question, "How does the distribution of external effects change the composition of the pool?" That is a question about selection. The diagram omits any explicit indication of the iterative nature of evolutionary change; that must be taken for granted. The reason that both proximate and ultimate explanations are necessary for a full understanding is that the two explanations address different parts of a whole process—mechanical connection and feedback, immediate causation and history.

Genetic evolutionary process

Figure 22.2B shows the general diagram applied to genetic evolution. The gene pool offers a distribution (*V*) of genotypes. Ontogeny or development transforms *V* into a distribution (*F*) of phenotypes. Environmental contingencies specify a distribution (*E*) of environmental effects, such as rates of obtaining resources, of predation, or of mating. Reproduction, which may be thought of as gene copying, depends on *E*. Sometimes it is thought of as simple transmission of genes from parents to offspring, but it may also be complex, because it may be the locus of other forms of selection besides natural selection—e.g., sexual selection. Its outcome *R*, the distribution of copying rates (fitnesses), in turn changes or maintains the composition of the gene pool. Proximate explanations focus on processes of ontogeny and development. Ultimate explanations focus on the feedback, the way the environmental contingencies and reproduction achieve selection among genotypes.

As an example, we may apply Figure 22.2B to the evolution of culture—that is, the transition from a non-cultural species to a cultural species as a result of genetic evolution (as opposed to cultural evolution, our second process). Ontogeny converted the distribution of genotypes *V* into a distribution of phenotypes (*F*) that varied in ability and propensity for culture—that is, in the mechanisms that cause a group of people to have a culture (group-level behavior transmitted from member to member). Elsewhere I have argued that three mechanisms would suffice: sensory specializations, imitation, and social reinforcers (Baum, 2000, 2005). Whether or not these are the answer, they suggest the sort of mechanisms that would constitute a proximate explanation for the development of culture in children and newcomers to a group. The distribution *F* might be called "variations on culture." A highly variable environment made culture beneficial to the hominids that possessed even its rudiments (Boyd & Richerson, 1985; Richerson & Boyd, 2000). It would, for example, solve many problems related to the obtaining of resources and mates. Those variants favored in the distribution of environmental effects (*E*) would be more represented in distribution *R*, with the result that the genes underlying the mechanisms for culture would increase in the gene pool. That feedback would be the ultimate explanation for the existence of culture.

Nesting cultural evolution within genetic evolution

Figure 22.4 diagrams the relationship between genetic evolution and cultural evolution and the different focuses of their proximate and ultimate explanations. It shows two feedback loops, an outer loop symbolizing genetic evolution and an inner loop (enclosed in the box) symbolizing cultural evolution (cf. Burgos, 1997). The inner loop takes the place

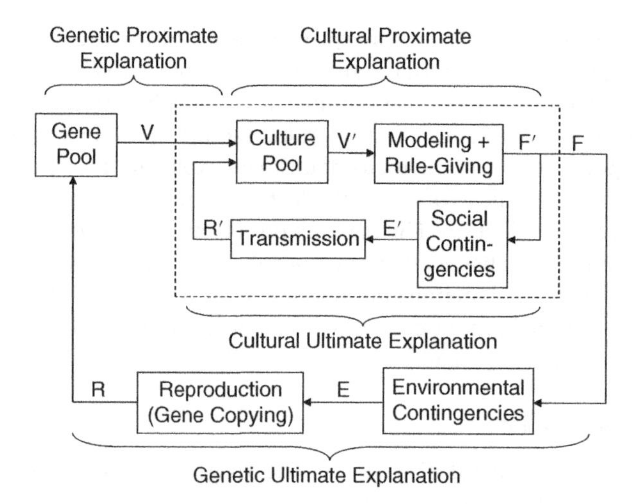

Figure 22.4 The cultural evolutionary process as an example of the general process (Figure 22.2A) and as nested within the genetic evolutionary process
Note: The rectangle of broken lines demarcates the cultural process. A distribution of all cultural variants (practices) available *V′* is transformed into a distribution of actually occurring variants *F′* by induction (e.g., modeling) and rule-giving ("instruction"). The distribution *F′* comes into contact with social contingencies, which transform it into a distribution of social effects (reinforcement and punishment), *E′*. The distribution of effects is transformed by transmission into a distribution of recurrences of the practices *R′*. The distribution of genetic variants that interact with culture *V* enters the culture pool as input, thus affecting *V′*. The distribution of actually occurring practices *F′* corresponds also to the distribution *F* that is the operand for environmental contingencies that form the context for the cultural process. The braces indicate the focuses of the different proximate and ultimate explanations.

of ontogeny in Figure 22.2B or expression in Figure 22.2A. The culture pool may be thought of as all the customs present in the group—the potential variants, resulting in a distribution *V'*. One might be tempted to think of *V'* as a distribution of memes. Our ignorance of the workings of the brain suggests that that temptation is best resisted (Baum, 2000; Simon & Baum, 2011). The mechanism of expression, labeled "Modeling + Rule-Giving," embodies the vagaries of the social environment that transform *V'* into a distribution *F'* of actual cultural variants (customs). Models and rules are stimuli that may induce imitation and rule-following, behavior that may be reinforced or punished by group members (Baum, 2000). Alternatively, one might think of the culture pool as directly producing *F'*. (As before, *F'* could also represent the relative frequency of a particular custom.) The structure of the social environment, particularly social contingencies, results in a distribution *E'* of social consequences. For example, food sharing might result in approval, whereas food hoarding might result in disapproval or aggression. From *E'*, transmission results in a distribution *R'* of recurrences that feeds back to change or maintain the composition of the culture pool. The means of transmission are imitation and instruction (Baum, 2000, 2005). They constitute the locus where Boyd and Richerson's (1985) indirect and frequency-dependent biases operate.

Figure 22.4 depicts the way in which cultural evolution and genetic evolution may interact. The distribution *F'* interacts as *F* in the external loop (genetic evolution) with environmental contingencies to produce distribution *E*—that is, cultural practices may affect genetic evolution (Boyd & Richerson, 1985). If, for example, group members who shared food were more likely to marry, then any genes that promoted tendency to share food would be selected.

The braces in Figure 22.4 indicate the different focuses of proximate and ultimate explanations in genetic and cultural evolution. Genetic proximate explanation focuses on the mechanisms by which genetic variants (distribution *V*; shown as input to the cultural pooling process) affect cultural evolution. Cultural proximate explanation focuses on the mechanisms (possibly in the nervous system, but specifically related to behavior in groups) by which cultural variants (distribution *F'*) are expressed. Modeling and exhortation, for example, might induce food sharing. Cultural ultimate explanation focuses on the iterative feedback from social consequences that shapes the culture pool over time. Once induced, for example, food sharing might be reinforced by approval, status, or reciprocation. Genetic ultimate explanation focuses on the feedback from environmental consequences of culture on the composition of the gene pool. It represents the feedback of culture on genes that prompts Boyd and Richerson (1985) to speak of "gene-culture co-evolution." Failure to keep these four different types of explanation distinct is likely to result in confusion. For example, development (genetic proximate explanation) might account for the presence of imitation and instruction, which underlie cultural ultimate explanation of cultural change over time. Cosmides and Tooby's (1992) idea that cultural variation can be explained by genetic expression probably arises from confusing these two types of explanation.

Proximate and ultimate operant explanations

In behavioral evolution (i.e., shaping), the distinction between proximate and ultimate explanations is the distinction between physiological mechanism and history of reinforcement. Advantageous behavior is defined by reinforcement and punishment. Explaining the occurrence of advantageous behavior may refer to events in the nervous system or to the history of advantage. As with the other two evolutionary processes, a question like, "Why does Liz brush her teeth before she goes to bed?" has two interpretations. The proximate interpretation focuses on the mechanism: "On any particular night, what causes Liz's tooth-brushing?" The proximate explanation would focus on stimuli that regularly precede going to bed and tooth-brushing and events in the nervous system that result from these stimuli and cause tooth-brushing at that time. The ultimate interpretation focuses on a history of selection: "How did it come about that Liz brushes her teeth before bed?" or "What advantage has Liz derived for brushing her teeth?" The ultimate explanation focuses on the differential consequences of tooth-brushing in Liz's life that selected tooth-brushing at bedtime over other behavior that might have occurred at bedtime. It would refer to bedtimes in Liz's childhood, her father's exhortations, reprimands, and approval, and the later incorporation of tooth-brushing into the pattern of behavior surrounding bedtime combined with events at the dentist's office. Different stories might be told, but they would all refer to the advantages of tooth-brushing over time.

Nesting of behavioral evolution within cultural evolution

Figure 22.5 parallels Figure 22.4, depicting the relationship between behavioral evolution and cultural evolution. The inner loop, behavioral evolution, contains the same elements as in Figure 22.2A. It stands in the place of "Modeling + Rule-Giving" in Figure 22.4 or "Expression" in Figure 22.2A. The behavior pool consists of all the individual's potential behavior—species-specific behavior plus all the behavior ever expressed in this individual. It results in a distribution *V"* of behavioral variants. The culture pool adds to the individual's behavior pool via the distribution *V'*. The mechanisms of stimulus control, including induction, transform *V"* into a distribution (or relative frequency) *F"* of

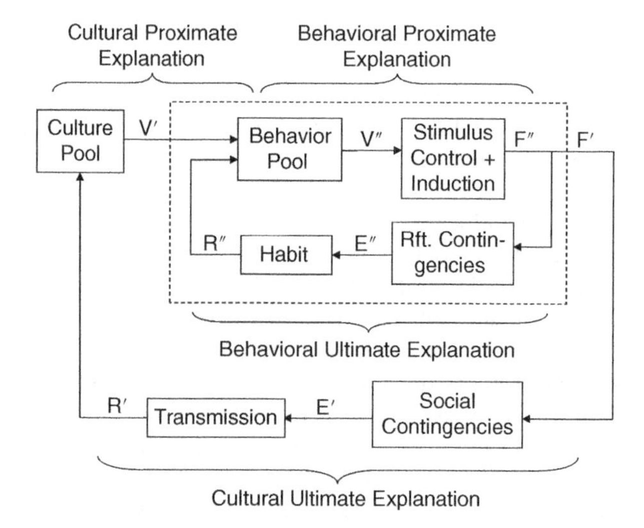

Figure 22.5 The operant (behavioral) evolutionary process as an example of the general process (Figure 22.2A) and as nested within the cultural evolutionary process
Note: The rectangle of broken lines demarcates the operant process. A distribution of all behavioral variants available *V″* is transformed into a distribution of actually occurring variants *F″* by stimulus control and induction (e.g., by releasers). The distribution *F″* comes into contact with reinforcement contingencies, which transform it into a distribution of effects (reinforcement and punishment), *E″*. The distribution of effects is transformed by habit (i.e., physiological mechanisms) into a distribution of recurrences of the behavioral variants *R″*. The distribution of cultural variants that interact with individual behavior *V′* enters the behavior pool as input, thus affecting *V″*. The distribution of actually occurring behavior *F″* corresponds also to the distribution *F′* that is the operand for social contingencies that form the context for the individual process. The braces indicate the focuses of the different proximate and ultimate explanations.

presently occurring behavior. Those mechanisms correspond to what Staddon and Simmelhag (1971) called "principles of variation." The structure of the environment, in the form of contingencies of reinforcement and punishment (or reinforcement and punishment feedback functions; Baum, 1973, 1989, 2012a), converts *F"* into a distribution *E"* of environmental effects, or consequences. These reinforcers and punishers constitute the relatively immediate tokens of ultimate reproductive success discussed earlier. Distributed differentially over behavioral variants, they result via mechanisms in the body, particularly in the nervous system, labeled "habit" in Figure 22.5, in a distribution *R"* of (relative) recurrences, which in turn changes or maintains the composition of the behavior pool. The behavioral proximate explanation focuses on the stimuli and physiological mechanisms that cause the behavior on particular occasions. The behavioral ultimate explanation refers to the selective effect that the token consequences have on the mechanism and behavior over time. A proximate explanation of sharing meat with other group members would point to the occasion of returning to one's village with a captured animal. The ultimate explanation would point to the history of exhortation in favor of sharing, social reinforcement for sharing, and punishment of hoarding. The effect of differential consequences on individuals' behavior provides the means for the cultural effects that Boyd and Richerson (1985) call guided variation and direct bias. Guided variation names the contribution of individual innovation to the culture pool. For example, the individual who makes a better arrowhead may catch more food personally, but the method also may be incorporated into the culture pool by imitation and instruction.

As we have already seen, direct bias is the effect of the token consequences on transmission of practices. The method of making the better arrowhead personally benefits anyone who uses it. The effect of behavioral evolution on cultural evolution is symbolized by the output of distribution *F"* as *F'* to cultural contingencies, which in turn produce *E'*, the distribution of social consequences.

Taken together, Figures 22.4 and 22.5 suggest two levels of nesting. Evolution of individual operant behavior is nested within cultural evolution, which is nested within genetic evolution. To understand human behavior, one needs to attend to all three levels and to all the various proximate and ultimate explanations they imply.

In a non-cultural species, the behavioral inner loop is directly nested within the genetic outer loop (cf. Burgos, 1997). In that situation, gene expression affects the mechanism that causes the behavior and is sensitive to feedback from the token consequences. As in cultural evolution, the token consequences affect gene copying, which feeds back on the composition of the gene pool, favoring alleles that promote the mechanism that allows behavioral evolution in the first place. If we ask why a laboratory rat presses a lever in an operant chamber, the answer might begin with an explanation of why the rat is so constructed that a contingency between presses and food pellets affects its behavior—i.e., the advantages of learning by reinforcement (Baum, 2005; Donahoe, 1999; Zeiler, 1999).

An example: Self-control

Behavior analysts have progressed on the problem of understanding self-control. Initially, it was analyzed as a conflict between deferred and immediate consequences (Ainslie, 1974). Self-control consists of choosing behavior that pays off in the long run over impulsiveness, behavior that pays off in the short run. For an alcoholic, acceptance of a drink is immediately reinforced, whereas abstinence, though it has a greater reward, is reinforced only at a great delay. The question remains, however, as to how the deferred consequences ever overcome the influence of the relatively immediate ones. How does self-control ever predominate? Rachlin and Green (1972) suggested one possibility: commitment. If one acts at an early time to prevent choosing impulsiveness (e.g., having part of one's salary deposited directly to a savings account before one can spend it), then self-control becomes more likely when the choice arises. Most situations calling for self-control, however, offer no opportunity for commitment, because occasions for impulsiveness are frequent and unpredictable. Some alcoholics take drugs that produce noxious effects when alcohol is in the bloodstream, and some dieters have their jaws wired shut, but most people rely on other means.

Rachlin (1995b) argued that self-control might be better understood if it were seen as a pattern of choices extended over time. Occasional impulsiveness ("defection" from the overall pattern) might occur without necessarily disrupting the extended pattern of "doing the right thing." The temporally extended pattern we call eating a healthy diet constitutes self-control even if it is peppered with minor defections like having an ice cream cone. The extended pattern, Rachlin argued, has a higher value—i.e., produces greater reinforcement—than any defection. Those greater and temporally extended consequences explain the persistence of the pattern.

Rachlin's (1995b) account stops, however, at an awkward place. Although it helps to frame the initial question, it leaves the question unanswered: How does the pattern of self-control ever emerge and persist in the face of frequent opportunities for gaining immediate reinforcement for impulsiveness? In comparison with the contingencies favoring impulsiveness—powerful because of the relative immediacy of the reinforcement—the contingencies favoring self-control are weak and vulnerable because of their temporal extendedness, as in our comparison to group selection. How could the weak selection ever prevail?

That question raises at least two others: (a) Why would an individual care about eating a healthy diet in the first place? and (b) Why would such a concern spread and persist among many members of a society? On the surface, the answers might seem obvious: quality of life and longevity. Quality of life, however, is a social construct, requiring explanation itself, and death cannot punish nor can longevity reinforce any subsequent operant behavior. Instead, the two questions may be interpreted as being about cultural and genetic evolution: (a) Why do other members of a group promote and enforce eating a healthy diet? and (b) How are genes selected that support acquisition and maintenance of behavior like eating a healthy diet? The answer to the first question requires identifying self-interest on the part of one member in other members' health. Such is near at hand, because practices concerning medicine and public health in advanced societies today are costly for the entire group (taxpayers, at least). The contingencies involved may be compared to those involved in so-called "altruism." We sacrifice to benefit others if greater benefit accrues to us eventually. The sacrificial behavior is dangerous, however, because of the possibility of cheating—i.e., someone's reaping benefits without paying the cost. When most members contribute to the cost of maintaining health in all, poor health is analogous to cheating. Customs promoting and enforcing health-enhancing behavior are selected in such a social setting (see "social contingencies" in Figures 22.4 and 22.5). The whole extended pattern of healthy eating is selected, in analogy to group selection, as in Equation 22.2.

A possible way that genes might be selected that support eating a healthy diet would be in response to variability in foods available from one environment to another. The advantages to imitation and easy instruction discussed earlier would apply to dietary habits. If you are an immigrant or child newly arrived in a group, the diet eaten by those around you is more likely to promote health than one you choose for yourself. Genes favoring imitation and easy instruction would be selected.

These suggested explanations are incomplete and may be incorrect, but they illustrate the sort of ultimate operant, cultural, and genetic explanations that an understanding of self-control requires. Many of the ideas are open to empirical study. They point to directions in research on behavior within an evolutionary framework.

Conclusion

Viewed from a sufficiently general perspective, such as diagrammed in Figure 22.2A, genetic, cultural, and behavioral evolution all may be seen as examples of the same sort of evolutionary process. Since all three allow the distinction between immediate causation of behavior and the historical origin of the causes, all three allow both proximate and ultimate explanations. Indeed, full understanding requires both proximate and ultimate explanations. The nesting of the three processes in the way suggested in Figures 22.4 and 22.5 means that the study of culture needs not only to take account of cultural evolution, but to be aware of the constraints imposed by genetic evolution and the contributions of operant behavior and its evolution to cultural evolution. Likewise, the study of human behavior needs to take account of the constraints imposed by both cultural evolution and genetic evolution. A complete explanation of a behavioral pattern needs to go beyond the processes that govern individual behavior. It needs to specify the provenance of behavior in cultural customs and in genetic effects on development and to explain the origins of effective stimuli and consequences in cultural and genetic evolution.

References

Ainslie, G., & Herrnstein, R. J. (1981). Preference reversal and delayed reinforcement. *Animal Learning & Behavior, 9,* 476–482.

Ainslie, G. W. (1974). Impulse control in pigeons. *Journal of the Experimental Analysis of Behavior, 21,* 485–489.

Alcock, J. (1993). *Animal behavior* (5th ed.). Sinauer Associates.

Alcock, J. (1998). *Animal behavior: An evolutionary approach* (6th ed.). Sinauer Associates.

Alessi, G. (1992). Models of proximate and ultimate causation in psychology. *American Psychologist, 47,* 1359–1370.

Allan, L. G., & Jenkins, H. M. (1983). The effect of representations of binary variables on judgment of influence. *Learning and Motivation, 14,* 381–405.

Allison, J. (1983). *Behavioral economics.* Praeger.

Allison, J., Miller, M., & Wozny, M. (1979). Conservation in behavior. *Journal of Experimental Psychology: General, 108,* 4–34.

Alloy, L. B., & Abramson, L. Y. (1979). Judgment of contingency in depressed and nondepressed students: Sadder but wiser? *Journal of Experimental Psychology: General, 108,* 441–485.

Alsop, B., & Elliffe, D. (1988). Concurrent-schedule performance: Effects of relative and overall reinforcer rate. *Journal of the Experimental Analysis of Behavior, 49,* 21–36.

Anger, D. (1963). The role of temporal discriminations in the reinforcement of Sidman avoidance behavior. *Journal of the Experimental Analysis of Behavior, 6,* 477–506.

Aparicio, C. F., & Baum, W. M. (2006). Fix and sample with rats in the dynamics of choice. *Journal of the Experimental Analysis of Behavior, 86,* 43–63.

Aparicio, C. F., & Baum, W. M. (2009). Dynamics of choice: Relative rate and amount affect local preference at three different time scales. *Journal of the Experimental Analysis of Behavior, 91,* 293–317.

Aparicio, C. F., & Cabrera, F. (2001). Choice with multiple alternatives: The barrier choice paradigm. *Mexican Journal of Behavior Analysis, 27,* 97–118.

Aparicio, C. F., Elcoro, M., & Alonso-Alvarez, B. (2015). A long-term study of the impulsive choices of Lewis and Fischer 344 rats. *Learning and Behavior, 43,* 251–271.

Aplin, L. M., Sheldon, B. C., & Morand-Ferron, J. (2013). Milk bottles revisited: Social learning and individual variation in the blue tit, *Cyanistes caeruleus. Animal Behaviour, 85,* 1225–1232.

Ashby, W. R. (1954). *Design for a brain.* Chapman & Hall.

Ashby, W. R. (1956). *An introduction to cybernetics.* Methuen.

Autor, S. M. (1969). *The strength of conditioned reinforcers as a function of frequency and probability of reinforcement.* [Doctoral dissertation, Harvard University], 1960. Reprinted in D. P. Hendry (Ed.), *Conditioned reinforcement* (pp. 127–162). Dorsey Press.

Science and Philosophy of Behavior: Selected Papers, First Edition. William M. Baum.

Azrin, N. H., & Holz, W. C. (1966). Punishment. In W. K. Honig (Ed.), *Operant behavior: Areas of research and application* (pp. 380–447). Appleton-Century-Crofts.
Baba, M. (1973). *God speaks* (2nd ed.). Dodd, Mead.
Baba, M. (1987). *Discourses* (7th ed.). Sheriar Press.
Baer, D. M. (1976). The organism as host. *Human Development, 19,* 87–98.
Bandura, A. (1989). Human agency in social cognitive theory. *American Psychologist, 44,* 1175–1184.
Barnard, C. J. (1980). Flock feeding and time budgets in the house sparrow (*Passer domesticus*). *Animal Behavior, 28,* 295–309.
Barnes-Holmes, D. (2000). Behavioral pragmatism: No place for reality and truth. *The Behavior Analyst, 23,* 191–202.
Baron, A., Kaufman, A., & Fazzini, D. (1969). Density and delay of punishment of free-operant avoidance. *Journal of the Experimental Analysis of Behavior, 12,* 1029–1037.
Baum, W. M. (1973). The correlation-based law of effect. *Journal of the Experimental Analysis of Behavior, 20,* 137–153.
Baum, W. M. (1974a). On two types of deviation from the matching law: bias and undermatching. *Journal of the Experimental Analysis of Behavior, 22,* 231–242.
Baum, W. M. (1974b). Choice in free-ranging wild pigeons. *Science, 185,* 78–79.
Baum, W. M. (1976). Time-based and count-based measurement of preference. *Journal of the Experimental Analysis of Behavior, 26,* 27–35.
Baum, W. M. (1979). Matching, undermatching, and overmatching in studies of choice. *Journal of the Experimental Analysis of Behavior, 32,* 269–281.
Baum, W. M. (1981a). Discrimination of correlation. In M. L. Commons & J. A. Nevin (Eds.), *Quantitative analyses of behavior: Discriminative properties of reinforcement schedules* (Vol. I, pp. 247–256). Ballinger.
Baum, W. M. (1981b). Optimization and the matching law as accounts of instrumental behavior. *Journal of the Experimental Analysis of Behavior, 36,* 387–403.
Baum, W. M. (1986). Performance on ratio and interval schedules: Some data and some theory. Invited address at the convention of the Association for Behavior Analysis, Milwaukee.
Baum, W. M. (1987). Random and systematic foraging, experimental studies of depletion, and schedules of reinforcement. In A. C. Kamil, J. R. Krebs, & H. R. Pulliam (Eds.), *Foraging behavior* (pp. 587–607). Plenum Press.
Baum, W. M. (1988). Selection by consequences is a good idea. *Behavioral and Brain Sciences, 11,* 447–448.
Baum, W. M. (1989). Quantitative prediction and molar description of the environment. *The Behavior Analyst, 12,* 167–176.
Baum, W. M. (1992). In search of the feedback function for variable-interval schedules. *Journal of the Experimental Analysis of Behavior, 57,* 365–375.
Baum, W. M. (1993a). Performances on ratio and interval schedules of reinforcement: Data and theory. *Journal of the Experimental Analysis of Behavior, 59,* 245–264.
Baum, W. M. (1993b). The status of private events in behavior analysis. *Behavioral and Brain Sciences, 16,* 644.
Baum, W. M. (1994a). John B. Watson and behavior analysis: Past, present, and future. In J. T. Todd & E. K. Morris (Eds.), *Modern perspectives on John B. Watson and classical behaviorism* (pp. 133–140). Greenwood Press.
Baum, W. M. (1994b). *Understanding behaviorism: Science, behavior, and culture.* HarperCollins.

Baum, W. M. (1995a). Introduction to molar behavior analysis. *Mexican Journal of Behavior Analysis, 21*, 7–25.
Baum, W. M. (1995b). Rules, culture, and fitness. *The Behavior Analyst, 18*, 1–21.
Baum, W. M. (1995c). Radical behaviorism and the concept of agency. *Behaviorology, 3*, 93–106.
Baum, W. M. (1997). The trouble with time. In L. J. Hayes & P. M. Ghezzi (Eds.), *Investigations in behavioral epistemology* (pp. 47–59). Context Press.
Baum, W. M. (1998). Why not ask, "Does the chimpanzee have a soul?". *Behavioral and Brain Sciences, 21*, 116.
Baum, W. M. (2000). Being concrete about culture and cultural evolution. In N. S. Thompson & F. Tonneau (Eds.), *Perspectives in ethology: Evolution, culture, and behavior* (Vol. 13, pp. 181–212). Kluwer Academic/Plenum.
Baum, W. M. (2001). Molar versus molecular as a paradigm clash. *Journal of the Experimental Analysis of Behavior, 75*, 338–341.
Baum, W. M. (2002). From molecular to molar: A paradigm shift in behavior analysis. *Journal of the Experimental Analysis of Behavior, 78*, 95–116.
Baum, W. M. (2003a). The molar view of behavior and its usefulness in behavior analysis. *The Behavior Analyst Today, 4*, 78–81.
Baum, W. M. (2003b). *Understanding behaviorism: Science, behavior, and culture*. Oxford: Blackwell Publishing. (Originally published in 1994.)
Baum, W. M. (2004). Molar and molecular views of choice. *Behavioural Processes, 66*, 349–359.
Baum, W. M. (2005). *Understanding behaviorism: Behavior, culture, and evolution* (2nd ed.). Blackwell Publishing.
Baum, W. M. (2010). Dynamics of choice: A tutorial. *Journal of the Experimental Analysis of Behavior, 94*, 161–174.
Baum, W. M. (2011a). What is radical behaviorism? A review of Jay Moore's *Conceptual foundations of radical behaviorism. Journal of the Experimental Analysis of Behavior, 95*, 119–126.
Baum, W. M. (2011b). Evasion, private events, and pragmatism: A reply to Moore's response to my review of *Conceptual foundations of radical behaviorism. Journal of the Experimental Analysis of Behavior, 95*, 141–144.
Baum, W. M. (2011c). Behaviorism, private events, and the molar view of behavior. *The Behavior Analyst, 34*, 185–200.
Baum, W. M. (2012a). Rethinking reinforcement: Allocation, induction, and contingency. *Journal of the Experimental Analysis of Behavior, 97*, 101–124.
Baum, W. M. (2012b). Mathematics and theory in behavior analysis: Remarks on Catania (1981), "The flight from experimental analysis." *European Journal of Behavior Analysis, 13*, 177–179.
Baum, W. M. (2012c). Extinction as discrimination. *Behavioural Processes, 90*, 101–110.
Baum, W. M. (2013). What counts as behavior: The molar multiscale view. *The Behavior Analyst, 36*, 283–293.
Baum, W. M. (2015). The role of induction in operant schedule performance. *Behavioural Processes, 114*, 26–33.
Baum, W. M. (2016a). Driven by consequences: The multiscale molar view of choice. *Decision and Managerial Economics, 37*, 239–248.
Baum, W. M. (2016b). On the impossibility of mental causation: Comments on Burgos' Antidualism and antimentalism in radical behaviorism. *Behavior and Philosophy, 44*, 1–5.

Baum, W. M. (2017a). *Understanding behaviorism: Behavior, culture, and evolution* (3rd ed.). Wiley Blackwell Publishing.
Baum, W. M. (2017b). Ontology for behavior analysis: Not realism, classes, or objects, but individuals and processes. *Behavior and Philosophy, 45*, 63–78.
Baum, W. M. (2017c). Selection by consequences, behavioral evolution, and the Price equation. *Journal of the Experimental Analysis of Behavior, 107*, 321–342.
Baum, W. M. (2018a). Three laws of behavior: Allocation, induction, and covariance. *Behavior Analysis: Research and Practice, 18*, 239–251.
Baum, W. M. (2018b). Multiscale behavior analysis and molar behaviorism: An overview. *Journal of the Experimental Analysis of Behavior, 110*, 302–322.
Baum, W. M. (2019). Relativity in hearing and stimulus discrimination. *Perspectives on Behavioral Science, 42*, 283–289.
Baum, W. M. (2020). Avoidance, induction, and the illusion of reinforcement. *Journal of the Experimental Analysis of Behavior, 114*, 116–141.
Baum, W. M. (2021). Matching, induction, and covariance with mixed response-contingent food and non-contingent food. *Journal of the Experimental Analysis of Behavior, 116*, 21–43.
Baum, W. M., & Aparicio, C. F. (1999). Optimality and concurrent variable-interval variable-ratio schedules. *Journal of the Experimental Analysis of Behavior, 71*, 75–89.
Baum, W. M., & Aparicio, C. F. (2020). Response-reinforcer contiguity versus response-rate-reinforcer-rate covariance in rats' lever pressing: Support for a multiscale view. *Journal of the Experimental Analysis of Behavior, 113*, 530–548.
Baum, W. M., & Davison, M. (2004). Choice in a variable environment: Visit patterns in the dynamics of choice. *Journal of the Experimental Analysis of Behavior, 81*, 85–127.
Baum, W. M., & Davison, M. (2014a). Choice with frequently-changing food rates and food ratios. *Journal of the Experimental Analysis of Behavior, 101*, 246–274.
Baum, W. M., & Davison, M. (2014b). Background activities, induction, and behavioral allocation in operant performance. *Journal of the Experimental Analysis of Behavior, 102*, 213–230.
Baum, W. M., & Grace, R. C. (2020). Matching theory and induction explain operant performance. *Journal of the Experimental Analysis of Behavior, 113*, 390–418.
Baum, W. M., & Heath, J. L. (1992). Behavioral explanations and intentional explanations in psychology. *American Psychologist, 47*, 1312–1317.
Baum, W. M., & Mitchell, S. H. (2000). Newton and Darwin: Can this marriage be saved? *Behavioral and Brain Sciences, 23*, 91–92.
Baum, W. M., & Rachlin, H. C. (1969). Choice as time allocation. *Journal of the Experimental Analysis of Behavior, 12*, 861–874.
Baum, W. M., Schwendiman, J. W., & Bell, K. E. (1999). Choice, contingency discrimination, and foraging theory. *Journal of the Experimental Analysis of Behavior, 71*, 355–373.
Belke, T. W. (1997). Running and responding reinforced by the opportunity to run: Effect of reinforcer duration. *Journal of the Experimental Analysis of Behavior, 67*, 337–351.
Belke, T. W., & Belliveau, J. (2001). The generalized matching law describes choice on concurrent variable-interval schedules of wheel-running reinforcement. *Journal of the Experimental Analysis of Behavior, 75*, 299–310.
Belke, T. W., Pierce, W. D., & Duncan, I. D. (2006). Reinforcement value and substitutability of sucrose and wheel running: Implications for activity anorexia. *Journal of the Experimental Analysis of Behavior, 86*, 131–158.
Bell, M. C., & Baum, W. M. (2017). Concurrent variable-interval variable-ratio schedules in a dynamic choice environment. *Journal of the Experimental Analysis of Behavior, 108*, 367–397.

Benedict, R. (1934). *Patterns of culture*. New American Library.
Bennett, M. R., & Hacker, P. M. S. (2003). *Philosophical foundations of neuroscience*. Blackwell Publishing.
Berens, N. M., & Hayes, S. C. (2007). Arbitrarily applicable comparative relations: Experimental evidence for a relational operant. *Journal of Applied Behavior Analysis, 40*, 45–71.
Berkeley, G. (1710/1939). Principles of human knowledge. In E. A. Burtt (Ed.), *The English philosophers from Bacon to Mill* (pp. 509–579). Random House.
Blakely, E., & Schlinger, H. (1987). Rules: Function-altering contingency-specifying stimuli. *The Behavior Analyst, 10*, 183–187.
Bloomfield, T. M. (1972). Reinforcement schedules: Contingency or contiguity? In R. M. Gilbert & J. R. Millenson (Eds.), *Reinforcement* (pp. 165–208). Academic Press.
Boren, J. J., & Sidman, M. A. (1957). Discrimination based upon repeated conditioning and extinction of avoidance behavior. *Journal of Comparative and Physiological Psychology, 50*, 18–22.
Boswell, J. (1791/2007). *Life of Samuel Johnson*. Penguin.
Boutros, N., Davison, M., & Elliffe, D. (2011). Contingent stimuli signal subsequent reinforcer ratios. *Journal of the Experimental Analysis of Behavior, 96*, 39–61.
Bower, G., McLean, J., & Meacham, J. (1967). Value of knowing when reinforcement is due. *Journal of Comparative and Physiological Psychology, 62*, 184–192.
Boyd, R., & Richerson, P. J. (1985). *Culture and the evolutionary process*. University of Chicago Press.
Boyd, R., & Richerson, P. J. (1992). Punishment allows the evolution of cooperation (or anything else) in sizable groups. *Ethology and Sociobiology, 13*, 171–195.
Brackney, R. J., & Sanabria, F. (2015). The distribution of response bout lengths and its sensitivity to reinforcement. *Journal of the Experimental Analysis of Behavior, 104*, 167–185.
Breland, K., & Breland, M. (1961). The misbehavior of organisms. *American Psychologist, 16*, 681–684.
Brown, P. L., & Jenkins, H. M. (1968). Auto-shaping of the pigeon's key peck. *Journal of the Experimental Analysis of Behavior, 11*, 1–8.
Brownstein, A. J., & Pliskoff, S. S. (1968). Some effects of relative reinforcement rate and changeover delay in response-independent concurrent schedules of reinforcement. *Journal of the Experimental Analysis of Behavior, 11*, 683–688.
Buege, D. J. (1997). An ecologically-informed ontology for environmental ethics. *Biology and Philosophy, 12*, 1–20.
Bullock, D. H., & Smith, W. C. (1953). An effect of repeated conditioning-extinction upon operant strength. *Journal of Experimental Psychology, 46*, 349–352.
Burgos, J. E. (1997). Evolving artificial neural networks in Pavlovian environments. In J. W. Donahoe & V. P. Dorsel (Eds.), *Neural-networks models of cognition* (pp. 58–79). Elsevier Science.
Burgos, J. E. (2016). Antidualism and antimentalism in radical behaviorism. *Behavior and Philosophy, 43*, 1–37.
Buzzard, J. H., & Hake, D. F. (1984). Stimulus control of schedule-induced activity in pigeons during multiple schedules. *Journal of the Experimental Analysis of Behavior, 42*, 191–209.
Byrd, L. D. (1969). Responding in the cat maintained under response-independent electric shock and response-produced electric shock. *Journal of the Experimental Analysis of Behavior, 12*, 1–10.

Caird, J. K., Willness, C. R., Steel, P., & Scialfa, C. (2008). A meta-analysis of the effects of cell phones on driver performance. *Accident Analysis and Prevention, 40,* 1282–1293.

Campbell, D. T. (1956). Adaptive behavior from random response. *Behavioral Science, 1,* 105–110.

Campbell, D. T. (1965). Variation and selective retention in socio-cultural evolution. In H. R. Barringer, G. I. Blanksten, & R. W. Mack (Eds.), *Social change in developing areas: A reinterpretation of evolutionary theory* (pp. 19–49). Schenkman.

Capra, F. (1982). *The turning point: Science, society, and the rising culture.* Simon & Schuster.

Capra, F. (1983). *The Tao of physics* (2nd ed.). Shambhala.

Carlson, R., & Shield, B. (Eds.) (1989). *Healers on healing.* Perigee Books.

Catania, A. C. (1963a). Concurrent performances: reinforcement interaction and response independence. *Journal of the Experimental Analysis of Behavior, 6,* 253–263.

Catania, A. C. (1963b). Concurrent performances: a baseline for the study of reinforcement magnitude. *Journal of the Experimental Analysis of Behavior, 6,* 299–300.

Catania, A. C. (1973). The concept of the operant in the analysis of behavior. *Behaviorism, 1,* 103–115.

Catania, A. C. (1981). The flight from experimental analysis. In C. M. Bradshaw, E. Szabadi, & C. F. Lowe (Eds.), *Quantification of steady-state operant behavior* (pp. 49–64). Elsevier North-Holland.

Catania, A. C. (2011). On Baum's public claim that he has no significant private events. *The Behavior Analyst, 34,* 227–236.

Catania, A. C., & Harnad, S. E. (1984). Canonical papers of B. F. Skinner. *Behavioral and Brain Sciences, 7,* 473–724.

Catania, A. C., & Reynolds, G. S. (1968). A quantitative analysis of the responding maintained by interval schedules of reinforcement. *Journal of the Experimental Analysis of Behavior, 11,* 327–383.

Cerutti, D. T. (1989). Discrimination theory of rule-governed behavior. *Journal of the Experimental Analysis of Behavior, 51,* 259–276.

Chatterji, M. M. (1960). *The Bhagavad Gitā.* Causeway Books.

Chiesa, M. (1992). Radical behaviorism and scientific frameworks: From mechanistic to relational accounts. *American Psychologist, 47,* 1287–1299.

Chiesa, M. (1994). *Radical behaviorism: The philosophy and the science.* Authors Cooperative.

Chopra, D. (1993). *Ageless body, timeless mind.* Harmony Books.

Chung, S., & Herrnstein, R. J. (1967). Choice and delay of reinforcement. *Journal of the Experimental Analysis of Behavior, 10,* 67–74.

Clark, F. C., & Hull, L. D. (1966). Free operant avoidance as a function of the response-shock = shock-shock interval. *Journal of the Experimental Analysis of Behavior, 9,* 641–647.

Córdoba-Salgado, O. (2017). Extended behavior-context relations: A molar view of functional analytic psychotherapy. *The Behavior Analyst, 40,* 257–273.

Cosmides, L., & Tooby, J. (1992). Cognitive adaptations for social exchange. In J. H. Barkow, L. Cosmides, & J. Tooby (Eds.), *The Adapted Mind: Evolutionary Psychology and the Generation of Culture* (pp. 163–228). Oxford University Press.

Courtney, K., & Perone, M. (1992). Reductions in shock frequency and response effort as factors in reinforcement by timeout from avoidance. *Journal of the Experimental Analysis of Behavior, 58,* 485–496.

Cowie, S., & Davison, M. (2016). Control by reinforcers across time and space: A review of recent choice research. *Journal of the Experimental Analysis of Behavior, 105,* 246–269.

Cowie, S., Davison, M., & Elliffe, D. (2011). Reinforcement: Food signals the time and location of future food. *Journal of the Experimental Analysis of Behavior, 96*, 63–86.
Dallery, J., McDowell, J. J., & Lancaster, J. S. (2000). Falsification of matching theory's account of single-alternative responding: Herrnstein's *k* varies with sucrose concentration. *Journal of the Experimental Analysis of Behavior, 73*, 23–43.
Davison, M. (1993). On the dynamics of behavior allocation between simultaneously and successively available reinforcer sources. *Behavioural Processes, 29*, 49–64.
Davison, M. (2004). Interresponse times and the structure of choice. *Behavioural Processes, 66*, 173–187. https://doi.org/10.1016/j.beproc.2004.03.003
Davison, M., & Baum, W. M. (2000). Choice in a variable environment: Every reinforcer counts. *Journal of the Experimental Analysis of Behavior, 74*, 1–24.
Davison, M., & Baum, W. M. (2006). Do conditional reinforcers count? *Journal of the Experimental Analysis of Behavior, 86*, 269–283.
Davison, M., & Baum, W. M. (2010). Stimulus effects on local preference: Stimulus-response contingencies, stimulus-food pairing, and stimulus-food correlation. *Journal of the Experimental Analysis of Behavior, 93*, 45–59. https://doi.org/10.1901/jeab.2010.93-45
Davison, M., & McCarthy, D. (1988). *The matching law: A research review*. Erlbaum Associates.
Davison, M. C. (1969). Preference for mixed-interval *versus* fixed-interval schedules. *Journal of the Experimental Analysis of Behavior, 12*, 247–252.
Dawkins, R. (1989a). *The selfish gene* (new ed.). Oxford University Press.
Dawkins, R. (1989b). *The extended phenotype: The long reach of the gene*. Oxford University Press.
Dennet, D. C. (1978). Skinner skinned. In *Brainstorms: Philosophical essays on mind and psychology* (pp. 53–70). MIT Press.
Dennett, D. C. (1984). *Elbow room: The varieties of free will worth wanting*. MIT Press.
Dews, P. B. (1960). Free-operant behavior under conditions of delayed reinforcement. I. CRF-type schedules. *Journal of the Experimental Analysis of Behavior, 3*, 221–234.
Dews, P. B. (1962). The effect of multiple SΔ periods on responding on a fixed-interval schedule. *Journal of the Experimental Analysis of Behavior, 5*, 369–374.
Dews, P. B. (1965). The effect of multiple SΔ periods on responding on a fixed-interval schedule: III. Effect of changes in pattern of interruptions, parameters, and stimuli. *Journal of the Experimental Analysis of Behavior, 8*, 427–435.
Diamond, J. M. (1992). Diabetes running wild. *Nature, 357*, 362–363.
Dickinson, A., Shanks, D., & Evenden, J. (1984). Judgement of act-outcome contingency: The role of selective attribution. *Quarterly Journal of Experimental Psychology, 36A*, 29–50.
Dinsmoor, J. A. (2001). Stimuli inevitably generated by behavior that avoids electric shock are inherently reinforcing. *Journal of the Experimental Analysis of Behavior, 75*, 311–333.
Donahoe, J. W. (1999). Edward L. Thorndike: The selectionist connectionist. *Journal of the Experimental Analysis of Behavior, 72*, 451–454.
Donahoe, J. W. (2012). Origins of the molar–molecular debate. *European Journal of Behavior Analysis, 2*, 195–200.
Duncan, B., & Fantino, E. (1970). Choice for periodic schedules of reinforcement. *Journal of the Experimental Analysis of Behavior, 14*, 73–86.
Eibl-Eibesfeldt, I. (1975). *Ethology: The biology of behavior* (2nd ed.). Holt, Rinehart, and Winston.
El Mouden, C., André, J. B., Morin, O., & Nettle, D. (2014). Cultural transmission and the evolution of human behaviour: a general approach based on the Price equation. *Journal of Evolutionary Biology, 27*, 231–241.

Epstein, R. (1984). Spontaneous and deferred imitation in the pigeon. *Behavioural Processes, 9*, 347–354.
Estes, W. K. (1943). Discriminative conditioning. I. A discriminative property of conditioned anticipation. *Journal of Experimental Psychology, 32*, 150–155.
Estes, W. K. (1948). Discriminative conditioning. II. Effects of a Pavlovian conditioned stimulus upon a subsequently established operant response. *Journal of Experimental Psychology, 38*, 173–177.
Ettinger, R. H., Reid, A. K., & Staddon, J. E. R. (1987). Sensitivity to molar feedback functions: A test of molar optimality theory. *Journal of Experimental Psychology: Animal Behavior Processes, 13*, 366–375.
Falk, J. L. (1971). The nature and determinants of adjunctive behavior. *Physiology and Behavior, 6*, 577–588.
Falk, J. L. (1977). The origin and functions of adjunctive behavior. *Animal Learning and Behavior, 5*, 325–335.
Fantino, E. (1968). Effects of required rates of responding upon choice. *Journal of the Experimental Analysis of Behavior, 11*, 15–22.
Farrington, B. (1944). *Greek science* (1980th ed.). Spokesman.
Felton, M., & Lyon, D. O. (1966). The post-reinforcement pause. *Journal of the Experimental Analysis of Behavior, 9*, 131–134.
Ferster, C. B., & Skinner, B. F. (1957). *Schedules of reinforcement*. Appleton-Century-Crofts.
Fleshler, M., & Hoffman, H. S. (1962). A progression for generating variable-interval schedules. *Journal of the Experimental Analysis of Behavior, 5*, 529–530.
Foxall, G. R. (2007). Intentional behaviorism. *Behavior and Philosophy, 35*, 1–55.
Galizio, M. (1999). Extinction of responding maintained by timeout from avoidance. *Journal of the Experimental Analysis of Behavior, 71*, 1–11.
Gallistel, C. R., King, A. P., Gottlieb, D., Balci, F., Papachristos, E. B., Szalecki, M., & Carbone, K. S. (2007). Is matching innate? *Journal of the Experimental Analysis of Behavior, 87*, 161–199.
Gallistel, C. R., Mark, T. A., King, A. P., & Latham, P. (2001). The rat approximates an ideal detector of changes in rates of reward: Implications for the law of effect. *Journal of Experimental Psychology: Animal Behavior Processes, 27*, 354–372.
Gardner, E. T., & Lewis, P. (1976). Negative reinforcement with shock-frequency increase. *Journal of the Experimental Analysis of Behavior, 25*, 3–14.
Gardner, R. A., & Gardner, B. T. (1988). Feedforward versus feedbackward: An ethological alternative to the law of effect. *Behavioral and Brain Sciences, 11*, 429–493.
Garrett, R. (1987). Practical reason and a science of morals. In S. Modgil & C. Modgil (Eds.), *B. F. Skinner: Consensus and controversy* (pp. 319–327). Falmer Press.
Ghiselin, M. T. (1981). Categories, life, and thinking. *Behavioral and Brain Sciences, 4*, 269–313.
Ghiselin, M. T. (1997). *Metaphysics and the origin of species*. State University of New York Press.
Ghiselin, M. T. (2018). B.F. Skinner and the metaphysics of Darwinism. *Perspectives on Behavior Science, 41*, 269–281.
Gilbert, R. M. (1970). Psychology and biology. *The. Canadian Psychologist, 11*, 221–238.
Gilbert, T. F. (1958). Fundamental dimensional properties of the operant. *Psychological Review, 65*, 272–285.
Glenn, S. S. (1988). Contingencies and metacontingencies: Toward a synthesis of behavior analysis and cultural materialism. *The Behavior Analyst, 11*, 161–179.

Glenn, S. S. (1991). Contingencies and metacontingencies: Relations among behavioral, cultural, and biological evolution. In P. A. Lamal (Ed.), *Behavioral analysis of societies and cultural practices* (pp. 39–73). Hemisphere.
Glenn, S. S., Ellis, J., & Greenspoon, J. (1992). On the revolutionary nature of the operant as a unit of behavioral selection. *American Psychologist, 47,* 1329–1336.
Glenn, S. S., & Field, D. P. (1994). Functions of the environment in behavioral evolution. *The Behavior Analyst, 17,* 241–259.
Gollub, L. R. (1970). Information on conditioned reinforcement. *Journal of the Experimental Analysis of Behavior, 14,* 361–372.
Green, L., & Freed, D. E. (1993). The substitutability of reinforcers. *Journal of the Experimental Analysis of Behavior, 60,* 141–158.
Green, L., & Myerson, J. (2013). How many impulsivities? A discounting perspective. *Journal of the Experimental Analysis of Behavior, 99,* 3–13.
Green, L., Myerson, J., Oliveira, L., & Chang, S. E. (2014). Discounting of delayed and probabilistic losses over a range of amounts. *Journal of the Experimental Analysis of Behavior, 101,* 186–200.
Green, L., & Rachlin, H. (1991). Economic substitutability of electrical brain stimulation, food, and water. *Journal of the Experimental Analysis of Behavior, 55,* 133–143.
Guerin, B. (1992). Behavior analysis and the social construction of knowledge. *American Psychologist, 47,* 1423–1432.
Guerin, B. (1997). How things get done: Socially, non-socially; with words, without words. In L. J. Hayes & P. M. Ghezzi (Eds.), *Investigations in behavioral epistemology* (pp. 219–235). Context Press.
Guerin, B. (1998). Religious behaviors as strategies for organizing groups of people: A social contingency analysis. *The Behavior Analyst, 21,* 53–72.
Hailman, J. P. (1969). How an instinct is learned. *Scientific American, 221,* 98–108.
Harris, M. (1980). *Cultural materialism.* Vintage Books.
Harris, M. (1987). Foodways: Historical overview and theoretical prolegomenon. In M. Harris & E. B. Ross (Eds.), *Food and evolution* (pp. 57–90). Temple University Press.
Harris, M., & Ross, E. B. (1987). *Death, sex, and fertility.* Columbia University Press.
Hayes, L. J., & Chase, P. N. (Eds.) (1991). *Dialogues on verbal behavior.* Context.
Hayes, S. C. (1984). Making sense of spirituality. *Behaviorism, 12*(2), 99–110.
Hayes, S. C. (Ed.) (1989). *Rule-governed behavior: Cognition, contingencies, and instructional control.* Plenum.
Henrich, J. (2004). Cultural group selection, coevolutionary processes and large-scale cooperation. *Journal of Economic Behavior and Organization, 53,* 3–35.
Herrnstein, R. J. (1961). Relative and absolute strength of response as a function of reinforcement. *Journal of the Experimental Analysis of Behavior, 4,* 267–272.
Herrnstein, R. J. (1964). Secondary reinforcement and rate of primary reinforcement. *Journal of the Experimental Analysis of Behavior, 7,* 27–36.
Herrnstein, R. J. (1969). Method and theory in the study of avoidance. *Psychological Review, 76,* 49–69.
Herrnstein, R. J. (1970). On the law of effect. *Journal of the Experimental Analysis of Behavior, 13,* 243–266.
Herrnstein, R. J. (1974). Formal properties of the matching law. *Journal of the Experimental Analysis of Behavior, 21,* 159–164.
Herrnstein, R. J. (1977). The evolution of behaviorism. *American Psychologist, 32,* 593–603.
Herrnstein, R. J. (1979). Derivatives of matching. *Psychological Review, 86,* 486–495.

Herrnstein, R. J., & Boring, E. G. (1966). *A source book in the history of psychology*. Harvard University Press.

Herrnstein, R. J., & Heyman, G. M. (1979). Is matching compatible with reinforcement maximization on concurrent variable interval, variable ratio? *Journal of the Experimental Analysis of Behavior, 31*, 209–223.

Herrnstein, R. J., & Hineline, P. N. (1966). Negative reinforcement as shock-frequency reduction. *Journal of the Experimental Analysis of Behavior, 9*, 421–430.

Herrnstein, R. J., & Loveland, D. H. (1964). Complex visual concept in the pigeon. *Science, 146*, 549–551.

Herrnstein, R. J., & Loveland, D. H. (1972). Food-avoidance in hungry pigeons, and other perplexities. *Journal of the Experimental Analysis of Behavior, 18*, 369–383.

Herrnstein, R. J., & Loveland, D. H. (1975). Maximizing and matching on concurrent ratio schedules. *Journal of the Experimental Analysis of Behavior, 24*, 107–116.

Herrnstein, R. J., Loveland, D. H., & Cable, C. (1976). Natural concepts in pigeons. *Journal of Experimental Psychology: Animal Behavior Processes, 2*, 285–302.

Heyes, C. M. (1998). Theory of mind in nonhuman primates. *Behavioral and Brain Sciences, 21*, 101–114.

Heyman, G. M. (1982). Is time allocation unconditioned behavior? In M. Commons, R. J. Herrnstein, & H. Rachlin (Eds.), *Quantitative analyses of behavior: Matching and maximizing accounts* (Vol. 2, pp. 459–490). Ballinger Press.

Heyman, G. M. (2009). *Addiction: A disorder of choice*. Harvard University Press.

Heyman, G. M., & Herrnstein, R. J. (1986). More on concurrent interval-ratio schedules: a replication and review. *Journal of the Experimental Analysis of Behavior, 46*, 331–351.

Heyman, G. M., & Tanz, L. (1995). How to teach a pigeon to maximize overall reinforcement rate. *Journal of the Experimental Analysis of Behavior, 64*, 277–297.

Hinde, R. A., & Stevenson-Hinde, J. (Eds.) (1973). *Constraints on learning*. Academic Press.

Hineline, P. (1995). The origins of environment-based psychological theory. In J. T. Todd & E. K. Morris (Eds.), *Modern perspectives on B. F. Skinner and contemporary behaviorism* (pp. 85–106). Greenwood Press.

Hineline, P. N. (1970). Negative reinforcement with shock reduction. *Journal of the Experimental Analysis of Behavior, 14*, 259–268.

Hineline, P. N. (1977). Negative reinforcement and avoidance. In W. K. Honig & J. E. R. Staddon (Eds.), *Handbook of operant behavior* (pp. 364–414). Prentice-Hall.

Hineline, P. N. (1978). Warmup in free-operant avoidance as a function of the response-shock = shock-shock interval. *Journal of the Experimental Analysis of Behavior, 30*, 281–291.

Hineline, P. N. (1980). The language of behavior analysis: Its community, its functions, and its limitations. *Behaviorism, 8(1)*, 67–86.

Hineline, P. N. (1984). Aversive control: A separate domain? *Journal of the Experimental Analysis of Behavior, 42*, 495–509.

Hineline, P. N. (2001). Beyond the molar-molecular distinction: We need multiscaled analyses. *Journal of the Experimental Analysis of Behavior, 75*, 342–347.

Hineline, P. N. (2011). Private versus inner in multiscaled interpretation. *The Behavior Analyst, 34*, 221–226.

Hineline, P. N., & Wanchisen, B. A. (1989). Correlated hypothesizing and the distinction between contingency-shaped and rule-governed behavior. In S. C. Hayes (Ed.), *Rule-governed behavior: Cognition, contingencies, and instructional control* (pp. 221–268). Plenum.

Hinson, J. M., & Staddon, J. E. R. (1983). Matching, maximizing, and hill-climbing. *Journal of the Experimental Analysis of Behavior, 40*, 321–331.
Hocutt, M. (2013). A behavioral analysis of morality and value. *The Behavior Analyst, 36*, 239–249.
Hocutt, M. (2018). George Berkeley Resurrected: A commentary on Baum's "Ontology for behavior analysis." *Behavior and Philosophy, 46*, 47–57.
Hoffman, H. S., Fleshler, M., & Chorny, H. (1961). Discriminated bar-press avoidance. *Journal of the Experimental Analysis of Behavior, 4*, 309–316.
Hollard, V., & Davison, M. C. (1971). Preference for qualitatively different reinforcers. *Journal of the Experimental Analysis of Behavior, 16*, 375–380.
Holt, E. B. (1965). *The Freudian wish and its place in ethics*. Johnson Reprint Corp. (Originally published in 1915 by Henry Holt and Co., New York.)
Holz, W. C. (1968). Punishment and rate of positive reinforcement. *Journal of the Experimental Analysis of Behavior, 11*, 285–292.
Honig, W. K., & Staddon, J. E. R. (Eds.) (1977). *Handbook of operant behavior*. Prentice-Hall.
Horne, P. J., & Lowe, C. F. (1996). On the origins of naming and other symbolic behavior. *Journal of the Experimental Analysis of Behavior, 65*, 185–241.
Houston, A. I., & McNamara, J. (1981). How to maximize reward rate on two variable-interval paradigms. *Journal of the Experimental Analysis of Behavior, 35*, 367–396.
Hull, D. L. (1988). *Science as a process: An evolutionary account of the social and conceptual development of science*. University of Chicago Press.
Huxley, A. (1945). *The perennial philosophy*. Harper & Row.
James, W. (1902). *The varieties of religious experience*. Longmans, Green, & Co.
James, W. (1974). *Pragmatism and four essays from The meaning of truth*. New American Library. (Reprinting of editions of 1907 and 1909.)
Jenkins, H. M., & Moore, B. R. (1973). The form of the auto-shaped response with food or water reinforcers. *Journal of the Experimental Analysis of Behavior, 20*, 163–181.
Jensen, G. (2014). Compositions and their application to the analysis of choice. *Journal of the Experimental Analysis of Behavior, 102*, 1–25.
Jensen, G., & Neuringer, A. (2009). Barycentric extension of generalized matching. *Journal of the Experimental Analysis of Behavior, 92*, 139–159.
Kelleher, R. T., & Morse, W. H. (1968). Schedules using noxious stimuli. III. Responding maintained with response-produced electric shocks. *Journal of the Experimental Analysis of Behavior, 11*, 819–838.
Keller, F. S., & Schoenfeld, W. N. (1950). *Principles of psychology*. Appleton-Century-Crofts.
Killeen, P. (1968). On the measurement of reinforcement frequency in the study of preference. *Journal of the Experimental Analysis of Behavior, 11*, 263–269.
Killeen, P. R. (1994). Mathematical principles of reinforcement. *Behavioral and Brain Sciences, 17*, 105–171.
Killeen, P. R., & Jacobs, K. W. (2017). Coal is not black, snow is not white, food is not a reinforcer: The roles of affordances and dispositions in the analysis of behavior. *The Behavior Analyst, 40*, 17–38.
Kimble, G. A. (1961). *Hilgard and Marquis' Conditioning and Learning* (2nd ed.). Appleton-Century-Crofts.
Kish, G. B. (1966). Studies of sensory reinforcement. In W. K. Honig (Ed.), *Operant behavior: Areas of research and application* (pp. 109–159). Appleton-Century-Crofts.
Koffka, K. (1935). *Principles of gestalt psychology*. Harcourt, Brace, & World. (Reprinted in 1963.)
Köhler, W. (1947). *Gestalt psychology*. The New American Library.

Krägeloh, C. U., Davison, M., & Elliffe, D. M. (2005). Local preference in concurrent schedules: The effects of reinforcer sequences. *Journal of the Experimental Analysis of Behavior, 84*, 37–64.

Krebs, J. R., & Davies, N. B. (1993). *An introduction to behavioural ecology* (3rd ed.). Blackwell Scientific.

Krebs, J. R., & Kacelnik, A. (1991). Decision-making. In J. R. Krebs & N. B. Davies (Eds.), *Behavioural ecology: An evolutionary approach* (3rd ed., pp. 105–136). Blackwell Scientific.

Kuhn, T. S. (1970). *The structure of scientific revolutions* (2nd ed.). University of Chicago Press.

Kuroda, T., Cançado, C. R. X., Lattal, K. A., Elcoro, M., Dickson, C. A., & Cook, J. E. (2013). Combinations of response-reinforcer relations in periodic and aperiodic schedules. *Journal of the Experimental Analysis of Behavior, 99*, 199–210.

Kuroda, T., & Lattal, K. A. (2018). Behavioral control by the response–reinforcer correlation. *Journal of the Experimental Analysis of Behavior, 110*, 185–200.

Lashley, K. S. (1961). The problem of serial order in behavior. In S. Saporta (Ed.), *Psycholinguistics: A book of readings* (pp. 180–198). Holt, Rinehart, and Winston. (Originally published in 1951.)

Lattal, K. A., & Gleeson, S. (1990). Response acquisition with delayed reinforcement. *Journal of Experimental Psychology: Animal Behavior Processes, 16*, 27–39.

Lee, V. L. (1983). Behavior as a constituent of conduct. *Behaviorism, 11*, 199–224.

Locey, M., & Rachlin, H. (2012). Commitment and self-control in a prisoner's dilemma game. *Journal of the Experimental Analysis of Behavior, 98*, 89–103.

Locey, M. L., & Rachlin, H. (2013). Shaping behavioral patterns. *Journal of the Experimental Analysis of Behavior, 99*, 245–259.

Logan, F. A. (1960). *Incentive*. Yale University Press.

Logue, A. W. (1978). Behaviorist John B. Watson and the continuity of the species. *Behaviorism, 6*, 71–79.

Logue, A. W. (1988). Research on self-control: An integrating framework. *Behavioral and Brain Sciences, 11*, 665–709.

Logue, A. W. (1994). Watson's behaviorist manifesto: Past positive and current negative consequences. In J. T. Todd & E. K. Morris (Eds.), *Modern perspectives on John B. Watson and classical behaviorism* (pp. 109–123). Greenwood.

Logue, A. W. (1995). *Self-control: Waiting until tomorrow for what you want today*. Prentice Hall.

Logue, A. W., & de Villiers, P. A. (1978). Matching in concurrent variable-interval avoidance schedules. *Journal of the Experimental Analysis of Behavior, 29*, 61–66.

Lubinski, D., & Thompson, T. (1993). Species and individual differences in communication based on private states. *Behavioral and Brain Sciences, 16*, 627–680.

Lumsden, C. J., & Wilson, E. O. (1981). *Genes, mind, and culture: The coevolutionary process*. Harvard University Press.

Mach, E. (1960). *The science of mechanics: A critical and historical account of its development*. Open Court. (Translation of the ninth German edition, 1933.)

Mahner, M., & Bunge, M. (1997). *Foundations of biophilosophy*. Springer-Verlag.

Malagodi, E. F., Gardner, M. L., Ward, S. E., & Magyar, R. L. (1981). Responding maintained under intermittent schedules of electric-shock presentation: "Safety" or schedule effects? *Journal of the Experimental Analysis of Behavior, 36*, 171–190.

Malagodi, E. F., & Jackson, K. (1989). Behavior analysts and cultural analysis: Troubles and issues. *The Behavior Analyst, 12*, 17–33.

Mallott, R. W. (2001). Moral and legal control. *Behavioral Development Bulletin, 1*, 1–7.
Mayr, E. (1959). Darwin and the evolutionary theory in biology. In B. J. Meggers (Ed.), *Evolution and anthropology: A centennial appraisal* (pp. 1–10). Anthropological Society of Washington.
de Villiers, P. A. (1972). Reinforcement and response rate interaction in multiple random-interval avoidance schedules. *Journal of the Experimental Analysis of Behavior, 18*, 499–507.
de Villiers, P. A. (1974). The law of effect and avoidance: A quantitative relationship between response rate and shock-frequency reduction. *Journal of the Experimental Analysis of Behavior, 21*, 223–235.
Tzu, L. (1955). *The way of life: Tao te ching.* New American Library.
Mayr, E. (1961). Cause and effect in biology. *Science, 134*, 1501–1506.
Mayr, E. (1970). *Populations, species, and evolution.* Harvard University Press.
Mazur, J. E. (1986). *Learning and behavior.* Prentice-Hall.
Mazur, J. E. (1997). Effects of rate of reinforcement and rate of change on choice behavior in transition. *Quarterly Journal of Experimental Psychology B, 50*, 111–128.
Mazur, J. E., & Logue, A. W. (1978). Choice in a "self-control" paradigm: Effects of a fading procedure. *Journal of the Experimental Analysis of Behavior, 30*, 11–17.
McDowell, J. J. (1986). On the falsifiability of matching theory. *Journal of the Experimental Analysis of Behavior, 45*, 63–74.
McDowell, J. J. (1987). A mathematical theory of reinforcer value and its application to reinforcement delay in simple schedules. In M. L. Commons, J. E. Mazur, J. A. Nevin, & H. Rachlin (Eds.), *Quantitative analyses of behavior, The effect of delay and of intervening events on reinforcement value* (Vol. 5, pp. 77–105). Erlbaum.
McElreath, R., & Boyd, R. (2007). *Mathematical models of social evolution: A guide for the perplexed.* University of Chicago Press.
McSweeney, F. K., Farmer, V. A., Dougan, J. D., & Whipple, J. E. (1986). The generalized matching law as a description of multiple-schedule responding. *Journal of the Experimental Analysis of Behavior, 45*, 83–101.
Mellitz, M., Hineline, P. N., Whitehouse, W. G., & Laurence, M. T. (1983). Duration-reduction of avoidance sessions as negative reinforcement. *Journal of the Experimental Analysis of Behavior, 40*, 57–67.
Miller, H. L. J. (1976). Matching-based hedonic scaling in the pigeon. *Journal of the Experimental Analysis of Behavior, 26*, 335–347.
Moerk, E. L. (1992). *A first language taught and learned.* Paul H. Brookes Publishing.
Moerk, E. L. (1996). Input and learning processes in first language acquisition. In H. W. Reese (Ed.), *Advances in child development and behavior* (Vol. 26, pp. 181–228). Academic Press.
Moore, J. (1995). Radical behaviorism and the subjective-objective distinction. *The Behavior Analyst, 18*, 33–49.
Moore, J. (2008). *Conceptual foundations of radical behaviorism.* Sloan Publishing.
Morgan, C. L. (1894). *An introduction to comparative psychology.* W. Scott Ltd.
Morse, W. H. (1966). Intermittent reinforcement. In W. K. Honig (Ed.), *Operant behavior: Areas of research and application* (pp. 52–108). Appleton-Century-Crofts.
Neuringer, A. J. (1967). Effects of reinforcement magnitude on choice and rate of responding. *Journal of the Experimental Analysis of Behavior, 10*, 417–424.
Neuringer, A. J. (1969). Animals respond for food in the presence of free food. *Science, 166*, 399–401.
Neuringer, A. J. (1970). Many responses per food reward with free food present. *Science, 169*, 503–504.

Nevin, J. A. (1974). Response strength in multiple schedules. *Journal of the Experimental Analysis of Behavior, 21*, 389–408.

Nevin, J. A. (1992). An integrative model for the study of behavioral momentum. *Journal of the Experimental Analysis of Behavior, 57*, 301–316.

Nevin, J. A., & Grace, R. C. (2000). Behavioral momentum and the law of effect. *Behavioral and Brain Sciences, 23*, 73–130.

Nevin, J. A., Tota, M. E., Torquato, R. D., & Shull, R. L. (1990). Alternative reinforcement increases resistance to change: Pavlovian or operant contingencies? *Journal of the Experimental Analysis of Behavior, 53*, 359–379.

Nicholson, D. J. (2018). Reconceptualizing the organism: From complex machine to flowing stream. In D. J. Nicholson & J. Dupré (Eds.), *Everything flows: Towards a processual philosophy of biology* (pp. 139–166). Oxford University Press.

Nisbett, R. E., & Wilson, T. D. (1977). Telling more than we can know: Verbal reports on mental processes. *Psychological Review, 84*, 231–259.

Nicholson, D. J., & Dupré, J. (2018). *Everything flows: Towards a processual philosophy of biology*. Oxford University Press.

Odum, A. L. (2011). Delay discounting: I'm a k, you're a k. *Journal of the Experimental Analysis of Behavior, 96*, 427–439.

Öhman, A., & Mineka, S. (2003). The malicious serpent: Snakes as a prototypical stimulus for an evolved module of fear. *Current Directions in Psychological Science, 12*, 5–9.

Ostlund, S. B., & Balleine, B. W. (2007). Selective reinstatement of instrumental performance depends on the discriminative stimulus properties of the mediating outcome. *Learning and Behavior, 35*, 43–52.

Palya, W. L. (1992). Dynamics in the fine structure of schedule-controlled behavior. *Journal of the Experimental Analysis of Behavior, 57*, 267–287.

Palya, W. L., & Zacny, J. P. (1980). Stereotyped adjunctive pecking by caged pigeons. *Animal Learning and Behavior, 8*, 293–303.

Pavlov, I. P. (1927). *Conditioned reflexes*. Dover.

Pavlov, I. P. (1960/1927). *Conditioned Reflexes: An Investigation of the Physiological Activity of the Cerebral Cortex*. Dover.

Pear, J. J. (1985). Spatiotemporal patterns of behavior produced by variable-interval schedules of reinforcement. *Journal of the Experimental Analysis of Behavior, 44*, 217–231.

Pear, J. J., Hemingway, M. J., & Keiser, P. (1978). Lever attacking and pressing as a function of conditioning and extinguishing a lever-press avoidance response in rats. *Journal of the Experimental Analysis of Behavior, 29*, 273–282.

Pear, J. J., Moody, J. E., & Persinger, M. A. (1972). Lever attacking by rats during free-operant avoidance. *Journal of the Experimental Analysis of Behavior, 18*, 517–523.

Petrovich, S. B., & Gewirtz, J. L. (1991). Imprinting and attachment: Proximate and ultimate considerations. In J. L. Gewirtz & W. M. Kurtines (Eds.), *Intersections with attachment* (pp. 69–93). Erlbaum.

Pinker, S. (1994). *The language instinct*. HarperCollins.

Pinker, S., & Bloom, P. (1992). Natural language and natural selection. In J. H. Barkow, L. Cosmides, & J. Tooby (Eds.), *The adapted mind: Evolutionary psychology and the generation of culture* (pp. 451–493). Oxford University Press.

Ploog, B. O., & Zeigler, H. P. (1997). Key-peck probability and topography in a concurrent variable-interval variable-interval schedule with food and water reinforcers. *Journal of the Experimental Analysis of Behavior, 67*, 109–129.

Poling, A., Edwards, T. L., Weeden, M., & Foster, M. T. (2011). The matching law. *Psychological Record, 61,* 313–322.
Powell, R. W., & Peck, S. (1969). Persistent shock-elicited responding engendered by a negative-reinforcement procedure. *Journal of the Experimental Analysis of Behavior, 12,* 1049–1062.
Prabhavananda, S., & Isherwood, C. (Eds.) (1944). *The song of God: Bhagavad-Gita.* Harper.
Premack, D. (1963). Rate differential reinforcement in monkey manipulation. *Journal of the Experimental Analysis of Behavior, 6,* 81–89.
Premack, D. (1965). Reinforcement theory. In D. Levine (Ed.), *Nebraska symposium on motivation* (pp. 123–180). University of Nebraska Press.
Premack, D. (1971). Catching up with common sense, or two sides of a generalization: reinforcement and punishment. In R. Glaser (Ed.), *The nature of reinforcement* (pp. 121–150). Academic Press.
Price, G. R. (1970). Selection and covariance. *Nature, 227,* 520–521.
Price, G. R. (1972). Extension of covariance selection mathematics. *Annals of Human Genetics, 35,* 485–490.
Pulliam, H. R. (1983). On the theory of gene-culture co-evolution in a variable environment. In R. L. Mellgren (Ed.), *Animal cognition and behavior* (pp. 427–443). Elsevier/North-Holland.
Rachlin, H. (1971). On the tautology of the matching law. *Journal of the Experimental Analysis of Behavior, 15,* 249–251.
Rachlin, H. (1976). *Behavior and learning.* Freeman.
Rachlin, H. (1985). Pain and behavior. *Behavioral and Brain Sciences, 8,* 43–83.
Rachlin, H. (1991). *Introduction to modern behaviorism* (3rd ed.). Freeman.
Rachlin, H. (1992). Teleological behaviorism. *American Psychologist, 47,* 1371–1382.
Rachlin, H. (1994). *Behavior and mind: The roots of modern psychology.* Oxford University Press.
Rachlin, H. (1995a). Self-control: Beyond commitment. *Behavioral and Brain Sciences, 18,* 109–159.
Rachlin, H. (1995b). The value of temporal patterns in behavior. *Current Directions in Psychological Science, 4,* 188–192.
Rachlin, H. (2000). *The science of self-control.* Harvard University Press.
Rachlin, H. (2002). Altruism and selfishness. *Behavioral and Brain Sciences, 25,* 239–296.
Rachlin, H. (2003). Privacy. In K. A. Lattal & P. N. Chase (Eds.), *Behavior theory and philosophy* (pp. 187–201). Kluwer Academis/Plenum.
Rachlin, H. (2014). *Escape of the mind.* Oxford University Press.
Rachlin, H. (2018). Skinner (1938) and Skinner (1945). *Behavior and Philosophy, 46,* 100–113.
Rachlin, H., & Baum, W. M. (1969). Response rate as a function of amount of reinforcement for a signaled concurrent response. *Journal of the Experimental Analysis of Behavior, 12,* 11–16.
Rachlin, H., & Baum, W. M. (1972). Effects of alternative reinforcement: Does the source matter? *Journal of the Experimental Analysis of Behavior, 18,* 231–241.
Rachlin, H., Battalio, R. C., Kagel, J. H., & Green, L. (1981). Maximization theory in behavioral psychology. *Behavioral and Brain Sciences, 4,* 371–417.
Rachlin, H., & Burkhard, B. (1978). The temporal triangle: Response substitution in instrumental conditioning. *Psychological Review, 85,* 22–47.
Rachlin, H., & Green, L. (1972). Commitment, choice, and self-control. *Journal of the Experimental Analysis of Behavior, 17,* 15–22.

Rachlin, H., Green, L., Kagel, J. H., & Battalio, R. C. (1976). Economic demand theory and psychological studies of choice. *Psychology of Learning and Motivation, 10*, 129–154.

Rachlin, H., Green, L., & Tormey, B. (1988). Is there a decisive test between matching and maximizing? *Journal of the Experimental Analysis of Behavior, 50*, 113–123.

Rachlin, H., & Herrnstein, R. J. (1969). Hedonism revisited: On the negative law of effect. In B. Campbell & R. M. Church (Eds.), *Punishment and aversive behavior* (pp. 83–109). Appleton-Century-Crofts.

Rachlin, H., & Krasnoff, J. (1983). Eating and drinking: An economic analysis. *Journal of the Experimental Analysis of Behavior, 39*, 385–404.

Rachlin, H., & Locey, M. (2011). A behavioral analysis of altruism. *Behavioural Processes, 87*, 25–33.

Rapport, D. J. (1980). Optimal foraging for complementary resources. *The American Naturalist, 116*, 324–346.

Rapport, D. J. (1981). Foraging of Stentor coeruleus: A microeconomic interpretation. In A. C. Kamil & T. D. Sargent (Eds.), *Foraging behavior* (pp. 77–93). Garland Press.

Reed, P., Hildebrandt, T., DeJongh, J., & Soh, M. (2003). Rats' performance on variable-interval schedules with a linear feedback loop between response rate and reinforcement rate. *Journal of the Experimental Analysis of Behavior, 79*, 157–173.

Reese, H. W. (1989). Rules and rule-governance: Cognitive and behavioristic views. In S. C. Hayes (Ed.), *Rule-governed behavior: Cognition, contingencies, and instructional control* (pp. 3–84). Plenum.

Reid, A. K., Bacha, G., & Moran, C. (1993). The temporal organization of behavior on periodic food schedules. *Journal of the Experimental Analysis of Behavior, 59*, 1–27.

Reid, R. L. (1958). The role of the reinforcer as a stimulus. *British Journal of Psychology, 49*, 202–209.

Rescorla, R. A. (1967). Pavlovian conditioning and its proper control procedures. *Psychological Review, 74*, 71–80.

Rescorla, R. A. (1968). Probability of shock in the presence and absence of CS in fear conditioning. *Journal of Comparative and Physiological Pyschology, 66*, 1–5.

Rescorla, R. A. (1988). Pavlovian conditioning: It's not what you think it is. *American Psychologist, 43*, 151–160.

Richerson, P. J., & Boyd, R. (2000). Built for speed: Pleistocene climate variation and the origin of human culture. In N. S. Thompson & F. Tonneau (Eds.), *Perspectives in ethology* (Vol. 13, pp. 1–45). Kluwer Academic/Plenum.

Richerson, P. J., Baldini, R., Bell, A., Demps, K., Frost, K., Hillis, V., . . . Zefferman, M. R. (2015). *Behavioral and Brain Sciences, 39*, E30.

Richerson, P. J., & Boyd, R. (2005). *Not by genes alone: How culture transformed human evolution*. University of Chicago Press.

Ringen, J. (1999). Radical behaviorism: B. F. Skinner's philosophy of science. In W. O'Donohue & R. Kitchener (Eds.), *Handbook of behaviorism* (pp. 160–178). Academic Press.

Rodewald, A. M., Hughes, C. E., & Pitts, R. C. (2010). Development and maintenance of choice in a dynamic environment. *Journal of the Experimental Analysis of Behavior, 94*, 175–195.

Roper, T. J. (1973). Nesting material as a reinforcer for female mice. *Animal Behaviour, 21*, 733–740.

Roper, T. J. (1978). Diversity and substitutability of adjunctive activities under fixed-interval schedules of food reinforcement. *Journal of the Experimental Analysis of Behavior, 30*, 83–96.

Rorty, R. (1979). *Philosophy and the mirror of nature*. Princeton University Press.
Rorty, R. (1989). *Contingency, irony, and solidarity*. Cambridge University Press.
Ryle, G. (1949). *The concept of mind*. University of Chicago Press.
Sanabria, F., Sitomer, M. T., & Killeen, P. R. (2006). Negative automaintenance omission training is effective. *Journal of the Experimental Analysis of Behavior, 86*, 1–10.
Schick, K. (1971). Operants. *Journal of the Experimental Analysis of Behavior, 15*, 413–423.
Schlinger, H. D., Jr. (2011). Introduction: Private events in a natural science of behavior. *The Behavior Analyst, 34*, 181–184.
Schneider, B. A. (1969). A two-state analysis of fixed-interval responding in the pigeon. *Journal of the Experimental Analysis of Behavior, 12*, 677–687.
Schneider, J. W. (1970). *Conditioned reinforcement and delay of reinforcement in concurrent-chain schedules*. [Doctoral dissertation, Harvard University].
Schneider, J. W. (1972). Choice between two-component chained and tandem schedules. *Journal of the Experimental Analysis of Behavior, 18*, 45–60.
Schneider, S. M., & Davison, M. (2005). Demarcated response sequences and generalized matching. *Behavioural Processes, 70*, 51–61.
Schoenfeld, W. N., & Cole, B. K. (1972). *Stimulus schedules: The t-τ systems*. Harper & Row.
Schoenfeld, W. N., & Farmer, J. (1970). Reinforcement schedules and the "behavior stream." In W. N. Schoenfeld (Ed.), *The theory of reinforcement schedules* (pp. 215–245). Appleton-Century-Crofts.
Schoneberger, T. (2016). Behavioral pragmatism: Making a place for reality and truth. *The Behavior Analyst, 39*, 219–242.
Schrödinger, E. (1961/1983). *My view of the world*. Ox Bow Press.
Schrödinger, E. (1956). *What is life? and other scientific essays*. Doubleday.
Schuster, R. H. (1969). A functional analysis of conditioned reinforcement. In D. P. Hendry (Ed.), *Conditioned reinforcement* (pp. 192–234). Dorsey Press.
Schuster, R., & Rachlin, H. (1968). Indifference between punishment and free shock: Evidence for the negative law of effect. *Journal of the Experimental Analysis of Behavior, 11*, 777–786.
Schwartz, B., & Williams, D. R. (1972). Two different kinds of key peck in the pigeon: Some properties of responses maintained by negative and positive response-reinforcer contingencies. *Journal of the Experimental Analysis of Behavior, 18*, 201–216.
Seeley, T. D. (1989). The honey bee colony as a superorganism. *American Scientist, 77*, 546–553.
Segal, E. F. (1972). Induction and the provenance of operants. In R. M. Gilbert & J. R. Millenson (Eds.), *Reinforcement: Behavioral Analyses* (pp. 1–34). Academic.
Seligman, M. E. P. (1970). On the generality of the laws of learning. *Psychological Review, 77*, 406–418.
Seligman, M. E. P., Maier, S. F., & Solomon, R. L. (1971). Unpredictable and uncontrollable aversive events. In F. R. Brush (Ed.), *Aversive conditioning and learning* (pp. 347–400). Academic Press.
Shahan, T. A. (2017). Moving beyond reinforcement and response strength. *The Behavior Analyst, 40*, 107–121.
Shahan, T. A., & Craig, A. R. (2017). Resurgence as choice. *Behavioural Processes, 141*, 100–127.
Shapiro, D. H. (1978). *Precision nirvana*. Prentice Hall.
Shapiro, M. M. (1961). Salivary conditioning in dogs during fixed-interval reinforcement contingent on lever pressing. *Journal of the Experimental Analysis of Behavior, 4*, 361–364.

Shimp, C. P. (2020). Molecular (moment-to-moment) and molar (aggregate) analyses of behavior. *Journal of the Experimental Analysis of Behavior, 114*, 394–429.
Shull, R. L., Gaynor, S. T., & Grimes, J. A. (2001). Response rate viewed as engagement bouts: Effects of relative reinforcement and schedule type. *Journal of the Experimental Analysis of Behavior, 75*, 247–274.
Shull, R. L., & Grimes, J. A. (2003). Bouts of responding from variable-interval reinforcement of lever pressing by rats. *Journal of the Experimental Analysis of Behavior, 80*, 159–171.
Shull, R. L., Grimes, J. A., & Bennett, J. A. (2004). Bouts of responding: The relation between bout rate and the rate of variable-interval reinforcement. *Journal of the Experimental Analysis of Behavior, 81*, 65–83.
Shull, R. L., & Pliskoff, S. S. (1967). Changeover delay and concurrent schedules: some effects on relative performance measures. *Journal of the Experimental Analysis of Behavior, 10*, 517–527.
Sidman, M. (1953). Two temporal parameters of the maintenance of avoidance behavior by the white rat. *Journal of Comparative and Physiological Psychology, 46*, 253–261.
Sidman, M. (1962). Reduction of shock frequency as reinforcement for avoidance behavior. *Journal of the Experimental Analysis of Behavior, 5*, 247–257.
Sidman, M. (1966). Avoidance behavior. In W. K. Honig (Ed.), *Operant behavior: Areas of research and application* (pp. 448–498). Appleton-Century-Crofts.
Sidman, M., Herrnstein, R. J., & Conrad, D. G. (1957). Maintenance of avoidance behavior by unavoidable shocks. *Journal of Comparative and Physiological Psychology, 50*, 553–557.
Simon, C., & Baum, W. M. (2011). Expelling the meme-ghost from the machine: An evolutionary explanation for the spread of cultural practices. *Behavior and Philosophy, 39(40)*, 127–144.
Simon, C., & Baum, W. M. (2017). Allocation of speech in conversation. *Journal of the Experimental Analysis of Behavior, 107*, 258–278.
Simon, H. A. (1990). A mechanism for social selection and successful altruism. *Science, 250*, 1665–1668.
Skinner, B. F. (1935/1961). The generic nature of the concepts of stimulus and response. In *Cumulative record* (Enlarged ed., pp. 347–366). Appleton-Century-Crofts. (Original work published 1935.)
Skinner, B. F. (1938). *Behavior of organisms.* Appleton-Century-Crofts.
Skinner, B. F. (1945). The operational analysis of psychological terms. *Psychological Review, 52*, 270–277.
Skinner, B. F. (1948). "Superstition" in the pigeon. *Journal of Experimental Psychology, 38*, 168–172.
Skinner, B. F. (1950). Are theories of learning necessary? *Psychological Review, 57*, 193–216.
Skinner, B. F. (1953). *Science and human behavior.* Macmillan.
Skinner, B. F. (1957). *Verbal behavior.* Appleton-Century-Crofts.
Skinner, B. F. (1961/1953). The analysis of behavior. In *Cumulative record* (Enlarged ed., pp. 70–76). Appleton-Century-Crofts. (Original work published 1953.)
Skinner, B. F. (1961/1957). The experimental analysis of behavior. In *Cumulative record* (Enlarged ed., pp. 100–131). Appleton-Century-Crofts. (Original work published 1957.)
Skinner, B. F. (1969a). An operant analysis of problem solving. In *Contingencies of reinforcement: A theoretical analysis* (pp. 133–171). Appleton-Century-Crofts.
Skinner, B. F. (1969b). The phylogeny and ontogeny of behavior. In *Contingencies of reinforcement: A theoretical analysis* (pp. 172–217). Appleton-Century-Crofts.

Skinner, B. F. (1969c). Behaviorism at fifty. In *Contingencies of reinforcement: A theoretical analysis* (pp. 221–268). Appleton-Century-Crofts.
Skinner, B. F. (1969d). Operant behavior. In *Contingencies of reinforcement: A theoretical analysis* (pp. 105–132). Appleton-Century-Crofts.
Skinner, B. F. (1971). *Beyond freedom and dignity*. Knopf.
Skinner, B. F. (1974). *About behaviorism*. Knopf.
Skinner, B. F. (1976). Farewell, my lovely! *Journal of the Experimental Analysis of Behavior, 25*, 218.
Skinner, B. F. (1981). Selection by consequences. *Science, 213*, 501–504.
Skinner, B. F. (1984). The phylogeny and ontogeny of behavior. *The Behavioral and Brain Sciences, 7*, 669–711.
Smith, R. F. (1974). Topography of the food-reinforced key peck and the source of 30-millisecond interresponse times. *Journal of the Experimental Analysis of Behavior, 21*, 541–551.
Solomon, R. L., & Wynne, L. C. (1954). Traumatic avoidance learning: The principles of anxiety conservation and partial irreversibility. *Psychological Review, 61*, 353–385.
Sommerville, C. J. (1982). *The rise and fall of childhood*. Sage.
Staddon, J. E. R. (1973). On the notion of cause, with applications to behaviorism. *Behaviorism, 1*, 25–63.
Staddon, J. E. R. (1977). Schedule-induced behavior. In W. K. Honig & J. E. R. Staddon (Eds.), *Handbook of operant behavior* (pp. 125–152). Prentice-Hall.
Staddon, J. E. R. (1980). Optimality analyses of operant behavior and their relation to optimal foraging. In J. E. R. Staddon (Ed.), *Limits to action* (pp. 101–141). Academic Press.
Staddon, J. E. R. (1982). Behavioral competition, contrast and matching. In M. L. Commons, R. J. Herrnstein, & H. Rachlin (Eds.), *Quantitative Analyses of Behavior, Matching and Maximizing Accounts* (Vol. II, pp. 243–261). Ballinger.
Staddon, J. E. R. (1983). *Adaptive behavior and learning*. Cambridge University Press.
Staddon, J. E. R. (1993). *Behaviorism: Mind, mechanism, and society*. Duckworth.
Staddon, J. E. R., & Simmelhag, V. L. (1971). The "superstition" experiment: A reexamination of its implications for the principles of adaptive behavior. *Psychological Review, 78*, 3–43.
Stephens, D. W., & Krebs, J. R. (1986). *Foraging theory*. Princeton University Press.
Sultan, S. E. (2017). Developmental plasticity: re-conceiving the genotype. *Interface Focus, 7*, 20170009.
Sultan, S. E. (2019). Genotype-environment interaction and the unscripted reaction norm. In T. Uller & K. N. Laland (Eds.), *Evolutionary causation: Biological and philosophical reflections*. (Chapter 6, pp. 109–126). MIT Press.
Suzuki, D. T. (1964). *An introduction to Zen Buddhism*. Grove Press.
Taylor, R. J. (1984). *Predation*. Chapman and Hall.
Teitelbaum, P. (1977). Levels of integration of the operant. In W. K. Honig & J. E. R. Staddon (Eds.), *Handbook of operant behavior* (pp. 7–27). Prentice-Hall.
Ten Eyck, R. L. (1970). Effects of rate of reinforcement-time upon concurrent operant performance. *Journal of the Experimental Analysis of Behavior, 14*, 269–274.
Thomas, G. (1981). Contiguity, reinforcement rate and the law of effect. *Quarterley Journal of Experimental Psychology, 33B*, 33–43.
Thompson, T. (2007). Relations among functional systems in behavior analysis. *Journal of the Experimental Analysis of Behavior, 87*, 423–440.
Thorndike, E. L. (1911). *Animal intelligence*. Macmillan.
Thorndike, E. L. (2000/1911). *Animal intelligence*. Hafner Publishing Co. (Originally published in 1911 by Macmillan, New York.)

Thorndike, E. L. (2012/1911). *Animal intelligence: Experimental studies.* Forgotten Books.
Timberlake, W. (1993). Behavior systems and reinforcement: An integrative approach. *Journal of the Experimental Analysis of Behavior, 60,* 105–128.
Timberlake, W., & Allison, J. (1974). Response deprivation: An empirical approach to instrumental performance. *Psychological Review, 81,* 146–164.
Timberlake, W., & Lucas, G. A. (1989). Behavior systems and learning: From misbehavior to general principles. In S. B. Klein & R. R. Mowrer (Eds.), *Contemporary learning theories: Instrumental conditioning theory and the impact of biological constraints on learning* (pp. 237–275). Erlbaum.
Tinbergen, N. (1963). On aims and methods of ethology. *Zeitschrift für Tierpshychologie, 20,* 410–433.
Todd, J. T. (1994). What psychology has to say about John B. Watson: Classical behaviorism in psychology textbooks, 1920–1989. In J. T. Todd & E. K. Morris (Eds.), *Modern perspectives on John B. Watson and classical behaviorism* (pp. 75–107). Greenwood.
Tooby, J., & Cosmides, L. (1992). The psychological foundations of culture. In J. H. Barkow, L. Cosmides, & J. Tooby (Eds.), *The adapted mind: Evolutionary psychology and the generation of culture* (pp. 19–136). Oxford University Press.
Trapold, M. A., & Overmier, J. B. (1972). The second learning process in instrumental learning. In A. H. Black & W. F. Prokasy (Eds.), *Classical conditioning. II. Current research and theory* (pp. 427–452). Appleton-Century-Crofts.
Turvey, M. T. (1977). Contrasting orientations to the theory of visual information processing. *Psychological Review, 84,* 67–88.
Urcuioli, P. J. (2005). Behavioral and associative effects of different outcomes in discrimination learning. *Learning and Behavior, 33,* 1–21.
Vaughan, W., Jr. (1988). Formation of equivalence sets in pigeons. *Journal of Experimental Psychology: Animal Behavior Processes, 14,* 36–42.
Vaughan, W. J., & Miller, H. L. J. (1984). Optimization versus response-strength accounts of behavior. *Journal of the Experimental Analysis of Behavior, 42,* 337–348.
Waldrop, M. M. (1992). *Complexity: The emerging science at the edge of order and chaos.* Simon & Schuster.
Wallace, A. F. C. (1965). Driving to work. In M. E. Spiro (Ed.), *Context and meaning in cultural anthropology* (pp. 277–292). Free Press.
Wasserman, E. A., DeVolder, C. L., & Coppage, D. J. (1992). Nonsimilarity-based conceptualization in pigeons via secondary or mediated generalization. *Psychological Science, 3,* 374–379.
Watkins, M. J. (1990). Mediationism and the obfuscation of memory. *American Psychologist, 45,* 328–335.
Watson, J. B. (1913). Psychology as the behaviorist views it. *Psychological Review, 20,* 158–177.
Watson, J. B. (1930). *Behaviorism.* W. W. Norton.
Watts, A. W. (1965). The individual as man/world. In G. M. Weil, R. Metzner, & T. Leary (Eds.), *The psychedelic reader* (pp. 47–57). University Books.
White, K. G. (1978). Behavioral contrast as differential time allocation. *Journal of the Experimental Analysis of Behavior, 29,* 151–160.
White, K. G. (1985). Interresponse-time analysis of stimulus control in multiple schedules. *Journal of the Experimental Analysis of Behavior, 43,* 331–339.
Whorf, B. L. (1956). *Language, thought, and reality.* MIT Press.
Williams, D. R., & Williams, H. (1969). Automaintenance in the pigeon: Sustained pecking despite contingent nonreinforcement. *Journal of the Experimental Analysis of Behavior, 12,* 511–520.

Williamson, M. (1993). *A return to love: Reflections on the principles of A Course in Miracles.* HarperCollins.
Wilson, D. S. (1999). Evolutionary psychology: *The new science of the mind,* by David M. *Buss. Evolution and Human Behavior, 20,* 279–287.
Wilson, D. S., & Sober, E. (1994). Re-introducing group selection to the human behavioral sciences. *Behavioral and Brain Sciences, 17,* 585–654. (Includes commentary)
Wilson, E. O. (1975). *Sociobiology: The new synthesis.* Harvard University Press.
Wilton, R. N., & Clements, R. O. (1971). Observing responses and informative stimuli. *Journal of the Experimental Analysis of Behavior, 15,* 199–204.
Wolin, B. R. (1968). Difference in manner of pecking a key between pigeons reinforced with food and with water. In A. C. Catania (Ed.), *Contemporary research in operant behavior* (pp. 286). Scott, Foresman. (Original work published 1948.)
Wormersley, D. (2007). *Introduction to Boswell's life of Samuel Johnson.* Penguin.
Zeiler, M. D. (1999). An odyssey through learning and evolution. *Mexican Journal of Behavior Analysis, 25,* 259–272.
Zener, K. (1937). The significance of behavior accompanying conditioned salivary secretion for theories of the conditioned response. *The American Journal of Psychology, 50,* 384–403.
Zettle, R. D., & Hayes, S. C. (1982). Rule-governed behavior: A potential theoretical framework for cognitive-behavioral therapy. In P. C. Kendall (Ed.), *Advances in cognitive-behavioral research and therapy* (Vol. I, pp. 73–118). Academic Press.
Zuriff, G. E. (1979). Ten inner causes. *Behaviorism, 7,* 1–8.
Zuriff, G. (1987). Naturalist ethics. In S. Modgil & C. Modgil (Eds.), *B. F. Skinner: Consensus and controversy* (pp. 309–318). Falmer Press.

Index

f